TECHNICAL LIBRARY (MAL)
U. S. NAVAL WEAPONS LAB.
DAHLGREN, VIRGINIA 22448

PARAMETRIC AMPLIFIERS

Philips Technical Library

Parametric Amplifiers

J. C. Decroly
L. Laurent
J. C. Lienard
G. Marechal
J. Vorobeitchik

A HALSTED PRESS BOOK

John Wiley & Sons
New York · Toronto

© N.V. Philips' Gloeilampenfabrieken, Eindhoven 1973

All rights reserved. No part of this publication may be reproduced or transmitted, in any form or by any means, without permission.

First published 1973 in English by
THE MACMILLAN PRESS LIMITED
London and Basingstoke
Associated companies in New York,
Melbourne, Dublin, Johannesburg and Madras

Published in the U.S.A. and Canada by
HALSTED PRESS
a Division of John Wiley & Sons, Inc., New York

Parametric amplifiers.
 (Philips technical library)
'A Halsted Press book.'

 1. Parametric amplifiers. I. Decroly, J. C. II. Lienard, J. C., ed. III. Series: Philips' Gloeilampenfabrieken N.V., Eindhoven. Technical library.
TK7871.24.P37   1973   621.3815'35   72–6547

ISBN 0–470–20065–0

 PHILIPS

Trademarks of N.V. Philips' Gloeilampenfabrieken

Printed in Great Britain

To the memory of our colleague J. Vorobeitchik, whose contribution to this book was much greater than it may seem, thanks to his fruitful criticism of the whole work.

CONTENTS

Introduction   xvii

Part I: Background Noise in Parametric Amplification

1 Background noise   **J. Vorobeitchik**   3
  1. Background noise in telecommunications   3
    1.1. General   3
    1.2. Noise functions   7
    1.3. Narrow-band signal with high signal-to-noise ratio   12
  2. Noise factor and effective noise temperature of an amplifier   19
    2.1. Equivalent circuits of a two-port containing internal noise sources   20
    2.2. Noise factor of a two-port   25
    2.3. Effective spectral noise temperature of a passive two-port at a uniform temperature   28
    2.4. Noise factor and effective noise temperature of a chain consisting of several two-ports   30
    2.5. Effective spectral noise temperature of a frequency converter   32
    2.6. Operational temperature of a system   40
  3. Sources of background noise   42
    3.1. Internal noise sources   42
    3.2. External noise sources   44

Part II: Parametric Amplifiers for Microwave Frequencies

2 Classification of parametric amplifiers and converters   **L. Laurent**   53
  1. Manley and Rowe relations   53
  2. Properties derived from the Manley and Rowe relations   56

3.	Application to parametric amplification and conversion	56
4.	Three-frequency converters	57
	4.1. Sum-frequency converters	57
	4.2. Difference-frequency converters	59
5.	Four-frequency converters	61
	5.1. Sum-frequency converters	61
	5.2. Difference-frequency converters	64
6.	Negative-resistance amplifiers	64
7.	Classification of three- and four-frequency parametric systems	66
8.	Relation between the power gain and the transducer gain	66
9.	Four-frequency, double-conversion converters	67

3 Small-signal analysis   L. Laurent   70

1.	Small-signal theory for an ideal junction	70
2.	Introduction of terminal immittances in the equations	75
3.	Introduction of the varactor losses in the equations	77
4.	Introduction of reduced variables	79
5.	Special case of three-frequency systems	80
	5.1. Sum-frequency converter	80
	5.2. Difference-frequency converter	81
	5.3. Reflection amplifier	81
	5.2. Degenerate amplifiers	82

4 Three-frequency non-inverting converters   L. Laurent   83

1.	Sum-frequency converter	83
	1.1. Equivalent circuit	83
	1.2. Transducer gain	84
	1.3. Instability factor of the gain	85
	1.4. Effective noise temperature	85
	1.5. Optimization of the gain	86
	1.6. Optimization of the intrinsic noise temperature	91
	1.7. Optimization of the total noise temperature	94
	1.8. Optimization of the stability	97
	1.9. Choice of $n_{(+)}$	97
	1.10. Bandwidth	98
	1.11. Conclusions	99
2.	Difference-frequency non-inverting converters	99

5	The three-frequency inverting converter and reflection amplifier		L. Laurent	101
	1.	Basic equations		101
		1.1.	Equivalent circuit	101
		1.2.	Circulators and isolators	103
		1.3.	Transducer gain of a negative-resistance amplifier without circulator	109
		1.4.	Transducer gain of a negative-resistance amplifier with circulator	111
		1.5.	Comparison between amplifiers with and without circulator	112
		1.6.	Use of circulators with four and five ports	112
		1.7.	Use of isolators in two port amplifiers	114
		1.8.	The six transducer gains of difference-frequency systems	115
	2.	The inverting converter		121
		2.1.	Transducer gain	121
		2.2.	Instability coefficient	122
		2.3.	Effective noise temperature	122
		2.4.	Part played by the instability factor in optimizations	124
		2.5.	Optimization of the gain for a given instability factor	125
		2.6.	Optimization of the intrinsic noise temperature for a given instability factor	126
		2.7.	Optimization of the total noise temperature for a given instability factor	131
		2.8.	High-gain bandwidth	133
	3.	The reflection amplifier		134
		3.1.	Transducer gain	134
		3.2.	Instability factor	136
		3.3.	Effective noise temperature	136
		3.4.	Optimization of the gain for a given instability factor	138
		3.5.	Optimization of the noise temperature for a given gain	138
		3.6.	Optimization of the noise temperature for a given instability factor	139
		3.7.	Minimization of the total noise temperature	140
		3.8.	High-gain bandwidth	140
	4.	Conclusion		141

6 Four-frequency converters and amplifiers — L. Laurent — 143

1. Introduction — 143
2. Equivalent diagram and general equations for four-frequency converters — 143
3. Calculation of the total reduced impedances of the three branches — 144
    - 3.1. Calculation of the total reduced input impedance $\bar{z}_e$ — 144
    - 3.2. Calculation of the equivalent circuit of the branch at $\omega_{(+)}$ — 145
    - 3.3. Calculation of the equivalent circuit of the branch at $\omega_{(-)}$ — 146
4. Introduction of new reduced variables — 147
5. Calculation of the transducer gains — 147
    - 5.1. Transducer gain of the sum-frequency converter for $f_p > f_e$ — 147
    - 5.2. Transducer gain of the sum-frequency converter for $f_p < f_e$ — 149
    - 5.3. Transducer gain of the difference-frequency converter for $f_p > f_e$ — 149
    - 5.4. Transducer gain of the difference-frequency converter for $f_p < f_e$ — 149
    - 5.5. Transducer gain of the four-frequency reflection amplifier — 150
6. Summary of the formulae when S_2 is negligible — 150
7. Stability conditions for the four-frequency converter when S_2 is negligible — 151
8. Applications — 153
    - 8.1. The sum-frequency converter — 153
    - 8.2. Deterioration in the performance of the reflection amplifier due to power consumption at $\omega_{(+)}$ — 155

7 Degenerate amplifiers — J. C. Lienard — 157

1. General study and classification — 157
    - 1.1. Classification of difference-frequency converters and amplifiers — 157
    - 1.2. Practical form of the three-port — 160
    - 1.3. Calculation of the transfer functions for the input signal and noise — 160
    - 1.4. Calculation of the transfer functions for the noise of R_s — 160
    - 1.5. Calculation of the impedances seen by the varactor — 161

	1.6.	Expressions for the output signals and noise	161
	1.7.	Classification of degenerate amplifiers	164
2.	Single-band reception		164
	2.1.	The three methods of operation	164
	2.2.	Amplifier X_e^e	165
	2.3.	Amplifier $X_{(-)}^e$	175
	2.4.	Amplifier $X_{e(-)}^e$	176
	2.5.	Comparison between the three amplifiers and the reflection amplifier	179
3.	Double-band reception, correlated signals		184
	3.1.	The two methods of operation	184
	3.2.	Amplifier $X_{e(-)}^e$	185
	3.3.	Amplifier $X_{e(-)}^{e(-)}$	187
	3.4.	Conclusions	188
4.	Double-band reception, random signals (radiometry)		189
	4.1.	The two methods of operation	189
	4.2.	Amplifier $X_e^{e(-)a}$	190
	4.3.	Amplifier $X_{e(-)}^{e(-)a}$	191
	4.4.	Conclusions	193

8 Pumping of parametric diodes for microwave frequencies
J. C. Lienard   194

1.	General		194
	1.1.	Description of the pumped varactor	194
	1.2.	Maximization of $|S_1|$	195
	1.3.	Discussion: waveform of the pumping voltage	198
2.	Voltage pumping		198
	2.1.	General formulae	198
	2.2.	Calculation of the $|C_k|$'s	202
	2.3.	Calculation of the $|S_k|$'s	205
	2.4.	Calculation of the pumping power	206
	2.5.	Special cases	207
3.	Current pumping		210
	3.1.	General formulae	210
	3.2.	Calculation of the $|C_k|$'s	213
	3.3.	Calculation of the $|S_k|$'s	213
	3.4.	Calculation of the bias	214
	3.5.	Special cases	215
4.	Tables and charts for voltage pumping		218
	4.1.	Introduction	218
	4.2.	Mean value of the elastance	219
	4.3.	First harmonic of the elastance	222
	4.4.	Higher harmonics of the elastance	224

	4.5.	Polarization	225
	4.6.	Pumping power	226
5.	Tables and charts for current pumping	228
	5.1.	Introduction	228
	5.2.	Mean value of the elastance	228
	5.3.	First harmonic of the elastance	231
	5.4.	Higher harmonics of the elastance	233
	5.5.	Polarization	233
	5.6.	Pumping power	236
6.	Pumping optimization	237
	6.1.	General. Pumping of negative-resistance amplifiers and sum-frequency converters	237
	6.2.	Square-wave pumping	238
	6.3.	Comparison between voltage and current pumping	239
	6.4.	Comparison of varactors	242

9 Measurement of parametric amplifiers — J. C. Lienard — 246

1.	The circle diagram and its interpretation		247
	1.1.	Properties of the Smith chart	247
	1.2.	Analysis of the pump circle diagram	249
	1.3.	Circle diagrams of small-signal circuits	252
	1.4.	Sum-frequency converter adjusted for maximum gain	255
	1.5.	Sum-frequency converter adjusted for minimum intrinsic noise	256
	1.6.	The inverting converter adjusted for minimum intrinsic noise	257
	1.7.	The reflection amplifier adjusted for minimum noise	258
2.	Adjustment of a parametric amplifier		258
	2.1.	Adjustment of the pump circuit	258
	2.2.	Adjustment of a reflection amplifier	260
3.	Measurement of varactors		261
	3.1.	Equivalent circuit of a varactor	261
	3.2.	Characteristics specified by the manufacturer	263
	3.3.	Measurement of the total capacitance	264
	3.4.	Measurement of the quality factor Q	264
	3.5.	Measurement of L_c	266
	3.6.	Measurement of C_c and of the non-linear capacitance	266
4.	Equivalent circuits of passive two-ports		267
	4.1.	Representation of the circuit elements	267

	4.2. Chain matrix of elementary two-ports	268
	4.3. Discussion: practical importance of the various equivalent circuits	269
	4.4. Study of the usual equivalent circuits	269
	4.5. Measurement of equivalent circuits by continuous displacement of a short circuit	269

10 Measurement of the noise temperature of an amplifier
                                                                                    L. Laurent 276

1. Equipment common to all methods of measurement — 276
2. Direct measurement of the noise temperature — 277
3. Measurement by comparison with a 'monochromatic' signal generator — 278
4. Hot and cold resistance method — 278
5. Use of a thermionic diode — 281
6. Use of a gas-discharge tube — 285
7. Automatic measurement of noise temperature — 288
8. Use of a directional coupler and a cooled termination — 291

11 Technical aspects   L. Laurent 297

1. Reactive networks including a varactor — 297
2. Construction of reactive networks corresponding to the different branches — 299
3. Filters — 301
   3.1. Types of filters most frequently used — 301
   3.2. Measurement of response curves and short-circuit planes — 302
4. Location of impedance transformers — 303
5. Description of a sum-frequency converter — 303
6. Description of a reflection amplifier — 306
   6.1. The parametric amplifier — 306
   6.2. Auxiliary circuits of the reflection amplifier — 307

Part III: Low-Frequency Parametric Amplifiers

12 Parametric amplification of low-frequency signals
                                                    J. C. Decroly 311

1. Introduction — 311
2. General equations — 314
3. Analysis of the operation for $\omega_e = 0$ — 319
   3.1. Phase and amplitude of the output signal of the parametric amplifier — 320

	3.2. Admittance Y_{in_0}	321
	3.3. Discussion of the relations (12.29) and (12.30)	321
	3.4. Operation of the parametric amplifier when $R_e Y_{in_0} \ll 1$	322
	3.5. Synchronous detection of the output signal	324
4.	Analysis of the operation when $\omega_e \neq 0$	326
	4.1. Phase and amplitude of the components of frequency f_+ and f_- of the voltage at the output of the parametric amplifier	327
	4.2. Instantaneous voltage energizing the detector	329
	4.3. Synchronous detection of the output signal	329
	4.4. Bandwidth at 3 dB	331
	4.5. Input admittance	332
5.	Background noise	334
	5.1. Internal noise sources in parametric amplifiers	334
	5.2. Noise voltage at the input of the detector	334
	5.3. Spectral density n_d of the parasitic signal at the output of the detector	337
	5.4. Noise factor	339
6.	Special case in which $\alpha = 0$; tuned-load modulator	340

13 Decoupling and methods of adjustment G. Marechal 342

1. Balancing of the bridge 342
   1.1. Preliminary conditions for satisfactory balancing 342
   1.2. Conditions for equilibrium 352
   1.3. Sensitivity of the trimming components 358
   1.4. Method of adjustment 364
   1.5. Special type of amplifier for d.c. signals 365
2. Auxiliary circuits 366
   2.1. Overall block diagram 366
   2.2. Conventional amplifier 367
   2.3. Pump oscillator 368
   2.4. Demodulator 368
   2.5. Internal correction networks 368
   2.6. Automatic gain control 369
   2.7. Anti-saturation circuit 369
3. Technical limitations 369
4. Examples 379
5. Details of amplification of d.c. and low-frequency signals by the parametric varactor method 381
   5.1. Input impedance 381
   5.2. Noise 382
   5.3. Gain 382

	5.4. Transformation of d.c. or low-frequency signals into a.c. high-frequency signals	382
	5.5. Bandwidth	383
	5.6. Remarks	383

14 The pumping of parametric diodes for low frequencies
G. Marechal 387

1. General 387
2. Cissoidal voltage pumping 390
   2.1. Calculation of the C_k's 390
   2.2. Calculation of the G_k's 403
   2.3. Calculation of the I_k's 405
3. Rectangular voltage pumping 406
   3.1. Calculation of the C_k's 406
   3.2. Calculation of the G_k's 412
   3.3. Calculation of the I_k's 412
   3.4. Recurrence formulae of C_{2k+1}, G_{2k+1}, I_{2k+1} ($k \geqslant 2$) 412
4. Specially important case: $I_0 = 0$ 413
   4.1. Relation between V_0 and $2V_p$ for $I_0 = 0$ 413
   4.2. Value of G_0 for $I_0 = 0$ 413
   4.3. Influence of temperature 414
   4.4. Nomographs 415
   4.5. Practical example 416
5. Reduced variables and important ratios 418
6. Comparison between cissoidal and rectangular pumping 422
   6.1. Comparison of the C_1/C_0's with β_v given 423
   6.2. Comparison for equal losses 423
   6.3. Comparison of losses for given V_0 and $2V_p$ 426
   6.4. Conclusion 426

15 Measurement of the parameters of the varactor G. Marechal 429

1. Measurement of I_s, a, G_s, R_s 430
   1.1. Measurement of the static characteristic 430
   1.2. Determination of a and I_s 431
   1.3. Determination of G_s 432
   1.4. Determination of R_s 433
2. Measurement of C_{0v}, ϕ, n, C_c 433
   2.1. Measurement of $C = C(v)$ and of $dC(v)/dv$ 433
   2.2. Determination of C_{0v}, ϕ, n, C_c 434
3. Cut-off frequency—quality factor 436

Index 439

INTRODUCTION

It has been known since the last century that a mechanical system or an electrical circuit in which there is a parameter which varies periodically with time, may oscillate under certain conditions, but it is only quite recently that it has been suggested that use should be made of this periodic variation for the design of amplifiers. Landon described an amplifier of this type in 1949. Several years later, Buhl suggested the use of two circuits coupled by a variable gyromagnetic effect, for a microwave-frequency amplifier. But real progress in the field of 'parametric amplifiers' dates from the development by Uhlir of new semiconductor diodes in which the barrier capacitance can be modulated at microwave frequencies by acting on the biasing (inverse) of the diodes. Since then, there has been a continual increase in the amount of theoretical and experimental work devoted to parametric amplifiers, providing evidence of great interest shown in this type of amplifier for both *microwave frequencies* and *low frequencies*.

The chief advantage of parametric amplifiers lies in the possibility of obtaining a low level of background noise. This feature is of particular interest in the case of transmissions in which the level of the noise, due to external sources, is very low (e.g. space telecommunications) and where, consequently, the sensitivity of the receiver depends chiefly on its intrinsic noise sources. In the case of parametric amplifiers using semiconductor diodes, with which this work is concerned, there is a further advantage, namely the possibility of having a very high input impedance at low frequencies, which opens a wide field of application for instrumentation to this type of amplifier.

Why does recourse to parametric amplification allow one to obtain noise levels lower than with the usual type of amplifier? In order to answer this question, let us remember first that an amplifier consists of both passive elements and active energy sources, the whole being terminated by two ports, one receiving the signal to be amplified and the other delivering the amplified signal. In the ordinary type of amplifier, the energy sources consist of d.c. sources which are incorporated into a network of passive elements which may be linear or non-linear. Now, it can be shown† that the essential condition

† R. J. Duffin, 'Impossible behaviour of nonlinear networks', *Journal of Applied Physics*, vol. 26, pp. 603–605, May 1955.

for such a system to act efficiently as an amplifier is; that at least one of the passive elements of the network should consist of a *resistance in which the current-voltage characteristic should not be of the quasi-linear type*, as defined by the following conditions.

(a) The relation between the current i through the resistance and the applied voltage v is biunivocal and is independent of any other voltage.

(b) The differential resistance dv/di is always positive.

It will thus be seen that ordinary amplifiers necessarily include a non-quasilinear element. Examples are: a triode or a transistor (which do not satisfy condition (a), since the relation $i = f(v)$ depends on the voltage of the control electrode), or a tunnel diode (which does not satisfy condition (b)).

Unfortunately, in spite of the improvements made in these devices, their operation is accompanied by background noise which is an inevitable result of their dissipative character. In order to obtain an amplifier with a low noise level, the energy must no longer be supplied to the network direct from d.c. sources, and so other methods of transfer must be envisaged. One of these is to employ a.c. sources. It can then be shown† that if a *non-linear reactance* is incorporated in the network, a system of this kind may be used efficiently as an amplifier. This method is much more interesting than the previous one, since a non-linear reactance does not generate any noise in the ideal state (in which it would be free from loss). Now, parametric diodes or varactors which are produced, in practice, behave as almost ideal non-linear reactances, both at microwave and at low frequencies. On the other hand, since the susceptance of varactors is very low at low frequencies, their use in parametric amplifiers makes it possible to obtain a very high input impedance at low frequencies. It is thus easy to appreciate the increasing importance which parametric amplifiers of the varactor type, are assuming in fields as different as those involving microwaves and very low frequencies.

This book is divided into three parts.

Part I

The first part (Chapter 1) deals with background noise. First of all the problem of background noise in telecommunications is discussed.

† J. M. Manley and H. E. Rowe, 'Some general properties of nonlinear elements; Part I, General energy relations', *Proc. I.R.E.*, vol. 44, pp. 904–913, July 1956.

This is followed by generalized definitions of the noise factor and the effective and operational noise temperatures. Next general expressions are derived for the parameters of the different types of parametric amplifiers and converters studied in this book. The chapter ends with a short discussion on the properties of the chief sources of noise, both internal and external.

Part II

In the second part we discuss parametric amplifiers for microwave frequencies and the Manley and Rowe relations are dealt with in Chapter 2. These relations are taken as the basis for the classification of parametric amplifiers and converters, and this classification is shown in a table at the end of the chapter. In Chapter 3 we introduce the small-signal hypothesis which enables the basic equations of a parametric system to be linearized. Chapter 4 deals with the calculation of the transducer gain of the effective noise temperature, and of the instability coefficient of non-inverting converters. After the general formulae have been established and discussed, the optimum operating conditions are determined, the results obtained being set out in the numerical tables. Chapter 5 deals with inverting converters and reflection amplifiers. After having given the basic equations common to these two types of amplifier, and having reviewed briefly the properties of non-reciprocal elements, we pass on to the determination of the gain, the noise temperature, the instability coefficient, the bandwidth and the optimum operating conditions. In Chapter 6 we give the general expressions for the four-frequency amplifier gains. Chapter 7 concerns degenerate amplifiers. First we establish a classification for this type of amplifier, and then consider in detail the gain, the effective noise temperature, the instability coefficient and the bandwidth. Chapter 8 is devoted to the theory of the pumping of parametric diodes, dealing with voltage and current pumping. Numerous tables and curves enable the results to be used immediately. Chapter 9 treats of the measurements required for the calculation, adjustment, and testing of parametric amplifiers, except for noise measurement which is dealt with in Chapter 10. Finally, Chapter 11 discusses some technological features peculiar to parametric amplifiers for microwave frequencies, and illustrates them by means of practical examples.

Part III

The third part deals with low-frequency parametric amplifiers. In Chapter 12 are given the general equations governing the operation of a parametric amplifier for low-frequency signals. Particular emphasis is laid on the necessity of giving this type of amplifier a differential structure by the use of two semiconductor diodes. Chapter 13 deals with the manufacture and adjustment of these amplifiers. First a study is made of the conditions for equilibrium of the differential structure introduced in Chapter 12, and a method of achieving this is given. Next there is an example of optimization, taking into account the technological limitations. Chapter 14 deals with the pumping of low-frequency diodes. Here again, the results obtained are given in the form of tables and curves. Finally, Chapter 15 describes the methods of measurement of these diodes.

This book is based partly on work done at M.B.L.E. under the auspices of the *Institut pour l'Encouragement de la Recherche Scientifique dans l'Industrie et l'Agriculture* (I.R.S.I.A.). The authors wish to thank the *Institut* for their permission to publish this section of the study. They are grateful also to their colleagues in the *Laboratoire de Recherches de la M.B.L.E.* for the computation of the tables included in this book.

J. C. DECROLY[*]
L. LAURENT[**]
J. C. LIENARD[**]
G. MARECHAL[**]
J. VOROBEITCHIK[***]

[*] Formerly with M.B.L.E., now with Disment Boort, Brussels.
[**] M.B.L.E., Brussels.
[***] Deceased, formerly with M.B.L.E.

Part I

BACKGROUND NOISE IN PARAMETRIC AMPLIFICATION

1

BACKGROUND NOISE

by J. VOROBEITCHIK

1. Background noise in telecommunications

1.1. General

In all telecommunication systems the transmission of a message consists, in principle, of the following three phases:

(1) Injection of the message into the carrier signal.
(2) Propagation of the signal through the medium connecting the transmitter and receiver.
(3) Reconstruction of the message from the received signal.

The transmission of a signal by telecommunication involves the reproduction at a distance of the excitations or physical signs representing the message (sound or luminous oscillations, graphical symbols, etc.). When these excitations or signs are transmitted they are transformed, by means of a suitable conversion device, into values of a certain variable m (modulating signal). Depending on the nature of the message and the type of conversion, the variations of m may be continuous (analogue input) or discontinuous (numerical coder). In either case, as the message proceeds, it determines the value of m at each instant t, and it is in the form of the function $m(t)$ that the information contained in the message modulates the carrier signal. When it is required to transmit several messages simultaneously on the same carrier signal, by means of the technique of cyclic sampling (time multiplex), each excitation-sample is converted into a value of m, and it is still in the form of the function of $m(t)$, that the information to be transmitted is injected into the carrier signal.

This carrier signal is, in most telecommunication systems, of the form:

$$s(t) = A \cos(\omega_c t + \theta)$$

A signal of this nature is characterized by two parameters: the amplitude A and the angle θ. Each of these parameters may be used in order to introduce the function $m(t)$ into the carrier signal, resulting in two types of modulation: amplitude modulation and angle modulation, the latter being subdivided into two types: phase modulation and frequency modulation.

At the receiver, a detection device, followed by a low-frequency (LF) filter, extracts from the received signal, (with amplification if required) the instantaneous value of the information-carrying parameter (A, θ or $d\theta/dt$ according to the type of modulation used). The LF filter thus delivers a signal $m_d(t)$ which consists of a replica of the modulating signal $m(t)$, and from this replica a suitable reconversion device re-establishes, in physical form, the message or messages transmitted. In an ideal transmission system, the function $m_d(t)$ should be identical with the function $m(t)$ (disregarding any difference in scale and also the inherent delay in the propagation of the signal). In practice this identity is never realized. There are several reasons for the difference between $m_d(t)$ and $m(t)$. Depending on the nature of the causes, these differences are predictable or unpredictable.

The divergences, due to the linearity of a chain of amplification (in front of the detector), are predictable in the sense that they can be precisely determined if we know the waveform of the input signal and the transfer characteristics of the amplification chain. On the other hand, the divergences due to the parasitic signals generated by noise sources, either external to the receiver (external background noise) or intrinsic to the receiver (internal background noise), are unpredictable, because of the random nature of parasitic signals. In other words, given a transmission path and a function $m(t)$, it is not possible to determine what, at the instant t, will be the precise value of the divergence $m_d(t) - m(t)$.

This value depends on the behaviour of the parasitic signals during the transmission time, and this is known to us only in the form of a probability. It is for this reason that the background noise confers a randomly varying character to the value of $m_d(t)$ for a given value of t. More precisely, if the transmission of the function $m(t)$ took place simultaneously over a very large number, k, of systems having the same physical structure as the system being studied, the k values of $m_d(t)$ obtained for a given value of t would form a certain set $\{^r m_d(t)\}$

($r = 1, 2, \ldots, k$), and it is this set that defines the random variable $m_d(t)$.

This variable is characterized by two essential parameters:

the statistical average

$$\langle {}^r m_d(t) \rangle = \lim_{k \to \infty} \frac{\sum_{r=1}^{k} {}^r m_d(t)}{k} \tag{1.1}$$

and the variance

$$\sigma^2_{m_d(t)} = \lim_{k \to \infty} \frac{\sum_{r=1}^{k} [{}^r m_d(t) - \langle {}^r m_d(t) \rangle]^2}{k} \tag{1.2}$$

In principle these two parameters may be determined if we know the function $m(t)$, the type of modulation, the transfer characteristics of the path and of the LF filter, the intensity of the usable signal at the detector input, and the mean intensity and the statistical properties of the parasitic signal which is superimposed on the useful signal.

It was stated above that, at the receiver, a reconversion device re-establishes, from $m_d(t)$, the physical magnitudes by means of which the message is expressed. Because of the differences between $m_d(t)$ and $m(t)$, the message which has been reconstructed in this way may be subjected to errors (information errors). These errors may be continuous or discrete depending on the conversion method used; more precisely, in analogue conversion systems, they manifest themselves by a difference between the reconstituted value and the original value of the magnitude transmitted. This difference is an increasing function of the divergence $m_d(t) - m(t)$, whilst in numerical conversion systems they are manifested by the substitution of one discrete value by another. This substitution occurs only if the divergence $m_d(t) - m(t)$ exceeds a certain threshold value (detection threshold).

Since the errors in the information depend on the value of the divergences $m_d(t) - m(t)$ they can be estimated only statistically. To each value $x(i)$ of the original magnitude x (continuous or discrete depending on the type of conversion used) there is associated a set of values ${}^r x_d(i)$ ($r = 1, 2, \ldots, k$) of the reconstituted magnitude x_d. This set defines the random variable $x_d(i)$ (reconstituted value) corresponding to the value $x(i)$ considered. The characteristic parametric values of $x_d(i)$ are:

the statistical average

$$\langle {}^r x_d(i) \rangle = \lim_{k \to \infty} \frac{\sum_{r=1}^{k} {}^r x_d(i)}{k} \tag{1.3}$$

and the variance

$$\sigma^2_{x_d(i)} = \lim_{k \to \infty} \frac{\sum_{r=1}^{k} [{}^r x_d(i) - \langle {}^r x_d(i) \rangle]^2}{k} \tag{1.4}$$

We must now ask ourselves how, under these conditions, we can evaluate the fidelity of a telecommunication system. When the ratio of the intensity of the useful signal to the mean intensity of the parasitic signal superimposed on the useful signal is high, the accepted criterion is the *signal-to-noise ratio after reconversion*. The ratio is defined by:

$$\frac{S'}{N'} = \frac{\sum_i p(i) \langle {}^r x_d(i) \rangle^2}{\sum_i p(i) \sigma^2_{x_d(i)}} \tag{1.5}$$

where $p(i)$ is the probability of occurrence of the value $x(i)$ in the messages to be transmitted.

Much work has been done, and is still continuing, on the improvement of the ratio S'/N'. In addition to the straightforward increase in the power of the useful signal, the methods considered to combat the degradation of the information transmitted due to background noise may be divided into two categories:

(a) methods to reduce the effect of parasitic disturbances;
(b) methods to reduce the disturbances themselves.

Amongst the first of these classes an important place is taken by those methods which, by the introduction of the time dimension, use a wider spectrum for the transmission of information than that occupied by the signal $m(t)$. Examples of this are frequency modulation, pulse code modulation and the use of self-correcting codes. The methods in the second class may be subdivided into:

(a) those aiming at reducing the external background noise (armoured cables, highly directional radio beams);
(b) those aiming at reducing the internal background noise (low-noise amplifiers).

Parametric amplification, and especially that which makes use of the properties of semiconductor diodes, belongs to this last class.

How are we to appreciate, from the point of view of the fidelity of transmission, the quality of an amplifier incorporated in a given telecommunication chain? In other words, how are we to estimate the role played by the noise sources of this amplifier in the deterioration of the information transmitted by this chain (account being taken here of the departure from linearity of the amplifier)? If the fidelity of the transmission is expressed by the ratio S'/N', it seems logical to

BACKGROUND NOISE

estimate this role by a comparison of the value S'/N' obtained under actual conditions, with that which would be obtained if the amplifier were assumed to be ideal, i.e. without its source of noise.

It should, however, be noted that the ratio S'/N' depends on two factors:

(a) The value of the *signal-to-noise ratio at the output of the LF filter* defined by

$$\frac{S}{N} = \frac{\overline{\langle {}^r m_d(t) \rangle^2}}{\overline{\sigma^2_{m_d}(t)}} \quad (1.6)$$

in which the barring indicates that we are dealing with the time mean of the barred function; in (1.6) this mean plays the same role as the sum weighted by $p(i)$ in (1.5).

(b) The structure and especially the redundancy factor of the code used. (This factor does not of course occur in analogue conversion systems.)

In numerical conversion systems, the factor (a) determines the value p of the probability that a figure produced will be incorrectly reconstituted. The probability p becomes smaller as S/N increases. The factor (b) affects the value of S'/N' as a function of p. For a given value of p, S'/N' increases with the degree of redundancy of the code used. Redundancy is paid by a reduction in the number of 'words' that can be transmitted in unit time (for equal widths of the bandwidth of the LF filter). On the other hand, for a given code structure, the value of S'/N' increases as p decreases, i.e. the ratio S/N of the path used is larger. Also, in order to become independent of the choice of the code structure, it is usual to accept as the criterion for the quality of an amplifier the ratio S/N instead of S'/N'. (In analogue conversion systems these two ratios are generally of the same value.) The criterion adopted is the *noise factor*, defined by:

$$F = \frac{[S/N]_{\text{ideal}}}{[S/N]_{\text{actual}}} \quad (1.7)$$

Since $[S/N]_{\text{actual}}$ and $[S/N]_{\text{ideal}}$ depend on noise sources exterior to the amplifier, it is usual to make the definition (1.7) agree with certain standardized conventions, which will be discussed later. Now that the noise factor has been defined, we can pass to the main subject of this chapter, namely the study of this factor. First, let us recall briefly the basic properties of noise functions.

1.2. Noise functions

In telecommunications the parasitic disturbances produced by noise sources (noise functions) may usually be expressed by random functions possessing stationary and ergodic properties. This book is concerned only with disturbances of this type. (Hypothesis I.) The concept of the random function suggests a trial. In this case, it consists in choosing at random, one of the transmissions effected by the k systems of the same physical structure, which was discussed in section 1.1. A function $^r x(t)$ corresponds to a trial of index r and the set $\{^r x(t)\}$, $r = 1, 2, \ldots, k$, defines the random function $x(t)$. A group of this nature is said to be *stationary* (of the second order) and *ergodic* if it possesses the following properties:

(a) For a given value of t, the mean values of the set of the first and second orders

$$\langle ^r x(t) \rangle = \lim_{k \to \infty} \frac{\sum_{r=1}^{k} {}^r x(t)}{k} \qquad \text{(mean)} \quad (1.8)$$

$$\langle ^r x(t)^r x(t + \tau) \rangle = \lim_{k \to \infty} \frac{\sum_{r=1}^{k} {}^r x(t)^r x(t + \tau)}{k}$$
(correlation function) (1.9)

are independent of t (stationary state of the second order).

(b) For a given value of r, the time means

$$\overline{^r x(t)} = \lim_{T \to \infty} \frac{1}{2T} \int_{-T}^{T} {}^r x(t) \, dt \qquad \text{(mean)} \quad (1.10)$$

$$\overline{^r x(t)^r x(t + \tau)} = \lim_{T \to \infty} \frac{1}{2T} \int_{-T}^{T} {}^r x(t)^r x(t + \tau) \, dt$$
(correlation function) (1.11)

are independent of r and are equal to the corresponding statistical means (ergodicity).

In what follows, in addition to hypothesis I, we will take the mean of $x(t)$ as zero. (Hypothesis II.)

$$\langle ^r x(t) \rangle = \overline{^r x(t)} = 0 \qquad (1.12)$$

If we admit hypotheses I and II, what are the statistical properties of the group $\{^r x(t)\}$? In order to find an answer to this question, let us consider a transmission of duration $2T$ and let

$$^r x(t) = \sum_{n=1}^{\infty} {}^r a_n \cos n\omega t + \sum_{n=1}^{\infty} {}^r b_n \sin n\omega t \qquad (1.13)$$

with

$$\omega = \frac{2\pi}{2T} \qquad (1.14)$$

the development in Fourier series of the function $^r x(t)$ corresponding to the trial of index r.

We see from this development that the statistical properties of the random function $x(t)$ are conditioned by the statistical properties of the coefficients a_n and b_n ($n = 1, 2, \ldots$), defined in terms of random variables respectively, by the sets:

$$\{^r a_n\} \quad \text{and} \quad \{^r b_n\} \quad (r = 1, 2, \ldots, k)$$

As the development (1.14) should satisfy hypotheses I and II, it can be shown [1] that these hypotheses impose the following conditions on the random variables a_n and b_n:

$$\langle ^r a_n \rangle = \langle ^r b_n \rangle = 0 \qquad \text{(zero means)} \quad (1.15)$$

$$\left. \begin{array}{l} \langle ^r a_n {}^r a_m \rangle = \langle ^r b_n {}^r b_m \rangle = 0 \quad \text{for } n \neq m \\ \langle ^r a_n {}^r b_m \rangle = 0 \end{array} \right\}$$

$$\text{(mutually independent variables)} \quad (1.16)$$

$$\langle ^r a_n^2 \rangle = \langle ^r b_n^2 \rangle \qquad \text{(equal variances)} \quad (1.17)$$

$$^r a_n^2 + {}^r b_n^2 = c_n^2 \qquad \text{(amplitudes independent of } r\text{)} \quad (1.18)$$

It should be noted that the last of these relations results from the property (b) (ergodicity) and for this reason is only valid, strictly speaking, at the limit for $T \to \infty$.

From (1.13) to (1.18) we derive:

$$\langle ^r a_n^2 \rangle = \langle ^r b_n^2 \rangle = \tfrac{1}{2} \langle ^r a_n^2 + {}^r b_n^2 \rangle = \frac{c_n^2}{2} \quad (1.19)$$

and

$$\langle ^r x^2(t) \rangle = \overline{^r x^2(t)} = \sum_{n=1}^{\infty} \frac{c_n^2}{2} \quad \text{(variance)} \quad (1.20)$$

We may therefore write (1.13) in the form

$$^r x(t) = \sum_{n=1}^{\infty} {}^r c_n \cos(n\omega t - {}^r \psi_n) \quad (1.21)$$

with

$$^r c_n^2 = {}^r a_n^2 + {}^r b_n^2 \quad (1.22)$$

$$\left. \begin{array}{l} \cos {}^r \psi_n = \dfrac{^r a_n}{^r c_n} \\[6pt] \sin {}^r \psi_n = \dfrac{^r b_n}{^r c_n} \end{array} \right\} \quad (1.23)$$

When $T \to \infty$, only the angles $^r \psi_n$ maintain their random character, whilst the amplitudes $^r c_n$ tend to become constant by virtue of

(1.18). From (1.15) to (1.18) it can be deduced that the random variables ψ_n, $n = 1, 2, \ldots$, defined respectively by the sets $\{^r\psi_n\}$, are mutually independent and that each set is distributed uniformly in the interval $(0 \cdot 2\pi)$. Finally, given a noise function $x(t)$, we see that in accordance with (1.20), its variance *is a function only of the amplitudes* c_n which characterize $x(t)$. We must ask ourselves how these amplitudes vary when the function $x(t)$ undergoes a certain transformation, but first let us define what is meant by a transformation of a random function. Consider a set $\{r_x(t)\}$ and cause each function belonging to this set to undergo the same transformation, which will be named A, and which transforms $^rx(t)$ into $^rx_A(t)$. We thus obtain a new set $\{^rx_A(t)\}$. The random function $x_A(t)$ defined by this group is the *transform* of $x(t)$.

We give below, for several usual transforms, the relations between c_{n_A} and c_n, established from (1.21):

Symbol	Nature of the transforms $x(t) \to x_A(t)$	Relation between c_{n_A} and c_n	
F	Filtering (amplitude gain: $H(n\omega)$)	$c_{n_F} = H(n\omega)c_n$	(1.24)
D	Differentiation with respect to time	$c_{n_D} = n\omega c_n$	(1.25)
M	Multiplication by $\cos(p\omega t + \phi)$ (modulation; p a whole number)	$c_{n_M}^2 = \tfrac{1}{4}(c_{n-p}^2 + c_{n+p}^2)$	(1.26)

The relation (1.26) calls for some comment. Whilst the sets $\{^rx_F(t)\}$ and $\{^rx_D(t)\}$ are stationary and ergodic, the same does not hold for the set $\{^rx_M(t)\}$ which results from the multiplication of $x(t)$ by $\cos(p\omega t + \phi)$: $\langle ^rx_M(t)^rx_M(t+\tau)\rangle$ *depends not only* on τ, but also on t and on ϕ; $\overline{^rx_M(t)^rx_M(t+\tau)}$ depends on r and ϕ, and also on τ. In order to resolve this difficulty and to take account of the fact that the angle ϕ is unknown, we consider not the correlation functions $\langle ^rx_M(t)^rx_M(t+\tau)\rangle$ and $\overline{^rx_M(t)^rx_M(t+\tau)}$, but their means in relation to ϕ, considered as a random variable, distributed in a uniform manner in the time interval $(0 \cdot 2\pi)$. It can be shown that the mean correlation functions thus defined are equal and dependent only on τ. The amplitudes c_{n_M} which appear in (1.26) are the amplitudes of a conventional random function with stationary and ergodic properties, and whose correlation function is exactly equal to the mean correlation function just defined. It should be noted that the mean of this conventional function may be different from zero ($c_{0_M}^2 = \tfrac{1}{2} c_p^2$).

It was stated earlier that the relation (1.18) is valid only up to the limit $T \to \infty$, that is for $\omega \to 0$, where ω is the frequency interval between the successive harmonics of the development (1.13). Let us take this development further:

$$^r x(t) = \sum_{n=1}^{\infty} {}^r a_n \cos n\omega t + \sum_{n=1}^{\infty} {}^r b_n \sin n\omega t \qquad (1.13)$$

with
$$\omega = 2\pi/2T \qquad (1.14)$$

$$\left.\begin{aligned} {}^r a_n &= \frac{1}{T} \int_{-T}^{T} {}^r x(t) \cos n\omega t \, dt \\ {}^r b_n &= \frac{1}{T} \int_{-T}^{T} x(t) \sin n\omega t \, dt \end{aligned}\right\} \qquad (1.27)$$

Let
$$n\omega = 2\pi f_n \qquad (1.28)$$

$$\omega = 2\pi \, \Delta f_n; \qquad \Delta f_n = \frac{1}{2T} \qquad (1.29)$$

$$\left.\begin{aligned} {}^r A(f_n) &= \frac{{}^r a_n}{\Delta f_n} = 2\pi \int_{-T}^{T} {}^r x(t) \cos 2\pi f_n t \, dt \\ {}^r B(f_n) &= \frac{{}^r b_n}{\Delta f_n} = 2\pi \int_{-T}^{T} {}^r x(t) \sin 2\pi f_n t \, dt \end{aligned}\right\} \qquad (1.30)$$

(1.22) and (1.30) give us:

$$^r c_n^2 = [{}^r A^2(f_n) + {}^r B^2(f_n)] \Delta f_n^2 = \frac{{}^r A^2(f_n) + {}^r B^2(f_n)}{2T} \Delta f_n$$

or
$$\frac{{}^r c_n^2}{2 \Delta f_n} = \frac{{}^r A^2(f_n) + {}^r B^2(f_n)}{4T}$$

If we pass to the limit for $\Delta f_n \to 0$ ($T \to \infty$) and make:

$$n(f) = \lim_{\Delta f_n \to 0} \frac{{}^r c_n^2}{2 \Delta f_n}$$
$$= \lim_{T \to \infty} \frac{\left[\int_{-T}^{T} {}^r x(t) \cos 2\pi f t \, dt\right]^2 + \left[\int_{-T}^{T} {}^r x(t) \sin 2\pi f t \, dt\right]^2}{T} \qquad (1.31)$$

(1.20) may be written in the form

$$\langle {}^r x^2(t) \rangle = \overline{{}^r x^2(t)} = \int_0^{\infty} n(f) \, df \qquad \text{(variance)} \quad (1.32)$$

This formula shows that the variance of a noise function $x(t)$ depends only on the function $n(f)$ which characterizes $x(t)$. The function $n(f)$ is termed the *spectral noise density* of the function $x(t)$. The following relations may be established by a similar reasoning.

Filtering:
$$n_F(f) = H^2(f)n(f) \qquad (1.33)$$

Differentiation
$$n_D(f) = (2\pi f)^2 n(f) \qquad (1.34)$$

Multiplication by $\cos(\omega_c t + \phi)$:
$$n_M(f) = \tfrac{1}{4}[n(f - f_c) + n(f + f_c)] \qquad (1.35)$$

Note, in connection with formula (1.35), that in accordance with definition (1.31), we have:
$$n(-f) = n(f) \qquad (1.36)$$

and that consequently
$$\int_0^\infty n(f)\,\mathrm{d}f = \tfrac{1}{2}\int_{-\infty}^\infty n(f)\,\mathrm{d}f$$

Since, on the other hand,
$$\int_{-\infty}^\infty n(f - f_c)\,\mathrm{d}f = \int_{-\infty}^\infty n(f + f_c)\,\mathrm{d}f = \int_{-\infty}^\infty n(f)\,\mathrm{d}f$$

it is easy to see that
$$\int_0^\infty n_M(f)\,\mathrm{d}f = \tfrac{1}{2}\int_0^\infty n(f)\,\mathrm{d}f \qquad (1.37)$$

which shows that the variance of $x_M(t)$ is equal to half that of $x(t)$. If one of the two side bands of $x_M(t)$ is suppressed by filtering, the variance of the remaining band is equal to one-quarter of the variance of $x(t)$.

After this short review of the basic properties of noise functions, in which it has been thought useful for the sake of clarity to use semi-intuitive rather than strictly proved concepts [2], we may now proceed to apply the formulae thus obtained to a special and important example.

1.3. Narrow-band signal with high signal-to-noise ratio

The following conditions are satisfied in a large number of actual cases. (Hypothesis III.)

(a) The width of the frequency band occupied by the modulated signal is small compared with the centre frequency.

(b) The ratio between the square of the minimum effective instantaneous amplitude of the useful signal, and the variance of the parasitic signal superimposed on the useful signal is high.

This being so, what, under these conditions, are the relations between the parameters characterizing the random variable $m_a(t)$ for a given value of t, and the parameters characterizing the total signal reaching the detector, that is, the signal flowing from the output stage of the amplification chain before the detector?

Two cases will be considered:

(1) *Amplitude modulation*

Let

$$^r x(t) = \sum_{n=1}^{\infty} {^r a_n} \cos n\omega t + \sum_{n=1}^{\infty} {^r b_n} \sin n\omega t \tag{1.13}$$

$$s(t) = A(1 + m \cos qt) \cos \omega_c t \tag{1.38}$$

represent the parasitic and useful signals respectively emanating from the amplification chain, and let $(f_c - B_1, f_c + B_1)$ represent the bandwidth of this chain. (Condition (a).)

Let

$$\left. \begin{aligned} \omega_c &= n_c \omega \\ \nu &= n_c - n \\ 2\pi B_1 &= \nu_1 \omega \end{aligned} \right\} \tag{1.39}$$

(n_c being assumed to be an integer) and

$$A(t) = A(1 + m \cos qt) \tag{1.40}$$

(1.13) and (1.38) become

$$^r x(t) = \sum_{\nu = -\nu_1}^{\nu_1} {^r a_{n_c + \nu}} \cos (n_c + \nu)\omega t + \sum_{\nu = -\nu_1}^{\nu_1} {^r b_{n_c + \nu}} \sin (n_c + \nu)\omega t \tag{1.41}$$

$$s(t) = A(t) \cos n_c \omega t \tag{1.42}$$

The summation in (1.41) is limited to the harmonics of frequency between $(n_c - \nu_1)\omega$ and $(n_c + \nu_1)\omega$.

Let us write (1.41) in the form

$$^r x(t) = {^r x_c}(t) \cos n_c \omega t - {^r x_s}(t) \sin n_c \omega t \tag{1.43}$$

with

$$\left.\begin{array}{l}{}^{r}x_c(t) = \displaystyle\sum_{\nu=-\nu_1}^{\nu_1} {}^{r}a_{n_c+\nu} \cos \nu\omega t + \sum_{\nu=-\nu_1}^{\nu_1} {}^{r}b_{n_c+\nu} \sin \nu\omega t \\ {}^{r}x_s(t) = \displaystyle\sum_{\nu=-\nu_1}^{\nu_1} {}^{r}a_{n_c+\nu} \sin \nu\omega t - \sum_{\nu=-\nu_1}^{\nu_1} {}^{r}b_{n_c+\nu} \cos \nu\omega t\end{array}\right\} \quad (1.44)$$

At a given instant t, the complex amplitudes of the useful signal and of the components of the parasitic signal, in phase or in quadrature with the useful signal, are as shown in Fig. 1.1.

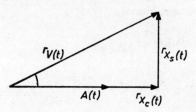

Figure 1.1

Taking condition (b) into account (high signal-to-noise ratio), it follows from Fig. 1.1 that the amplitude ${}^{r}V(t)$ of the resulting signal is almost exactly equal to

$${}^{r}V(t) = A(t) + {}^{r}x_c(t) \tag{1.45}$$

If we assume that the detecting unit extracts the value ${}^{r}V(t) - A$ from the energizing signal, and if we designate by $H(\nu\omega)$ and $\phi(\nu\omega)$ respectively the amplitude gain and the phase shift produced by the LF filter after the detector, we have, by virtue of (1.40), (1.44), (1.45):

$${}^{r}m_d(t) = H(q)mA \cos [qt + \phi(q)] + \sum_{\nu=-\nu_0}^{\nu_0} H(\nu\omega) {}^{r}a_{n_c+\nu} \cos [\nu\omega t + \phi(\nu\omega)]$$

$$+ \sum_{\nu=-\nu_0}^{\nu_0} H(\nu\omega) {}^{r}b_{n_c+\nu} \sin [\nu\omega t + \phi(\nu\omega)] \quad (1.46)$$

with

$$2\pi B = \nu_0 \omega \tag{1.47}$$

where $2B$ is the bandwidth of the LF filter ($\nu_0 \leqslant \nu_1$). In (1.46) the summation is effected from $-\nu_0$ to ν_0. Equation (1.46) gives us, taking hypothesis II into account,

$$\langle {}^{r}m_d(t)\rangle^2 = \{H(q)mA \cos [qt + \phi(q)]\}^2 \tag{1.48}$$

$$\sigma^2_{m_d(t)} = \langle [{}^{r}m_d(t) - \langle {}^{r}m_d(t)\rangle]^2 \rangle = \sum_{\nu=-\nu_0}^{\nu_0} H^2(\nu\omega) \frac{c^2_{n_c+\nu}}{2} \tag{1.49}$$

BACKGROUND NOISE 15

Due to (1.31) this last relation may be written:

$$\sigma^2_{m_d(t)} = \int_{-B}^{B} H^2(f) n(f_c + f) \, df \qquad (1.50)$$

where $n(f_c + f)$ is the spectral noise density of the parasitic signal emanating from the amplification chain. Relations (1.49) and (1.50) now enable us to clarify the ratio S/N in the case under consideration (the signal defined by (1.38) is adopted as the normalized useful signal):

$$\frac{S}{N} = \frac{\overline{\langle {^r}m_d(t) \rangle^2}}{\sigma^2_{m_d(t)}} = \frac{\overline{\{H(q)mA \cos[qt + \phi(q)]\}^2}}{\sigma^2_{m_d(t)}} = \frac{[H(q)mA]^2}{2\sigma^2_{m_d(t)}} \qquad (1.51)$$

($\sigma^2_{m_d(t)}$ is independent of t insofar, of course, as the condition (b) is satisfied in the dip of modulation.) From (1.51) we obtain the expression for the noise factor

$$F_{AM} = \frac{[S/N]_{\text{ideal}}}{[S/N]_{\text{actual}}} = \frac{[\sigma^2_{m(t)}]_{\text{actual}}}{[\sigma^2_{m_d(t)}]_{\text{ideal}}} \qquad (1.52)$$

Hence, following from (1.50)

$$F_{AM} = \frac{\int_{-B}^{B} H^2(f)[n(f_c + f)]_{\text{actual}} \, df}{\int_{-B}^{B} H^2(f)[n(f_c + f)]_{\text{ideal}} \, df} \qquad (1.53)$$

(2) *Frequency modulation*

Let

$$^r x(t) = \sum {^r}a_n \cos n\omega t + \sum {^r}b_n \sin n\omega t \qquad (1.14)$$

$$s(t) = A \cos(\omega_c t + \theta) \quad \text{with} \quad \frac{d\theta}{dt} = m\omega_D \cos qt \qquad (1.54)$$

represent the parasitic and useful signals respectively emanating from the amplification chain and let $(f_c - B_1, f_c + B_1)$ represent the bandwidth of this chain. At a given instant t and for a time interval Δt sufficiently short for the variation of $d\theta/dt$ during this interval to be ignored, (1.54) may be written

$$s(t') = A \cos(\omega_t t' + \theta_t) \qquad (1.55)$$

with

$$\left.\begin{array}{l} 0 < t' < \Delta t \\[4pt] \omega_t = \omega_c + m\omega_D \cos qt \\[4pt] \theta_t = \omega_c t + m\omega_D \displaystyle\int_0^t \cos qt \, dt \end{array}\right\} \qquad (1.56)$$

In order to simplify the writing of the expressions, we may ignore θ_t and take
$$s(t') = A \cos \omega_t t' \qquad (1.57)$$
Let
$$\left.\begin{array}{l} \omega_c = n_c \omega \\ \omega_t = n_t \omega \\ \nu = n - n_t \\ 2\pi B_1 = \nu_1 \omega \\ n_t - n_c = \nu_{tc} \\ \nu' = \nu_1 + \nu_{tc} \\ \nu'' = \nu_1 - \nu_{tc} \end{array}\right\} \qquad (1.58)$$

(1.13) and (1.57) become
$$^r x(t') = \sum_{\nu=-\nu'}^{\nu''} {}^r a_{n_t+\nu} \cos(n_t + \nu)\omega t' + \sum_{\nu=-\nu'}^{\nu''} {}^r b_{n_t+\nu} \sin(n_t + \nu)\omega t' \qquad (1.59)$$

$$s(t') = A \cos n_t \omega t' \qquad (1.60)$$

In (1.59) the summation is limited to the harmonics of which the frequency lies between $(n_c - \nu_1)\omega$ and $(n_c + \nu_1)\omega$ (the marking of the frequency axis in values of ν is shown in Fig. 1.2).

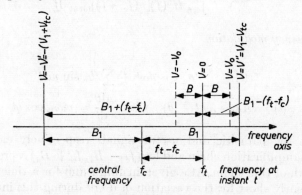

Figure 1.2

(1.59) may be written in the form
$$^r x(t') = {}^r x_c(t') \cos n_t \omega t' - {}^r x_s(t') \sin n_t \omega t' \qquad (1.61)$$
with
$$\left.\begin{array}{l} {}^r x_c(t') = \displaystyle\sum_{\nu=-\nu'}^{\nu''} {}^r a_{n_t+\nu} \cos \nu\omega t' + \sum_{\nu=-\nu'}^{\nu''} {}^r b_{n_t+\nu} \sin \nu\omega t' \\ {}^r x_s(t') = \displaystyle\sum_{\nu=-\nu'}^{\nu''} {}^r a_{n_t+\nu} \sin \nu\omega t' - \sum_{\nu=-\nu'}^{\nu''} {}^r b_{n_t+\nu} \cos \nu\omega t' \end{array}\right\} \qquad (1.62)$$

BACKGROUND NOISE   17

The complex amplitudes of the useful signal and of the components of the parasitic signal are as shown in Fig. 1.3. If we take condition (b) into account we see that according to Fig. 1.3, the phase shift $^r\alpha(t')$ of the resultant signal with respect to the usable signal is, to a good approximation, equal to:

$$^r\alpha(t') = \frac{^rx_s(t')}{A} \tag{1.63}$$

If we assume that the detecting device extracts from the energizing signal the value of $\omega_t + (d\alpha/dt') - \omega_c$ and if, as before, we represent

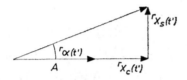

Figure 1.3

by $H(\nu\omega)$ and $\phi(\nu\omega)$ respectively, the amplitude gain and the phase shift produced by the LF filter after the detector, then by virtue of (1.56), (1.62), (1.63), we have:

$$^rm_a(t) = H(q)m\omega_D \cos[qt + \phi(q)]$$

$$+ \frac{1}{A} \sum_{\nu=-\nu_0}^{\nu_0} H(\nu\omega)\nu\omega\, ^ra_{n_t+\nu} \cos[\nu\omega t' + \phi(\nu\omega)]$$

$$+ \frac{1}{A} \sum_{\nu=-\nu_0}^{\nu_0} H(\nu\omega)\nu\omega\, ^rb_{n_t+\nu} \sin[\nu\omega t' + \phi(\nu\omega)] \tag{1.64}$$

In this formula the summation is carried out from $-\nu_0$ to ν_0, for, usually, $\nu_0 \leqslant \nu''$ (Fig. 1.2), even for 100% modulation, when $f_t - f_c$ reaches its maximum value f_D ($B \leqslant B_1 - f_D$). If we take account of hypothesis II, (1.64) gives us:

$$\langle ^rm_a(t)\rangle^2 = \{H(q)m\omega_D \cos[qt + \phi(q)]\}^2 \tag{1.65}$$

$$\sigma^2_{m_a(t)} = \frac{1}{A^2} \sum_{\nu=-\nu_0}^{\nu_0} H^2(\nu\omega)\nu^2\omega^2 \frac{C_{n_t+\nu}}{2} \tag{1.66}$$

By virtue of (1.64) this last equation may be written:

$$\sigma^2_{m_a(t)} = \frac{(2\pi)^2}{A^2} \int_{-B}^{B} H^2(f)f^2 n(f_t + f)\, df \tag{1.67}$$

(1.66) and (1.67) give us:

$$\frac{S}{N} = \frac{[H(q)m\omega_D]^2}{2\overline{\sigma^2_{m_a(t)}}} \tag{1.68}$$

Formula (1.67) shows that if, for $(f_c - B_1 < f_t + f < f_c + B_1)$, we have approximately:

$$n(f_t + f) = n(f_c)$$

the variance $\sigma^2_{m_a(t)}$ is independent of t. If this does not hold, it is usual in order to normalize the value of S/N to replace the time mean of $\sigma^2_{m_a(t)}$ by the value of $\sigma^2_{m_a(t)}$ for $f_t = f_c$. In these conditions the noise factor is given by:

$$F_{\rm FM} = \frac{\int_{-B}^{B} H^2(f) f^2 [n(f_c + f)]_{\rm actual}\, df}{\int_{-B}^{B} H^2(f) f^2 [n(f_c + f)]_{\rm ideal}\, df} \tag{1.69}$$

Note 1. Formulae (1.53) and (1.69) show that when the spectral density $n(f_c + f)$ is constant in the frequency range $(f_c - B, f_c + B)$, the noise factor of an amplifier inserted in a transmission chain does not depend, for a given spectral density $n(f_c)$, on the type of modulation used in this chain. In fact, when

$$n(f_c + f) = n(f_c) \quad \text{for} \quad -B < f < B$$

then from (1.53) and (1.69) we obtain:

$$F_{\rm AM} = F_{\rm FM} = \frac{n(f_c)_{\rm actual}}{n(f_c)_{\rm ideal}} \tag{1.70}$$

Note 2. Unlike the noise factor of an amplifier, the signal-to-noise ratio of a transmission chain depends to a great extent on the type of modulation used. We will consider as an example the improvement in S/N produced by frequency modulation over that with amplitude modulation. We will assume that in both cases the characteristics of the LF filter and the values of A, m and $n(f)$ respectively are the same. With this assumption we obtain from (1.50), (1.51) and from (1.67), (1.68):

$$\frac{(S/N)_{\rm FM}}{(S/N)_{\rm AM}} = \frac{f_D^2 \int_{-B}^{B} H^2(f) n(f_c + f)\, df}{\int_{-B}^{B} H^2(f) f^2 n(f_c + f)\, df} \tag{1.71}$$

(S/N ratios normalized).

When $-B < f < B$ we have, to a good approximation

$$\left. \begin{array}{l} H(f) = H(0) \\ n(f_c + f) = n(f_c) \end{array} \right\} \tag{1.72}$$

Relation (1.71) becomes:

$$\frac{(S/N)_{FM}}{(S/N)_{AM}} = 3\left(\frac{f_D}{B}\right)^2 \tag{1.73}$$

The factor $3(f_D/B)^2$ is called the *improvement factor*. It should be noted that the improvement obtained is at the expense of an increase in the width of the frequency band occupied by the modulated signal ($2B_1$), which is approximately equal to $2B$ for amplitude modulation and usually exceeds $2(f_D + B)$ for frequency modulation.

Note 3. As far as the distribution law of the random variables $[m_d(t)]_{AM}$ and $[m_d(t)]_{FM}$ is concerned, for a given value of t, it should be noted that if, in addition to conditions (a) and (b) certain other conditions are satisfied (validity of the central limit theory in the calculation of probabilities), which are generally true in practice, these variables are distributed approximately in accordance with a Gaussian distribution about their mean values, which are equal to $[H(q)mA \cos qt]$ and $[H(q)m\omega_D \cos qt]$ respectively.

The formulae established in this part of the book will enable us to pass to the analysis of the noise factor.

2. Noise factor and effective noise temperature of an amplifier

It was seen in the last section that for a narrow-band signal with a high signal-to-noise ratio, the noise factor of an amplifier inserted in a transmission chain is given by:

$$F_{AM} = \frac{\int_{-B}^{B} H^2(f)[n(f_c + f)]_{actual} \, df}{\int_{-B}^{B} H^2(f)[n(f_c + f)]_{ideal} \, df}$$
(amplitude modulation) (1.53)

$$F_{FM} = \frac{\int_{-B}^{B} H^2(f)f^2[n(f_c + f)]_{actual} \, df}{\int_{-B}^{B} H^2(f)f^2[n(f_c + f)]_{ideal} \, df}$$
(frequency modulation) (1.69)

where f_c is the central frequency,
$2B$ is the pass band of the LF filter after the detector,
$H(f)$ is the amplitude gain of the LF filter at frequency f,
$[n(f_c + f)]_{actual}$ is the spectral noise density of the parasitic signal produced at the detector input by the total noise sources (both external and internal) contained in the transmission chain,
$[n(f_c + f)]_{ideal}$ is the spectral noise density of the parasitic signal which would be obtained if the intrinsic noise sources of the amplifier concerned were assumed to be suppressed.

In order to determine the parasitic signal which the internal noise sources in the amplifier produce at the detector stage, it is usual to replace the amplifier by an *equivalent circuit*. This will be discussed in the next paragraph.

2.1. Equivalent circuits of a two-port containing internal noise sources

2.1.1. EQUIVALENT CIRCUIT USING TWO CORRELATED NOISE GENERATORS

Let us consider a two-port Q (Fig. 1.4) consisting of internal noise sources, and let

$$\left.\begin{aligned} V_1 &= Z_{11}I_1 + Z_{12}I_2 \\ V_2 &= Z_{21}I_1 + Z_{22}I_2 \end{aligned}\right\} \quad (1.74)$$

be the equations of this two-port when the internal noise sources are assumed to be suppressed $(Q)_{\text{ideal}}$. We will now proceed to show that

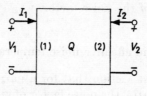

Figure 1.4

with the aid of Thévenin's theorem it is possible to replace the two-port Q by an equivalent circuit that takes account of the internal noise sources of the two-port concerned.

Let $x_1(t)$ and $x_2(t)$ represent the random voltages generated by these sources at the terminals of the ports (1) and (2) respectively of the two-port, when the ports are assumed to be open. In order to make Thévenin's theorem applicable we will pass from the temporal to the frequency domain.

Let us consider an infinitely narrow frequency range $(f, f + \Delta f)$ and let $e_{1_f}\sqrt{\Delta f}(e_{2_f}\sqrt{\Delta f})$ represent the complex effective amplitude of the corresponding spectral component of $x_1(t)$ (of $x_2(t)$).† Since the functions $x_1(t)$ and $x_2(t)$ are *random functions*, the magnitudes e_{1_f} and e_{2_f} corresponding to the value selected for f, are *random variables*. Consequently, the statistical properties of $x_1(t)$ and $x_2(t)$ may be expressed by the parameters which characterize the statistical distribu-

† See section 1.2.

tion of the values e_{1_f} and e_{2_f} for each frequency. From the point of view we are now concerned with, the essential parameters are:

(1) *The spectral noise densities*: $n_1(f) = \langle e_{1f} e_{1f}^* \rangle$ and $n_2(f) = \langle e_{2_f} e_{2_f}^* \rangle$ (real magnitudes), $\langle u \rangle$ and u^* which represent the statistical mean† and the conjugate magnitude of u respectively.

(2) *The mutual spectral noise density*: $n_{12}(f) = \langle e_{1_f} e_{2_f}^* \rangle$ (complex magnitude). The form of the functions $n_1(f)$, $n_2(f)$ and $n_{12}(f)$ depend on the structure of the two-port and on the nature of the internal noise sources. Depending on the case under consideration, these functions may be determined either theoretically or empirically. The introduction of the random spectral amplitudes $e_{1_f}\sqrt{\Delta f}$ and $e_{2_f}\sqrt{\Delta f}$ means that Thévenin's theorem is applicable. The resulting equivalent circuit is formed by the two-port Q_{ideal} to which two generators of random e.m.f. are connected in parallel externally: one in series with port (1) and the other in series with port (2) (Fig. 1.5), the spectral random amplitudes of these e.m.f.s being exactly $e_{1_f}\sqrt{\Delta f}$ and $e_{2_f}\sqrt{\Delta f}$ (for convenience in setting out we will, in what follows, adopt e or e' for the notation e_f).

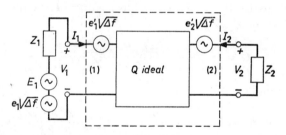

Figure 1.5

We will now connect in parallel with port (1) of the equivalent circuit of the two-port, a voltage generator having an internal impedance $Z_1 = R_1 + jX_1$, and with port (2) a load $Z_2 = R_2 + jX_2$ (Fig. 1.5). We will assume that the e.m.f. of the voltage generator has two components: one, of effective amplitude E_1, is the useful signal, and the other, of effective (random) spectral amplitude $e_1\sqrt{\Delta f}$, is the parasitic signal from the group of noise sources in front of the equivalent circuit (including Z_1). Ignoring the useful signal but taking account of (1.74), we find:

$$\left.\begin{array}{r}e_1 + e_1' = (Z_{11} + Z_1)i_1 + Z_{12}i_2 \\ e_2' = Z_{21}i_1 + (Z_{22} + Z_2)i_2\end{array}\right\} \quad (1.75)$$

† See Section 1.2.

Hence:
$$i_2 = \frac{-Z_{21}(e_1 + e_1') + (Z_{11} + Z_1)e_2'}{(Z_{11} + Z_1)(Z_{22} + Z_2) - Z_{12}Z_{21}} \tag{1.76}$$

The magnitude which interests us for the determination of the noise factor is the spectral density of the noise supplied to the load Z_2:
$$p = \langle i_2 \, i_2^* \rangle R_2 \tag{1.77}$$

If we take account of (1.76) and of the fact that e_1 is correlated neither with e_1' nor with e_2', ($\langle e_1 \, e_1'^* \rangle = \langle e_1 \, e_2'^* \rangle = 0$), we obtain:

$$p = \frac{|Z_{21}|^2(n_1 + n_1') + |Z_{11} + Z_1|^2 n_2' - 2\,\mathrm{Re}\, Z_{21}(Z_{11}^* + Z_1^*)n_{12}'}{|(Z_{11} + Z_1)(Z_{22} + Z_2) - Z_{12}Z_{21}|^2} R_2$$

(Re represents the real part) (1.78)

with:
$$\left. \begin{aligned} n_1 &= \langle e_1 \, e_1^* \rangle \\ n_1' &= \langle e_1' \, e_1'^* \rangle \\ n_2' &= \langle e_2' \, e_2'^* \rangle \\ n_{12}' &= \langle e_1' \, e_2'^* \rangle \end{aligned} \right\} \tag{1.79}$$

If we designate, by G, the transducer gain of the equivalent circuit, terminated by Z_1, Z_2:

$$G = \frac{4R_1 R_2 |Z_{21}|^2}{|(Z_{11} + Z_1)(Z_{22} + Z_2) - Z_{12}Z_{21}|^2}$$

and if we make

$$n' = n_1' + \left|\frac{Z_{11} + Z_1}{Z_{21}}\right|^2 n_2' - 2\,\mathrm{Re}\, \frac{Z_{11}^* + Z_1^*}{Z_{21}^*} n_{12}'$$

(1.78) may be written:
$$p = \frac{G}{4R_1}(n_1 + n_1') \tag{1.80}$$

It should be noted that, unlike the densities n_1', n_2' and n_{12}' which are independent of the terminations Z_1, Z_2, the density n' depends on Z_1, and consequently does not constitute a characteristic magnitude peculiar to the two-port.

2.1.2. EQUIVALENT CIRCUIT MAKING DIRECT USE OF THE NOISE GENERATORS OF THE TWO-PORT

Let us consider a two-port Q in which the internal noise sources resemble voltage generators of given internal impedance and e.m.f.

(more precisely, of noise spectral density of the latter). We will now show that, in this case, it is possible to obtain an equivalent circuit for the two-port Q directly from the network formed by the components of the two-port. Let us then remove all these generators of internal noise from the two-port Q. This operation transforms the two-port Q into an ideal multiport M. If we terminate each

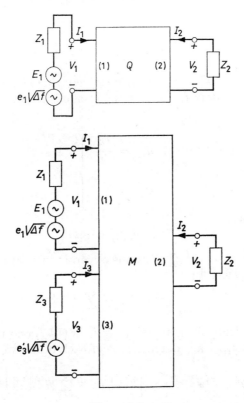

Figure 1.6

port of M, produced by the removal of a noise generator, by this same generator, we reconstruct the two-port Q. The equivalent circuit of Fig. 1.6 is the result. The circuit shown refers to the case of a single generator of internal noise; this generator is connected at port (3) of the multiport M.

Therefore, let:

$$\left.\begin{aligned} V_1 &= Z_{11}^M I_1 + Z_{12}^M I_2 + Z_{13}^M I_3 \\ V_2 &= Z_{21}^M I_1 + Z_{22}^M I_2 + Z_{23}^M I_3 \\ V_3 &= Z_{31}^M I_1 + Z_{32}^M I_2 + Z_{33}^M I_3 \end{aligned}\right\} \qquad (1.81)$$

represent the equations of the multiport M. If the useful signal is suppressed, we find:

$$\left.\begin{aligned} e_1 &= (Z_{11}^M + Z_1)i_1 + Z_{12}^M i_2 + Z_{13}^M i_3 \\ 0 &= Z_{21}^M i_1 + (Z_{22}^M + Z_2)i_2 + Z_{23}^M i_3 \\ e_3' &= Z_{31}^M i_1 + Z_{32}^M i_2 + (Z_{33}^M + Z_3)i_3 \end{aligned}\right\} \quad (1.82)$$

Hence:

$$i_2 = \frac{D_{(2)}^{(1)}}{D} e_1 + \frac{D_{(2)}^{(3)}}{D} e_3' \quad (1.83)$$

in which D and $D_{(2)}^{(1)}$, $D_{(2)}^{(3)}$ are respectively the determinant and the minors of the matrix formed by the coefficients of (1.82).

Let

$$\left.\begin{aligned} d_{(2)}^{(1)} &= \frac{D_{(2)}^{(1)}}{D} \\ d_{(2)}^{(3)} &= \frac{D_{(2)}^{(3)}}{D} \end{aligned}\right\} \quad (1.84)$$

(1.83) becomes

$$i_2 = d_{(2)}^{(1)} e_1 + d_{(2)}^{(3)} e_3' \quad (1.85)$$

When the two-port Q contains several noise generators, (1.85) should be replaced by

$$i_2 = d_{(2)}^{(1)} e_1 + \sum_{i>2} d_{(2)}^{(i)} e_1' \quad (1.86)$$

If we substitute the expression for i_2 in (1.77) and take account of the fact that e_1, e_i' are independent of each other, we obtain:

$$p = \langle i_2 i_2^* \rangle R_2 = \left[|d_{(2)}^{(1)}|^2 n_1 + \sum_{i>2} |d_{(2)}^{(i)}|^2 n_i' \right] R_2 \quad (1.87)$$

with

$$n_1 = \langle e_1 e_i^* \rangle$$
$$n_i' = \langle e_i' e_i'^* \rangle$$

Introducing the transducer gains of the multiport M : port $(r) \to$ port (2)

$$G_{(2)}^{(r)} = \frac{|d_{(2)}^{(r)}|^2 R_2 |e_r|^2}{|e_r|^2/4R_r} = 4R_2 R_r |d_{(2)}^{(r)}|^2$$

Hence

$$|d_{(2)}^{(r)}|^2 = \frac{G_{(2)}^{(r)}}{4R_2 R_r} \quad (1.88)$$

which gives, after substitution in (1.87)

$$p = \frac{G_{(2)}^{(1)}}{4R_1} n_1 + \sum_{i>2} \frac{G_{(2)}^{(i)}}{4R_i} n'_i$$

and noting that $G_{(2)}^{(1)} = G$,

$$p = \frac{G}{4R_1}\left[n_1 + \sum_{i>2} \frac{R_1}{R_i} \cdot \frac{G_{(2)}^{(i)}}{G} n'_i\right] \qquad (1.89)$$

Comparing (1.89) with (1.80), we see that in the present case the expression n' is:

$$n' = \sum_{i>2} \frac{R_1}{R_i} \frac{G_{(2)}^{(i)}}{G} n'_i \qquad (1.90)$$

and just as in the previous paragraph we obtain:

$$p = \frac{G}{4R_1} (n_1 + n') \qquad (1.80)$$

If we note that

$$[p]_{\text{actual}} = \frac{G}{4R_1} (n_1 + n') \qquad (1.91)$$

$$[p]_{\text{ideal}} = \frac{G}{4R_1} n_1$$

we are now in a position to clarify formulae (1.53) and (1.69) which give the noise factor.

2.2. Noise factor of a two-port

In order to facilitate our study, we propose from now on to confine our attention to amplitude modulation; frequency modulation can be treated in just the same way. Before explaining formula (1.53), let us replace $n(f_c + f)$ by $p(f_c + f)$, and $H^2(f)$ by the transducer gain $G_0(f)$ of the LF filter; we now obtain (omitting the index AM):

$$F = \frac{\int_{-B}^{B} G_0(f)[p(f_c + f)]_{\text{actual}}\, df}{\int_{-B}^{B} G_0(f)[p(f_c + f)]_{\text{ideal}}\, df} \qquad (1.92)$$

If we substitute in (1.92) the expressions (1.91) of p, we find:

$$F = 1 + \frac{\int_{-B}^{B} G(f_c + f) G_0(f) n'(f_c + f)\, df}{\int_{-B}^{B} G(f_c + f) G_0(f) n_1(f_c + f)\, df} \qquad (1.93)$$

2.2.1. NORMALIZATION OF THE NOISE FACTOR

Formula (1.93) shows that given a two-port and its terminations, the value of the noise factor F depends on n_1. Also, in order to normalize the expression for F, it is usual to take the case in which the parasitic signal at the input of the two-port is reduced to only the thermal noise of the impedance Z_1. It will be seen in section C that the spectral noise density of the e.m.f. generated by the thermal agitation in an impedance $Z = R + jX$, is independent of the frequency and is equal to:

$$n = \langle e\,e^* \rangle = 4kTR$$

where k is the Boltzmann constant and T the absolute temperature of the one-port (in degrees K). Finally, by a last convention let us take T as $T_0 = 290°K$. Thus, renormalized, the noise factor is given by:

$$F = 1 + \frac{\int_{-B}^{B} G(f_c + f)G_0(f)n'(f_c + f)\,\mathrm{d}f}{4kT_0 R_1 \int_{-B}^{B} G(f_c + f)G_0(f)\,\mathrm{d}f} \tag{1.94}$$

2.2.2. NOISE BANDWIDTH OF A TWO-PORT

The last formula may be written in a different form if we introduce the *noise bandwidth* of a two-port. This is defined by:

$$B_N = \frac{\int_{-B}^{B} G(f_c + f)G_0(f)\,\mathrm{d}f}{G(f_c)G_0(0)} \tag{1.95}$$

which enables us to write:

$$F = 1 + \frac{\int_{-B}^{B} G(f_c + f)G_0(f)n'(f_c + f)\,\mathrm{d}f}{4kT_0 R_1 G(f_c)G_0(0)B_N} \tag{1.96}$$

2.2.3. EFFECTIVE NOISE TEMPERATURE OF A TWO-PORT

A third expression for F may be obtained by the introduction of the *effective noise temperature* of a two-port. This is defined by:

$$T_{\text{eff}} = \frac{\int_{-B}^{B} G(f_c + f)G(f)n'(f_c + f)\,\mathrm{d}f}{4kR_1 \int_{-B}^{B} G(f_c + f)G_0(f)\,\mathrm{d}f} \tag{1.97}$$

which leads to:

$$F = 1 + \frac{T_{\text{eff}}}{T_0} \tag{1.98}$$

This formula shows us that the effective noise temperature, like the noise factor, may serve as a criterion for the quality of an amplifier. As will be seen in section 4.2 of Chapter 2, when frequency converters are studied, T_{eff} serves as a more convenient criterion than F. Figure 1.7 gives the relation between T_{eff} and F.

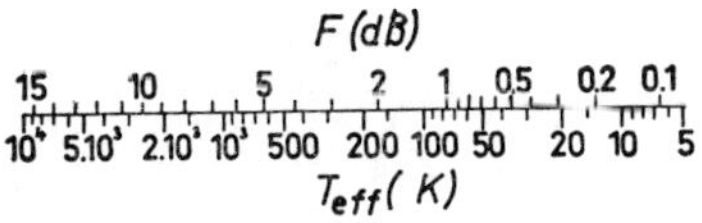

```
           F (dB)
 15    10     5    2   1  0.5   0.2 0.1
|·|·|·|·|·|·|·|·|·|·|·|·|·|·||·|·| ' |·|·|·|
10⁴ 5.10³ 2.10³ 10³ 500 200 100 50  20 10   5
              T_eff( K)
```

Figure 1.7

2.2.4. SPECTRAL NOISE FACTOR AND EFFECTIVE SPECTRAL NOISE TEMPERATURE OF A TWO-PORT

If, instead of considering the frequency band $(f_c - B, f_c + B)$, we confine ourselves to an infinitely narrow frequency band $(f_c + f, f_c + f + \Delta f)$, (1.94), (1.97) can be reduced to:

$$F(f_c + f) = 1 + \frac{n'(f_c + f)}{4kT_0 R_1} \tag{1.99}$$

$$T_{eff}(f_c + f) = \frac{n'(f_c + f)}{4kR_1} \tag{1.100}$$

These relations define respectively the *spectral noise factor* and the *effective spectral noise temperature* of a two-port. When, for $-B < f < B$, we have almost exactly:

$$n'(f_c + f) = n'(f_c) \tag{1.101}$$

we obtain from (1.94), (1.97)

$$\left. \begin{array}{l} F = F(f_c) \\ T_{eff} = T_{eff}(f_c) \end{array} \right\} \tag{1.102}$$

Formulae (1.99), (1.100) show that the spectral noise factor and the effective spectral noise temperature of a given two-port depend only on the impedance Z_1, when n' is independent of Z_2. Noting that by virtue of (1.91) we have:

$$n' = \frac{[p]_{actual} - [p]_{ideal}}{G} 4R_1 \tag{1.103}$$

(1.100) may be written in the form

$$T_{eff}(f_c + f) = \frac{[p(f_c + f)]_{actual} - [p(f_c + f)]_{ideal}}{kG} \tag{1.104}$$

2.2.5. EQUIVALENT SPECTRAL NOISE TEMPERATURE OF A ONE-PORT

When the equivalent circuit employed is of the type described in section 2.1.2, that is, when the spectral density n' is given by:

$$n' = \sum_{i>2} \frac{R_1}{R_i} \frac{G_{(2)}^{(i)}}{G} n'_i \qquad (1.90)$$

formula (1.100) may be written in a more convenient form, by the introduction of the *equivalent spectral noise temperature* of a noise generator. Let $Z = R + jX$ represent the internal impedance of a noise generator, and n the spectral noise density of its e.m.f. The equivalent spectral noise temperature of a one-port of this nature is defined by:

$$n(f_c + f) = 4kT_{\text{eq}}(f_c + f)R \qquad (1.105)$$

From (1.90), (1.100), (1.105) we obtain:

$$T_{\text{eff}}(f_c + f) = \sum_{i>2} \frac{G_{(2)}^{(i)}}{G} T_{\text{eq}_i}(f_c + f) \qquad (1.106)$$

with

$$n'_i(f_c + f) = 4kT_{\text{eq}_i}(f_c + f)R_i \qquad (1.107)$$

2.3. Effective spectral noise temperature of a passive two-port at a uniform temperature

Let us consider a passive two-port Q, for example an attenuator terminated by impedances Z_1, Z_2, and let us suppose that this two-port is at a uniform temperature T. The problem is to find the effective spectral noise temperature of this two-port. Since this temperature does not depend on the temperature of the impedances Z_1 and Z_2, we will assume that these impedances are also at a temperature T, which will enable us to base our search for $T_{\text{eff}}(f_c + f)$ on the second law of thermodynamics. This law may be described thus: in a passive system at a uniform temperature, no *net* transfer of energy can take place from one part of the system to another, and this applies to each of the spectral components of the energy, considered separately (principle of equilibrium of parts).

It follows that in our particular case, the spectral density $P_{Q \to Z_2}$ of the power of the noise transferred from Q to Z_2 is equal to the spectral density $P_{Z_2 \to Q}$ of the noise power transferred from Z_2 to Q. We then have:

$$P_{Q \to Z_2} = P_{Z_2 \to Q} = \frac{\langle e_2 e_2^* \rangle R_0}{|Z_2 + Z_0|^2} = \frac{4kTR_2R_0}{|Z_2 + Z_0|^2} \qquad (1.108)$$

where Z_0 is the output impedence of the two-port Q, and $e_2\sqrt{\Delta f}$ the spectral random amplitude of the e.m.f. generated by the thermal agitation in Z_2. Let G represent the transducer gain of the two-port Q, terminated by impedances Z_1, Z_2, and G_a the transducer gain of the same two-port when it is terminated by Z_1, Z_0^* (matching of the impedances at port (2)). We know that:

$$\frac{G}{G_a} = \frac{4R_2R_0}{|Z_2 + Z_0|^2} \qquad (1.109)$$

which enables us to write (1.108) in the form

$$[P_{Q\to Z_2}]_{\text{actual}} = \frac{G}{G_a} kT \qquad (1.110)$$

Now let us assume that, while maintaining Z_1 at temperature T,

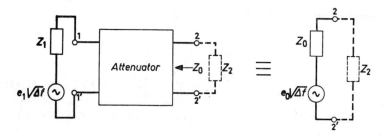

Figure 1.8

we suppress the noise sources of the two-port Q. In this case the spectral density $P_{Q\to Z_2}$ becomes:

$$[P_{Q\to Z_2}]_{\text{ideal}} = \frac{\langle e_1 e_1^* \rangle}{4R_1} G = \frac{4kTR_1}{4R_1} G = GkT \qquad (1.111)$$

where $e_1\sqrt{\Delta f}$ is the random spectral amplitude of the e.m.f. generated by the thermal agitation in Z_1. If we replace expressions (1.110) and (1.111) by $P_{Q\to Z_2}$ in (1.104) we find:

$$T_{\text{eff}}(f_c + f) = \left[\frac{1}{G_a(f_c + f)} - 1\right]T \qquad (1.112)$$

This formula will now be applied to the determination of the equivalent spectral noise temperature of the generator obtained if we place an attenuator at a temperature T (Fig. 1.8) after a noise generator of equivalent temperature T_1.

The spectral density p of the noise power that circuits I and II deliver to the load Z_2 is given by:

$$p = \frac{G}{4R_1}(n_1 + n') \quad \text{(circuit I)} \tag{1.80}$$

$$p = \frac{n_0 R_2}{|Z_2 + Z_0|^2} \quad \text{(circuit II)} \tag{1.113}$$

If we equate these two expressions for p, we find:

$$n_0 = \frac{|Z_2 + Z_0|^2}{4R_1 R_2} G(n_1 + n') \tag{1.114}$$

Now,

$$\left.\begin{array}{l} n_1 = 4kT_1 R_1 \\ n' = 4kT_{\text{eff}} R_1 \\ n_0 = 4kT_{\text{eq}} R_0 \end{array}\right\} \tag{1.115}$$

After substitution in (1.114) we obtain:

$$T_{\text{eq}} = \frac{|Z_2 + Z_0|^2}{4R_2 R_0} G(T_1 + T_{\text{eff}})$$

$$= \frac{|Z_2 + Z_0|^2}{4R_2 R_0}\left[GT_1 + \left(\frac{G}{G_a} - G\right)T\right] \tag{1.116}$$

When there is impedance matching at the output of the attenuator, (1.116) becomes:

$$T_{\text{eq}}(f_c + f) = G_a(f_c + f)T_1 + [1 - G_a(f_c + f)]T \tag{1.117}$$

2.4. Noise factor and effective noise temperature of a chain consisting of several two-ports

Let us consider a chain composed of two two-ports Q_1 and Q_2 connected in cascade (Fig. 1.9).

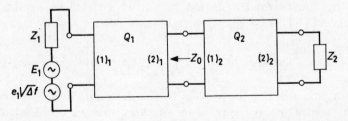

Figure 1.9

Let:

$Z_0(=R_0 + jX_0)$ be the impedance of Q_1 seen from the terminals of port $(2)_1$,

$G^{(r)_1}_{a(2)_1}$ - be the transducer gain, port $(r)_1 \to$ port $(2)_1$, of the multiport M_1 obtained from the two-port Q_1, the multiport M_1 being terminated by Z_1, Z_0^* (impedance matching at port $(2)_1$),

$G^{(r)_2}_{(2)_2}$ - be the transducer gain, port $(r)_2 \to (2)_2$, of the multiport M_2 obtained from the two-port Q_2, the multiport M_2 being terminated by Z_0, Z_2.

By a reasoning similar to that in section 2.1 we find that the spectral density of the noise power fed to the load Z_2 is given by:

$$p = \frac{G_{a_1} G_2}{4R_1}(n_1 + n_1') + \frac{G_2}{4R_0} n_2' \qquad (1.118)$$

with

$$\left. \begin{aligned} G_{a_1} &= G^{(1)_1}_{a(2)_1} \text{ (transducer gain of } Q_1 \text{ terminated by } Z_1, Z_0^*) \\ G_2 &= G^{(1)_2}_{(2)_2} \text{ (transducer gain of } Q_2, \text{ terminated by } Z_0, Z_2) \end{aligned} \right\} \qquad (1.119)$$

and

$$\left. \begin{aligned} n_1' &= \sum_{i>2} \frac{R_1}{R_i} \frac{G^{(i)_1}_{a(2)_1}}{G_{a_1}} n_{i_1}' \\ n_2' &= \sum_{i>2} \frac{R_0}{R_{i_2}} \frac{G^{(i)_2}_{(2)_2}}{G_2} n_{i_2}' \end{aligned} \right\} \qquad (1.120)$$

It follows that the normalized noise factor ($n_1 = 4kT_0 R_1$) of the chain Q_1, Q_2 is given by

$$F_{12} = 1 + \frac{\int_{-B}^{B} G_{a_1} G_2 G_0 n_1' \, df}{4kT_0 R_1 \int_{-B}^{B} G_{a_1} G_2 G_0 \, df} + \frac{\int_{-B}^{B} G_2 G_0 n_2' \, df}{4kT_0 R_0 \int_{-B}^{B} G_{a_1} G_2 G_0 \, df} \qquad (1.121)$$

If we assume that for $-B < f < B$, we have almost exactly

$$G_{a_1}(f_c + f) = G_{a_1}(f_c)$$

and

$$n_1'(f_c + f) = n_1'(f_c) \qquad (1.122)$$

which is generally the case (pass band of Q_1 considerably wider than $2B$), relation (1.121) becomes:

$$F_{12} = 1 + \frac{n'(f_c)}{4kT_0R_1} + \frac{1}{G_{a_1}(f_c)} \times \frac{\int_{-B}^{B} G_2 G_0 n_2' \, df}{4kT_0 R_0 \int_{-B}^{B} G_2 G_0 \, df} \quad (1.123)$$

If we designate by $F_1(F_2)$ the noise factor of the two-port $Q_1(Q_2)$ terminated by Z_1, $Z_0^*(Z_0, Z_2)$, we find:

$$F_{12} = F_1 + \frac{F_2 - 1}{G_{a_1}} \quad (1.124)$$

Hence

$$T_{\text{eff}_{12}} = T_{\text{eff}_1} + \frac{T_{\text{eff}_2}}{G_{a_1}} \quad (1.125)$$

When the chain contains r two-ports and when the bandwidth of the first $r - 1$ two-ports is considerably wider than $2B$, it can easily be shown that:

$$F_{12\ldots r} = F_1 + \frac{F_2 - 1}{G_{a_1}} + \cdots + \frac{F_r - 1}{G_{a_1} G_{a_2} \ldots G_{a_{r-1}}} \quad (1.124a)$$

$$T_{\text{eff}_{12\ldots r}} = T_{\text{eff}_1} + \cdots + \frac{T_{\text{eff}_r}}{G_{a_1} G_{a_2} \ldots G_{a_{r-1}}} \quad (1.125a)$$

We will now proceed to apply formulae (1.112) and (1.125) to determine the effective noise temperature of a chain formed by an amplifier A, preceded by an attenuator α. Let T_{eff_α}, T_{eff_A} and $T_{\text{eff}_{\alpha A}}$ respectively represent the effective noise temperatures of the attenuator α, of the amplifier A and of the chain α, A. From (1.123) we find:

$$T_{\text{eff}_{\alpha, A}} = T_{\text{eff}_\alpha} + \frac{T_{\text{eff}_A}}{G_{a_\alpha}}$$

whence, taking (1.112) into account:

$$T_{\text{eff}_{\alpha, A}} = \left(\frac{1}{G_{a_\alpha}} - 1\right)T + \frac{1}{G_{a_\alpha}} T_{\text{eff}_A} \quad (1.126)$$

2.5. Effective spectral noise temperature of a frequency converter

The definitions and relations discussed above apply only to linear two-ports. Nevertheless, they can be adapted for frequency converters, which involve the introduction of a nonlinear element. During the process of frequency changing, the incoming signal, of frequency f_1, combines in a non-linear element with a signal from the local oscillator, of frequency f_p, so as to create frequency components:

$$f = nf_p + mf_1 \quad (n, m \text{ integers}) \tag{1.127}$$

In parametric amplifiers used in practice only a small number of these components are used (filtering). We will limit our discussion to the case of two frequencies (in addition to f_p), namely:

sum-frequency converter $\quad\begin{cases} f_1 \\ f_2 = f_p + f_1 \end{cases}$

difference-frequency converter $\quad\begin{cases} f_1 \\ f_2 = f_p - f_1 \end{cases}$

Each of these converters may be represented schematically by a two-port C, as shown in Fig. 1.10, in which V_{11}, I_{11} and V_{21},

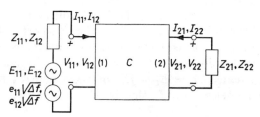

Figure 1.10

I_{21} (V_{12}, I_{12} and V_{22}, I_{22}) represent respectively the effective complex amplitudes of the frequency component $f_1(f_2)$ of the voltages and the currents at the ports (1) and (2) of the two-port C. As will be seen later, the 'small-signal' analysis of these converters shows that the two-port C may be represented by a system of linear equations. These equations, in their general form are:

$$\left.\begin{aligned}
V_{11} &= Z_{11}^{11} I_{11} + Z_{11}^{12} I_{12}^{(*)} + Z_{12}^{11} I_{21} + Z_{12}^{12} I_{22}^{(*)} \\
V_{12}^{(*)} &= Z_{11}^{21} I_{11} + Z_{11}^{22} I_{12}^{(*)} + Z_{12}^{21} I_{21} + Z_{12}^{22} I_{22}^{(*)} \\
V_{21} &= Z_{21}^{11} I_{11} + Z_{21}^{12} I_{12}^{(*)} + Z_{22}^{11} I_{21} + Z_{22}^{12} I_{22}^{(*)} \\
V_{22}^{(*)} &= Z_{21}^{21} I_{11} + Z_{21}^{22} I_{12}^{(*)} + Z_{22}^{21} I_{21} + Z_{22}^{22} I_{22}^{(*)}
\end{aligned}\right\} \tag{1.128}$$

in which $V^{(*)}$ signifies: replace V by V^* when the converter is of the spectrum inversion type (see Chapter 3). This being so, we can, with the aid of equations (1.128), once more apply the reasoning which, in section 2.1.2, gave us relation (1.86) which we repeat here, but replacing notations $d_{(2)}^{(r)}$ by d_2^r for the sake of simplicity.

$$i_2 = d_2^1 e_1 + \sum_{i>2} d_2^i e_i' \tag{1.86}$$

It can easily be shown that, in the case of a frequency converter, we find:

$$\left. \begin{array}{l} i_{21} = d_{21}^{11} e_{11} + d_{21}^{12} e_{12}^{(*)} + \sum_{i>2} (d_{21}^{i1} e_{i1}' + d_{21}^{i2} e_{i2}'^{(*)}) \\ i_{22} = d_{22}^{11} e_{11}^{(*)} + d_{22}^{12} e_{12} + \sum_{i>2} (d_{22}^{i1} e_{i1}'^{(*)} + d_{22}^{i2} e_{i2}') \end{array} \right\} \tag{1.129}$$

with (cf. 1.88)

$$|d_{2\beta}^{r\alpha}|^2 = \frac{G_{2\beta}^{r\alpha}}{4 R_{2\beta} R_{r\alpha}} \quad (\alpha, \beta = 1, 2) \tag{1.130}$$

where $G_{2\beta}^{r\alpha}$ represents the transducer gain, port (r) component $f_\alpha \to$ port (2) component f_β, of the ideal multiport M obtained by removing all the noise generators from the two-port C. It is assumed that the multiport M is terminated by impedances Z_1, Z_2 and by the internal noise generators of C. When the real part of the internal impedance $Z_i = R_i + jX_i$ of these sources has the same value for frequencies f_1 and f_2, that is, when $R_{i1} = R_{i2} = R_i$, we may write:

$$|d_{2\beta}^{i\alpha}|^2 = \frac{G_{2\beta}^{i\alpha}}{4 R_{2\beta} R_i} \quad (\alpha, \beta = 1, 2) \tag{1.131}$$

Depending on the relative values of the gains $G_{2\beta}^{r\alpha}$, we can distinguish between two types of converters, namely the degenerate and the non-degenerate converter.

2.5.1. NON-DEGENERATE CONVERTERS

In this type of converter the *currents* at the ports (1) and (2), but not necessarily at the other ports of the multiport M, contain only a single component, as a result of appropriate filtering. There are two versions to be considered:

(1) *Frequency-conversion amplifiers* (input signal of frequency f_1, output signal of frequency f_2). This type of amplifier is characterized by

$$\left. \begin{array}{l} i_{12} = 0 \quad (d_{22}^{12} = 0) \\ i_{21} = 0 \quad (d_{21}^{r\alpha} = 0) \end{array} \right\} \tag{1.132}$$

Relations (1.129) reduce to

$$\left.\begin{aligned} i_{21} &= 0 \\ i_{22} &= d_{22}^{11} e_{11}^{(*)} + \sum_{i<2} [d_{22}^{i1} e'_{i1} + d_{22}^{i2} e'_{i2}] \end{aligned}\right\} \quad (1.133)$$

The spectral density of the noise power fed to the load Z_2 is therefore equal to

$$p = \langle i_{22} i_{22}^* \rangle R_2$$

Substituting i_{22} in this equation by its expression (1.133) and taking account of the fact that e_{11}, e'_{i1}, e'_{i2} are independent of each other, we obtain:

$$p = |d_{22}^{11}|^2 n_{11} + \sum_{i>2} [|d_{22}^{i1}|^2 n'_{i1} + |d_{22}^{i2}|^2 n'_{i2}] R_2 \quad (1.134)$$

with:

$$\left.\begin{aligned} n_{11} &= \langle e_{11} e_{11}^* \rangle \\ n'_{i1} &= \langle e'_{i1} e'^*_{i1} \rangle \\ n'_{i2} &= \langle e'_{i2} e'^*_{i2} \rangle \end{aligned}\right\} \quad (1.135)$$

Assuming that:

$$\left.\begin{aligned} n_{11} &= 4kT_1 R_1 \\ n'_{i1} &= n'_{i2} = 4kT_i R_i \end{aligned}\right\} \quad (1.136)$$

we may write (1.134), by virtue of (1.131) in the form:

$$p = (G_{22}^{1N})_{nd} kT_1 + \sum_{i>2} G_{22}^{iN} T_i \quad (1.137)$$

with:

$$(G_{22}^{1N})_{nd} = G_{22}^{11} \quad (1.138)$$

$$G_{22}^{iN} = G_{22}^{i1} + G_{22}^{i2} \quad (1.139)$$

in which the subscript nd serves to remind us that we are dealing with a non-degenerate converter, (1.137) gives us:

$$[p]_{\text{ideal}} = (G_{22}^{1N})_{nd} kT_1 \quad (1.140)$$

$$[p]_{\text{actual}} - [p]_{\text{ideal}} = \sum_{i>2} G_{22}^{iN} kT_i \quad (1.141)$$

It should be noted that in the case we are considering we have:

$$(G_{22}^{1N})_{nd} = G(= G_{22}^{11}) \quad (1.142)$$

where G is the transducer gain of the converter. In these conditions, the effective temperature of this converter may be defined either by formula (1.104):

$$T_{\text{eff}} = \frac{[p]_{\text{actual}} - [p]_{\text{ideal}}}{kG} \quad (1.104)$$

or by the relation:

$$T_{\text{eff}} = \frac{[p]_{\text{actual}} - [p]_{\text{ideal}}}{kG^{1N}} \tag{1.143}$$

with:

$$[p]_{\text{ideal}} = G^{1N}kT_1 \tag{1.144}$$

When $G^{1N} \neq G$, the two definitions are not equivalent and it is only the definition (1.143) which is compatible in every case with formula (1.98).

$$F = 1 + \frac{T_{\text{eff}}}{T_0} \tag{1.98}$$

It is for this reason that we will use (1.143) as our general definition of the effective spectral noise temperature of an amplifier. (We will not discuss at this stage the modifications that this new definition of T_{eff} imposes, on the formulae relating to the connecting of the two-ports in cascade.) We may therefore deduce from (1.140) and (1.143):

$$(T_{\text{eff }\frac{1}{2}})_{nd} = \sum_{i>2} \frac{G_{22}^{iN}}{G_{22}^{11}} T_i \tag{1.145}$$

in which the indices 1 and 2 attached to T_{eff} remind us that we are dealing with the case, in which the input signal is of frequency f_1, and the output signal of frequency f_2.

(2) *Amplifiers without frequency conversion* (input signal and output signal of the same frequency f_1). This type of amplifier is characterized by:

$$\left. \begin{array}{ll} i_{12} = 0 & (d_{21}^{12} = 0) \\ i_{22} = 0 & (d_{22}^{r\alpha} = 0) \end{array} \right\} \tag{1.146}$$

which gives us:

$$\left. \begin{array}{l} (G_{21}^{1N})_{nd} = G_{21}^{11} = G \\ G_{21}^{iN} = G_{21}^{i1} + G_{21}^{i2} \\ (T_{\text{eff }1})_{nd} = \sum_{i>2} \frac{G_{21}^{iN}}{G_{21}^{11}} T_i \end{array} \right\} \tag{1.147}$$

2.5.2. DEGENERATE CONVERTERS

In this type of converter the currents at the ports (1) and (2) of the multiport M have two components, of frequencies f_1 and f_2 respectively. We will have to consider various methods of operation.

(1) SINGLE-BAND RECEPTION. Reception is called *single-band* when the useful incoming signal consists of only one component, say, of

frequency f_1. The corresponding output signal consists of two components, of frequencies f_1 and f_2, respectively. There are two cases to consider.

Case 1.1. A single component of the output signal is used, for example the component of frequency f_2. We then have:

$$p = \langle i_{22} i_{22}^* \rangle R_2$$

If we substitute expression (1.129) for i_{22} in this equation and if we take account of the fact that e_{11}, e_{12}, e'_{i1}, e'_{i2} are independent of each other, we obtain:

$$p = \{|d_{22}^{11}|^2 n_{11} + |d_{22}^{12}|^2 n_{12} + \sum_{i>2}[|d_{22}^{i1}|^2 n'_{i1} + |d_{22}^{i2}|^2 n'_{i2}]\} R_2 \quad (1.148)$$

If we assume that

$$\left.\begin{array}{l} n_{11} = n_{12} = 4kT_1 R_1 \\ n'_{i1} = n'_{i2} = 4kT_i R_i \end{array}\right\} \quad (1.149)$$

we may write (1.148), by virtue of (1.131), in the form

$$p = (G_{22}^{1N})_d kT_1 + \sum_{i>2} G_{22}^{iN} kT_i \quad (1.150)$$

with

$$(G_{22}^{1N})_d = G_{22}^{11} + G_{22}^{12} \quad (1.151)$$

the suffix d indicating that we are dealing with a degenerate converter. Equations (1.143) and (1.150) give us:

$$(T_{\text{eff }\frac{1}{2}})_d = \sum_{i>2} \frac{G_{22}^{iN}}{G_{22}^{1N}} T_i \quad (1.152)$$

On the other hand, we have:

$$[p]_{\text{ideal}} = (G_{22}^{1N})_d kT_1 \quad (1.153)$$

with

$$(G_{22}^{1N})_d > G(=G_{22}^{11}) \quad (1.154)$$

When the useful component of the output signal is of frequency f_1, we find:

$$\left.\begin{array}{l} (G_{21}^{1N})_d = G_{21}^{11} + G_{21}^{12} \\ (T_{\text{eff }1})_d = \sum_{i>2} \frac{G_{21}^{iN}}{G_{21}^{1N}} T_i \end{array}\right\} \quad (1.155)$$

$$(G_{21}^{1N})_d > G(=G_{21}^{11}) \quad (1.156)$$

Case 1.2. The two components of the output signal are used. To take advantage of the phase correlation between the two components, synchronous detection is generally used. As the effect of the correlation appears only in the detected signal, the effective temperature of the converter will be determined not by consideration of the powers fed into the *input* impedance Z_2 of the detector, but by the powers fed to the *terminating* load of the detector. In order to simplify the calculations, we will assume that the impedance of this load is Z_2. In order to study synchronous detection it will be necessary to derive the expressions for each of the components of the output signal, and these can be obtained by substituting e_{11} and e_{12} respectively in (1.129) by E_{11} and E_{12}, and by making $e_{i1} = e_{i2} = 0$. We then find:

$$\left. \begin{array}{l} I_{21} = d_{21}^{11} E_{11} + d_{21}^{12} E_{12}^{(*)} \\ I_{22} = d_{22}^{11} E_{11}^{*} + d_{22}^{12} E_{12} \end{array} \right\} \quad (1.157)$$

Let

$$2\pi f_1 = \omega' - \Delta$$

with

$$\left. \begin{array}{ll} \omega' = \omega_p/2 & (\omega_p \text{ frequency of the local oscillator}) \\ 2\pi f_2 = \omega' + \Delta & (\text{spectrum inversion converter}) \end{array} \right\} \quad (1.158)$$

In synchronous detection the output signal of the converter is modulated by a signal of frequency ω'.

$$i_2(t) = \frac{1}{\sqrt{2}} [(I_{21} + i_{21}) e^{j(\omega' - \Delta)t} + (I_{21}^* + i_{21}^*) e^{-j(\omega' - \Delta)t}$$
$$+ (I_{22} + i_{22}) e^{j(\omega' + \Delta)t} + (I_{22}^* + i_{22}^*) e^{-j(\omega' + \Delta)t}] \quad (1.159)$$

$$I(t) = I' e^{j\omega' t} + I'^* e^{-j\omega' t} \quad (1.160)$$

Only the component at frequency Δ is retained in the product of modulation (the quantity $I(t)$ is assumed to be dimensionless). It can easily be shown that the effective complex amplitude of this component is given by:

$$\left. \begin{array}{ll} \text{the useful signal:} & I_\Delta = I_{21}^* I' + I_{22} I'^* \\ \text{the parasitic signal:} & i_\Delta = i_{21}^* I' + i_{22} I'^* \end{array} \right\} \quad (1.161)$$

in which the subscript Δ indicates that we are dealing with detected signals.

Substituting in these relations the expressions I_{21}, I_{22} and i_{21}, i_{22} given by (1.157) and (1.129) respectively, and taking into account the fact that the converter is of the spectrum inversion type, we obtain:

$$I_\Delta = (d_{21}^{11*} I' + d_{22}^{11} I'^*) E_{11}^* + (d_{21}^{12*} I' + d_{22}^{12} I'^*) E_{12} \quad (1.162)$$

$$i_\Delta = (d_{21}^{11*} I' + d_{22}^{11} I'^*) e_{11}^* + (d_{21}^{12*} I' + d_{22}^{12} I'^*) e_{12}$$
$$+ \sum_{i>2} [(d_{21}^{i1*} I' + d_{22}^{i1} I'^*) e_{i1}'^* + (d_{21}^{i2*} I' + d_{22}^{i2} I'^*) e_{i2}'] \quad (1.163)$$

If it is assumed that we are dealing with single-band reception, equation (1.162) is reduced to

$$I_\Delta = (d_{21}^{11*}I' + d_{22}^{11}I'^*)E_{11}^* \qquad (1.164)$$

An appropriate choice of the initial phase of the signal $I(t)$ (frequency $f' = f_p/2$) with respect to that of the signal of the local oscillator of the receiver, frequency f_p, enables a maximum value to be obtained for $|I_\Delta|$ for a given $|E_{11}|$ (see Chapter 7). The effective complex amplitude I' having been chosen, we can determine with the aid of (1.164) the value of the transducer gain G_Δ^{11} of the converter, and with the aid of (1.163) the values of G_Δ^{1N}, G_Δ^{iN}. The final relations are:

$$T_{\text{eff}\,\Delta}^{\,1} = \sum_{i>2} \frac{G_\Delta^{iN}}{G_\Delta^{1N}} T_i \qquad (1.165)$$

$$[p]_{\text{ideal}} = G_\Delta^{1N} k T_1 \qquad (1.166)$$

$$G_\Delta^{1N} > G \quad (=G_\Delta^{11}) \qquad (1.167)$$

(2) DOUBLE-BAND RECEPTION. Reception is called *double-band* when the useful input signal consists of two components, of frequencies f_1 and f_2 respectively. We have to consider two cases:

Case 2.1. Only one of the components of the output signal is used; let the frequency of this signal be designated by f_1. We then find:

$$T_{\text{eff}\,1}^{\,1,2} = \sum_{i>2} \frac{G_{21}^{iN}}{G_{21}^{1N}} T_i \qquad (1.168)$$

The transducer gain $G_{21}^{1(1,2)}$ of the converter can be determined with the aid of the first part of the equation (1.157). It is to be noted that $G_{21}^{1(1,2)}$ depends on the ratio $|E_{11}|/|E_{12}|$ and on the initial phases of E_{11} and E_{12}.

When the components E_{11} and E_{12} are derived from the modulation of a carrier of frequency $f' = f_p/2$, a suitable choice of the initial phase of the signal of the local oscillator with respect to that of the carrier enables a maximum value to be obtained for $G_{21}^{1(1,2)}$ (see Chapter 7). It should be noted that this maximum value is higher than $(G_{21}^{1N})_d$; it follows that in the case under consideration:

$$(G_{21}^{1N})_d < G_{max}(=G_{21\,\text{max}}^{1(1,2)}) \qquad (1.169)$$

On the other hand, when the components E_{11} and E_{12} are without any phase correlation and when also, $|E_{11}| = |E_{12}|$, we have:

$$(G_{21}^{1N})_d = G$$

Case 2.2. The two components of the output signal are used (synchronous detection). We now have:

$$T_{\text{eff}\,\Delta}^{1,2} = \sum_{i>2} \frac{G_\Delta^{iN}}{G_\Delta^{1N}} T_i \qquad (1.170)$$

The transducer gain $G_\Delta^{1(1,2)}$ of the converter is determined with the aid of equation (1.162). When the components E_{11} and E_{12} have been obtained by modulation of a carrier wave, a suitable choice of the initial phases of the signal given by the local oscillator of the receiver and of the signal $I(t)$ with respect to the initial phase of the carrier wave, enables us to bring the value of $G_\Delta^{1(1,2)}$ to a maximum (see Chapter 7).

2.6. Operational temperature of a system

According to the definition of the effective noise temperature:

$$T_{\text{eff}} = \frac{[p]_{\text{actual}} - [p]_{\text{ideal}}}{kG^{1N}} \qquad (1.143)$$

this defines the contribution of all the internal noise sources of the amplifier to the background noise computed as an equivalent input noise amplified by the gain G^{1N}. However, when we are studying a telecommunication system from the point of view of its fidelity of transmission we have to consider the signal-to-noise ratio S/N which constitutes our criterion, and this depends on the total background noise $[p]_{\text{actual}}$. By analogy with (1.143), this noise is calculated at the amplifier input, but this time with the intervention of the transducer gain G, which is expressed by a temperature T_{op}, called *the operational temperature of the system*. This is defined by:

$$T_{\text{op}} = \frac{[p]_{\text{actual}}}{kG} \qquad (1.171)$$

Let this relation be written in the form

$$T_{\text{op}} = \frac{[p]_{\text{ideal}}}{kG} + \frac{[p]_{\text{actual}} - [p]_{\text{ideal}}}{kG} \qquad (1.172)$$

and note that, in accordance with the definition given in the above paragraph, we have:

$$[p]_{\text{ideal}} = G^{1N} k T_1$$
$$[p]_{\text{actual}} - [p]_{\text{ideal}} = G^{1N} k T_{\text{eff}} \qquad (1.173)$$

After substitution in (1.171) we obtain:

$$T_{\text{op}} = \frac{G^{1N}}{G} (T_1 + T_{\text{eff}}) \qquad (1.174)$$

This relation is the generalization of the classic formula

$$T_{op} = T_1 + T_{eff} \qquad (1.175)$$

to which it is equivalent when $G^{1N} = G$. Formula (1.174) shows that the use of low-noise amplifiers, that is those with a low value of T_{eff}, is justified only if T_1 is at the very most of the same order of magnitude as T_{eff}. On the other hand, when $G^{1N} \neq G$, the value of T_1 is the determining factor in the choice of the parameters of the amplifier so as to cause the value of T_{op} to be a minimum. It should be noted that when T_1 is small and when the chief contribution to T_{op} comes from the term $(G^{1N}/G)T_{eff}$, we may be satisfied by taking for the criterion of optimization the *operational effective temperature* of the amplifier

$$T_{eff_{op}} = \frac{G^{1N}}{G} T_{eff} = \frac{[p]_{actual} - [p]_{ideal}}{kG} = \sum_{i>2} \frac{G^{iN}}{G} T_i \qquad (1.176)$$

General notes

(1) During the determination of the spectral density p of the noise power fed into the load Z_2, which terminates a given amplification chain, we have not so far mentioned the contribution made by the intrinsic noise of Z_2. This contribution is equivalent to (with the notations in section 2.3):

$$[p]_{Z_2} = \frac{\langle e_2 e_2^* \rangle R_2}{|Z_e + Z_0|^2} = \frac{4k R_2^2 T_2}{|Z_2 + Z_0|^2}$$

The increase in the resulting effective operational temperature is given by:

$$(T_{eff_{op}})_{Z_2} = \frac{[p]_{Z_2}}{kG} = \frac{1}{G} \frac{4R_2}{|Z_2 + Z_0|^2} T_2 \qquad (1.177)$$

where G is the overall transducer gain of the amplification chain. Formula (1.177) shows us that when the value of G is high and when the real part of Z_0 is positive, the contribution made by the intrinsic noise of Z_2 to the effective operational temperature of the chain is almost negligible. When the real part of Z_2 is negative, this contribution may again be made negligible by the use of circulators (see Chapter 5).

(2) In this section we have discussed voltage generators and impedance matrices. It is clear that identical treatment with dual representations would give quite similar results.

The group of formulae established enables us to determine T_{eff} and T_{op} if we know the structure of the amplifier used, as well as the equivalent temperatures T_1 and T_i; these depend on the physical nature of the noise sources.

The last section of this chapter deals briefly with a number of noise sources and introduces the concept of the *radiation temperature* of an aerial.

3. Sources of background noise

3.1. Internal noise sources

3.1.1. JOHNSON EFFECT (thermal noise)

When we consider a conductor in thermal equilibrium it is observed that there is a fluctuating e.m.f. at its terminals. This phenomenon, foretold by Einstein in 1906, is due to the thermal agitation of the charge carriers (electrons). This effect was studied experimentally by Johnson (1927) and expressed in theoretical form by Nyquist (1928). Nyquist showed that the spectral noise density of the random e.m.f. e which exists between the terminals of the conductor, may be expressed by:

$$n = \langle e\,e^* \rangle = \overline{e\,e^*} = \frac{4Rhf}{e^{hf/kT} - 1} \tag{1.178}$$

where n is the spectral noise density, V^2 s;
  h is Planck's constant, $6 \cdot 63 \times 10^{-34}$ J s;
  k is Boltzmann's constant, $1 \cdot 374 \times 10^{-23}$ J/°K;
  T is the absolute temperature of the conductor, °K;
  R is resistance of the conductor, Ω;
  f is the frequency, Hz.

At the temperatures and frequencies commonly met with in telecommunication techniques, hf/kT is much less than unity and we have almost exactly:

$$n = 4kTR \tag{1.179}$$

which shows that the spectral noise density due to thermal agitation may be considered as independent of frequency ('white noise'). If we apply the second law of thermodynamics we can determine the spectral density of the thermal-agitation noise of any type of passive one-port. Let us therefore consider a passive one-port having an impedance $Z = R(f) + jX(f)$ and a resistance R (Fig. 1.11) connected across its terminals.

If we assume that Z and R are at the same temperature T, we have, by virtue of the second law (cf. section 2.2.4):

$$\underset{R \to Z}{p} = \underset{Z \to R}{p}$$

which gives:

$$\frac{\langle e\,e^*\rangle R(f)}{|R+Z|^2} = \frac{\langle e_{Z_{th}} e^*_{Z_{th}}\rangle R}{|R+Z|^2}$$

Hence, taking account of (1.179):

$$n_{Z_{th}} = \langle e_{Z_{th}} e^*_{Z_{th}}\rangle = 4kTR(f) \qquad (1.180)$$

Thus, from the point of view of thermal agitation noise, a passive one-port of impedance Z is equivalent to a voltage generator of internal impedance Z. The spectral noise density of the e.m.f. of this generator is given by relation (1.180). This relation shows that $n_{Z_{th}}$ depends only on T and on the real part of Z. Thus, pure reactances do not

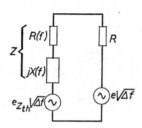

Figure 1.11

give rise to any thermal-agitation noise. When the passive one-port is characterized by its admittance $Y = C + jB$, it is more convenient to represent it by an equivalent current generator of internal admittance C. The spectral noise density of the current i supplied by this generator is then given by:

$$n = \langle i\,i^*\rangle = \overline{i\,i^*} = 4kTG \qquad (1.181)$$

3.1.2. EQUIVALENT SPECTRAL NOISE TEMPERATURE OF A ONE-PORT

Thermal agitation is not the sole cause of noise. In vacuum tubes the shot effect, the flicker effect and the fluctuation of the current distribution between the positive electrodes give rise to noise; similar effects occur in semiconductors [3]. The spectral density of these noises is not necessarily constant as in the case of thermal-agitation noise. When the noise source is a one-port of impedance $Z = R + jX$, it is convenient to characterize this one-port, as was done in the last section, by its equivalent spectral temperature T_{eq}. This is defined by:

$$n(f) = \langle e_Z(f)e^*_Z(f)\rangle = 4kT_{eq}(f)R \qquad (1.182)$$

where $n(f)$ is the spectral density of the noise produced by the one-port.

The equivalent spectral temperature of the noise $T_{eq}(f)$ of a one-port is the temperature at which a resistance, of value equal to that of the one-port, would have a thermal noise spectral density equal to $n(f)$.

3.1.3. SPECTRAL DENSITY OF THE AVAILABLE NOISE POWER OF A ONE-PORT

Let us consider a one-port of impedance $Z = R + jX$ and let T_{eq} be its equivalent noise temperature. Connect a load of impedance $Z^* = R - jX$ to the terminals of this one-port. The spectral density of the noise power that the one-port Z supplies to Z^* (available power) equals:

$$\underset{Z \to Z^*}{p} = \frac{\langle e_Z e_Z^* \rangle R}{|Z + Z^*|^2}$$

Whence, by virtue of (1.182):

$$\underset{Z \to Z^*}{p} = kT_{eq} \qquad (1.183)$$

Conversely, if the spectral density of the available noise power of a one-port is equal to kT_1, its equivalent noise temperature is equal to T_1. This principle will be made use of in section 3.2.2.

3.2. External noise sources

Background noise is produced on the one hand by the dissipative elements of the aerial and the transmission lines which connect it to the receiver, and on the other hand, by the noise sources which the aerial picks up. The effect of losses in the transmission lines and in the aerial are equivalent to an attenuator, at ambient temperature T, connected between the aerial, assumed to be loss-free (ideal aerial), and the receiver. If we assume that there is impedance matching at the input of the receiver, the equivalent noise temperature of the generator, formed by the ideal aerial followed by the attenuator, is given by the formula (1.117).

$$T_{eq} = G_a T_1 + (1 - G_a)T \qquad (1.117)$$

The equivalent noise temperature T_1 of the generator feeding the attenuator appears in this formula. What value must be given to T_1 when the parasitic signal at the input of the attenuator consists of noises picked up by the ideal aerial? Before answering this question, let us say a few words on the sources that give rise to these noises. Parasitic signals of industrial origin being outside the scope of this study, we can classify these sources in three categories:

(a) extra-terrestrial noises,
(b) the atmosphere,
(c) the earth.

BACKGROUND NOISE

These sources emit random signals of very wide spectrum and of a spectrum distribution which is usually not uniform. Just as we took the thermal-agitation noise as reference for the study of internal noise, so we will take as reference for our present purposes, the radiation from a black body, and this calls for a few definitions.

3.2.1. A REMINDER OF SOME DEFINITIONS OF RADIATION PHENOMENA

(1) *Spectral density of brightness*

The spectral density of brightness B_f in a given direction Ox of a radiating surface element ds situated at O, is defined by

$$B_f = \frac{\mathrm{d}P}{\mathrm{d}f \, \mathrm{d}\Omega_0 \, \mathrm{d}\sigma} \tag{1.184}$$

where dP is the power emitted in the frequency band $(f, f + \mathrm{d}f)$ by

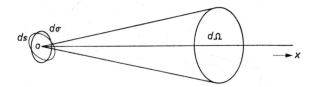

Figure 1.12

the element ds in the solid angle dΩ_0 originating at O and surrounding Ox, and dσ is the projection of ds on the surface normal to Ox (Fig. 1.12).

(2) *Black-body radiation*

The spectral distribution of the brightness of the black body does not depend on the direction Ox (Lambert's Law). It is a function only of the temperature T and of the frequency f. The expression for it is given by Planck's law.

$$B_f = \frac{2hc}{\lambda^3}(e^{hc/k\lambda T} - 1)^{-1} \tag{1.185}$$

where c is the velocity of light and λ the wavelength corresponding to frequency f.

As we saw when discussing formula (1.178):

$$\frac{hf}{kT} = \frac{hc}{k\lambda T} \ll 1$$

In these conditions, Planck's law reduces to the Rayleigh–Jeans law, in the much simpler form

$$B_f = \frac{2kT}{\lambda^2} \quad (1.186)$$

(3) *Spectral flux density per unit of surface*

The spectral flux density per unit of surface at a point A in space, following a given path Ax, is expressed by:

$$S_f = \frac{\mathrm{d}P}{\mathrm{d}f\,\mathrm{d}s'} \quad (1.187)$$

where $\mathrm{d}P$ is the power in the frequency band $(f', f + \mathrm{d}f)$ passing

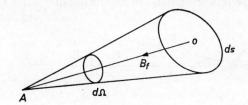

Figure 1.13

through an element of surface $\mathrm{d}s'$ situated at A and normal to Ax. It can be shown that the elementary spectral flux density $\mathrm{d}S_f$ that an element of radiating surface $\mathrm{d}s$ at a point O of the space produces at another point A, is given by

$$\mathrm{d}S_f = B_f \cdot \mathrm{d}\Omega \quad (1.188)$$

where B_f is the spectral density of the brightness of $\mathrm{d}s$ in the direction OA, and $\mathrm{d}\Omega$ the solid angle subtended by $\mathrm{d}s$ at A (Fig. 1.13).

(4) *Brightness temperature of a source of radiation*

We defined the equivalent noise temperature of a one-port earlier by comparing its noise with the thermal-agitation noise of a resistance. We will define similarly the noise radiated by a source at a given frequency by comparing it with the radiation of a black body:

The brightness temperature T_{bf}, at frequency f, of a source of radiation is the temperature of a black body having the same spectral density of brightness at that frequency.

3.2.2. RADIATION TEMPERATURE OF AN IDEAL AERIAL

Let it be assumed that an aerial receives an extended source of radiation, and let $T_{bf}(\theta, \psi)$ be the temperature of the brightness of the element of the source seen by the aerial in the direction defined by the co-ordinates (θ, ψ). By virtue of (1.184), (1.186) and (1.188) the elementary flux per unit of surface at the aerial in the frequency band $(f, f + df)$, due to the element of the source seen in the direction (θ, ψ) in the solid angle $d\Omega$, is given by:

$$dS_f . df = \frac{2kT_{bf}(\theta, \psi)}{2} d\Omega\, df \quad (1.189)$$

The total noise power in the frequency band $(f, f + df)$ is then equal to:

$$P_u = \tfrac{1}{2} \iint A(\theta, \psi)\, dS_f\, df \quad (1.190)$$

where:

$$A(\theta, \psi) = \frac{\lambda^2}{4\pi} G(\theta, \psi) \quad (1.191)$$

is the effective section of the aerial. The factor $\tfrac{1}{2}$ appears because the aerial is sensitive to only one polarization component of the field.

Equations (1.189) and (1.190) give us:

$$P_u = \frac{k\, df}{4\pi} \iint T_{bf}(\theta, \psi) G(\theta, \psi)\, d\Omega \quad (1.192)$$

In accordance with the definition of $G(\theta, \psi)$:

$$\iint G(\theta, \psi)\, d\Omega = 4\pi \quad (1.193)$$

When the brightness temperature of the source of radiation is uniform (at least in the solid angle corresponding to the main lobe of the aerial), (1.192) reduces to

$$P_u = kT_{bf}\, df \quad (1.194)$$

This relation shows that the spectral density of the available noise power of the generator, formed by the ideal aerial receiving the noise emitted by a temperature source of brightness T_{bf}, is equal to kT_{bf}. By virtue of the proposition discussed at the end of section 3.1.3, the equivalent spectral noise temperature of this generator is equal to T_{bf}: this is the *radiation temperature* T_a of the ideal aerial.

When the brightness temperature is not uniform, for example

when the aerial is exposed to several sources, the radiation temperature is given by the general formula

$$T_a = \frac{1}{4\pi} \iint T_{bf}(\theta, \psi) G(\theta, \psi) \, d\Omega \qquad (1.195)$$

3.2.3. EXTRA-TERRESTRIAL NOISE SOURCES

The radiation of extra-terrestrial origin comes either from localized sources or from scattered sources. Amongst the chief localized sources may be mentioned the sun, the moon, Cassiopeia A, Cygnus A, Taurus A, Virgo A.† It is chiefly the first two which give rise to a

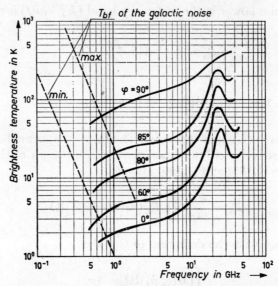

Figure 1.14

flux density large enough to cause a considerable increase in the radiation temperature of the aerial when they are in its main lobe or even in its side lobes. The contribution made to the radiation temperature of the aerial by the other radio sources, which are to all intents and purposes point sources, would be important only for highly directional aerials; but in this case, their presence in the main lobe of the aerial would be temporary and accidental. The source of galactic noise is of a scattered nature. Its brightness temperature is

† This type of nomenclature is used for the chief radio sources. For example, Cygnus A signifies the first radio source discovered in the constellation of Cygnus, Taurus B the second radio source discovered in the constellation of Taurus, etc.

maximum along the galactic plane and is minimum in the direction of the galactic poles. It decreases as the frequency increases. The maximum and minimum brightness temperatures as functions of the frequency are shown in Fig. 1.14.

3.2.4. ATMOSPHERIC NOISE

The thermal radiation of the atmosphere makes an important contribution to the brightness temperature of the sky. This contribution varies with frequency and with the inclination of the aerial. Figure 1.14 shows both the brightness temperature due to the atmosphere as a function of frequency and of the angle of the axis of the main lobe of the aerial from the vertical. It will be seen that this temperature is a minimum when the aerial is directed towards the zenith.

3.2.5. THERMAL-AGITATION NOISE OF THE EARTH

The earth can be compared to a black body at the ambient temperature. The contribution of the earth to the radiation temperature of the aerial can be calculated by means of formula (1.195). The noise from the earth increases in importance as the aerial becomes less directional and is directed low on the horizon. In order to eliminate this effect adequately it is not enough to give the aerial a high gain, but it is necessary also that the secondary lobes and especially those directed towards the earth should be negligible.

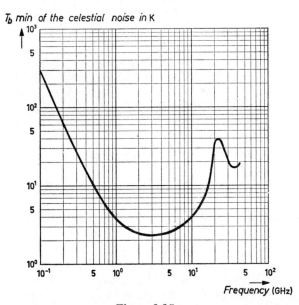

Figure 1.15

To terminate this short discussion on noise sources it should be mentioned that the increase in atmospheric noise and the decrease in galactic noise with increasing frequency are illustrated in the curve (Fig. 1.15) in which the minimum occurs between 0·4 and 18 GHz. It is thus in this 'window' that we must choose the carrier frequency of a transmission path so as to derive appreciable benefit from the use of a low-noise amplifier.

References

[1] Bendat, J. S., *Principles and Applications of Random Noise Theory*, John Wiley and Sons, Inc., New York, 1958.
[2] Davenport, W. B. and Root, W. L., *An Introduction to the Theory of Random Signals and Noise*, McGraw-Hill, New York, 1958.
[3] Grivet, P. and Blaquière, A., *Le Bruit de fond*, Masson et Cie, Paris, 1958.
[4] Laurent, L., 'Le Bruit de fond', *Revue M.B.L.E.*, vol. V (No. 4), December 1962.

*Part II*

PARAMETRIC AMPLIFIERS FOR MICROWAVE FREQUENCIES

2

CLASSIFICATION OF PARAMETRIC AMPLIFIERS AND CONVERTERS

by L. LAURENT

Our study of background noise has shown that one of the methods of achieving very low-noise amplification consists in the use of a purely reactive amplifier, namely the parametric amplifier. Depending on the sign of the reactance used, one can expect that there would be two classes of parametric amplifiers: those with variable capacitance, and those with variable inductance. However, for technical reasons, the first are the only ones in practical use, whilst the second are still in the laboratory stage. In this chapter we will deal with the Manley and Rowe equations. These equations, in which the influence of the losses in the circuits and in the variable reactance are ignored, give to some extent the 'conditions for the existence of a parametric amplifier' and enable us to classify them qualitatively into different categories.

1. Manley and Rowe relations

Instead of giving proofs of the Manley and Rowe relation we refer the reader to the original article [1], and we will confine ourselves to establishing their origins and to a discussion of their applications. Let us consider series circuits of any number, placed in parallel with a capacitance of a value depending on the voltage applied across it. This capacitance, as shown in Fig. 2.1, will be assumed to be lossless. Each of the first two circuits consists of a generator and a filter having zero impedance at the generator frequency and presenting an open circuit to all other frequencies. The other circuits consist of an impedance Z_i and a filter of zero impedance at frequency f_i, presenting an open circuit to all other frequencies. It is assumed that the

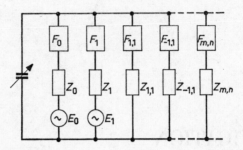

Figure 2.1. Schematic of a parallel-type parametric device

frequencies of the generators f_0 and f_1 are non-commensurable and positive, and that the frequencies f_1 are of the form

$$f_i = mf_1 + nf_0 = f_{mn} \tag{2.1}$$

where m and n are positive, negative or zero integers.

In the series-type configuration the parametric device is as shown in Fig. 2.2: a number of parallel circuits are connected in series with a capacitance which varies as a function of the voltage applied across it. The first two circuits consist of a current generator and a filter of zero admittance at the generator frequency and infinite at all other frequencies. The other circuits consist of an admittance Y_i in parallel with a filter of zero admittance at f_i and infinite at all other frequencies. Since the Manley and Rowe relations are applicable to both configurations, we will confine our discussions to the parallel configuration. In order to prove that this device is capable of amplifying a signal, it is necessary to know the relations between the powers in the different branches. The relation between the voltage across the non-linear capacitance and its charge may be written in the most general form:

$$v = f(q) \tag{2.2}$$

The only restriction is that $f(q)$ should be biunivocal, i.e. that its non-linear capacitance should be free from hysteresis.

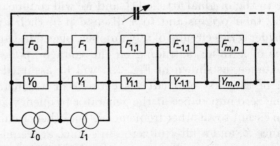

Figure 2.2. Schematic of a series-type parametric device

PARAMETRIC AMPLIFIERS AND CONVERTERS 55

Since the charge q is a doubly periodic function of time, it can be developed as a double Fourier series:

$$q = \sum_{m=-\infty}^{+\infty} \sum_{n=-\infty}^{+\infty} Q_{m,n} e^{j(m\omega_1 + n\omega_0)t} \tag{2.3}$$

In order to find the current through the non-linear capacitance, we will differentiate the two members of (2.3) with respect to time:

$$i = \frac{dq}{dt} = \frac{d}{dt}\left[\sum_{m=-\infty}^{+\infty} \sum_{n=-\infty}^{+\infty} Q_{m,n} e^{j(m\omega_1 + n\omega_0)t} \right] \tag{2.4}$$

As the series is convergent,

$$i = \sum_{m=-\infty}^{+\infty} \sum_{n=-\infty}^{+\infty} \frac{d}{dt}[Q_{m,n} e^{j(m\omega_1 + n\omega_0)t}] \tag{2.5}$$

or

$$i = \sum_{m=-\infty}^{+\infty} \sum_{n=-\infty}^{+\infty} I_{m,n} e^{j(m\omega_1 + n\omega_0)t} \tag{2.6}$$

with

$$I_{m,n} = j(m\omega_1 + n\omega_0)Q_{m,n} \tag{2.7}$$

The same method can be used to examine v by a double series development. By combining the expression obtained in this way with the preceding equations, we obtain relations connecting the mean powers $P_{m,n}$ entering the non-linear capacitance

$$\sum_{m=0}^{+\infty} \sum_{n=-\infty}^{+\infty} \frac{mP_{m,n}}{mf_1 + nf_0} = 0 \tag{2.8}$$

$$\sum_{m=-\infty}^{+\infty} \sum_{n=0}^{+\infty} \frac{nP_{m,n}}{mf_1 + nf_0} = 0 \tag{2.9}$$

in which $P_{m,n}$ is negative when the non-linear reactance supplies, and positive when it receives energy at frequency $f_{m,n}$.

These are the Manley and Rowe relations, which are quite general; we have made no assumption about the law governing the variation of the non-linear capacitance except for the absence of hysteresis and freedom from losses. As a result of this last condition it is clear that

$$\sum_{m=0}^{+\infty} \sum_{n=-\infty}^{+\infty} P_{m,n} = 0 \tag{2.10}$$

This relation is obtained from (2.8) and (2.9).

2. Properties derived from the Manley and Rowe relations

(1) If the number of filters F_i is limited, only those components whose frequencies correspond to the filters used will appear among the components $I_{m,n}$, and consequently $P_{m,n}$.

(2) Each of the equations (2.8) or (2.9) can only involve the power corresponding to one of the generators. In fact, (2.8) contains $P_{i,0}$ but not $P_{0,1}$ which is multiplied by zero. Similarly, (2.9) contains $P_{0,1}$ but not $P_{1,0}$

(3) When all the $P_{m,n}$'s $(m, n \neq 0)$ correspond to sum-frequencies (m and n are of the same sign), factors $m/(mf_1 + nf_0)$ and $n/(mf_1 + nf_0)$ are positive. As the branches corresponding to the $P_{m,n}$'s other than $P_{0,1}$ and $P_{1,0}$ do not contain a generator, the values of these $P_{m,n}$'s can only be zero or negative. Hence, because of the property (2), $P_{1,0}$ and $P_{0,1}$ are both positive: the two generators supply energy and the system is unconditionally stable.†

(4) When all the $P_{m,n}$'s $(m, n \neq 0)$ correspond to difference frequencies (m and n of different signs), then either the factor $m/(mf_1 + nf_0)$ or the factor $n/(mf_1 + nf_0)$ is negative. Since the $P_{m,n}$'s can be only negative or zero, (2) tells us that either $P_{1,0}$ or $P_{0,1}$ is negative: one of the generators receives energy and the system is potentially unstable.†

(5) When the system contains sum-frequency and difference-frequency circuits, no general conclusions can be drawn. The system may be either unconditionally stable, or potentially unstable depending on the ratios of the powers consumed at the sum and difference frequencies respectively.

3. Application to parametric amplification and conversion

The properties just mentioned are independent of the respective levels of the signals present. If we assume that the amplitude of one

---

† Adopting the sign conventions used earlier, when the power exchanged between a source and a one-port is positive, the one-port is passive. That is, the real part of its immittance is positive; the system is unconditionally stable. When the power exchanged between a source and a one-port is negative (the source receives energy from the one-port) the one-port is necessarily active, the real part of its immittance being negative. We know that in this case spontaneous oscillations may occur in the circuit. The instability of this circuit depends on the immittance of the source: the system is potentially unstable. This is true for all negative-resistance amplifiers (parametric or tunnel diode amplifiers).

PARAMETRIC AMPLIFIERS AND CONVERTERS 57

of these signals (called the pump) is much larger than that of all the other signals, we may classify parametric devices other than multipliers ($m = 0$) in groups possessing special properties. It should be noted that this assumption is justified in practice: parametric amplifiers and converters are almost always used as amplifiers for weak signals. The pump generator then acts as the local oscillator in frequency-changing circuits and constitutes the energy source of the amplifier, as we will shortly see. Amplifiers or converters used in practice employ only a small number of components $f_{m,n}$. The powers $P_{m,n}$ corresponding to the other frequencies $f_{m,n}$ are then zero.

Let f_p represent the pumping frequency,
f_e the frequency of the input signal,
$f_{(+)}$ ($=f_p + f_e$) the first sum-frequency
$|f_{(-)}|$ the first difference frequency†, with
$$f_{(-)} = f_p - f_e \qquad (2.12)$$
(2.11)

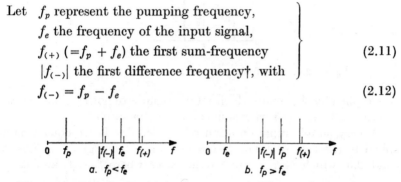

Figure 2.3. Distribution of the first sum and difference frequencies

According to whether the pumping frequency is lower or higher than the frequency of the input signal, the frequency distribution will be as shown in Figs. 2.3a or 2.3b. The signal f_e has been assigned to a sideband in order to make clear that in the first case the spectrum is not changed for $f_{(+)}$ or for $|f_{(-)}|$, whilst in the second case the spectrum is reversed for $f_{(-)}$.

4. Three-frequency converters

4.1. Sum-frequency converters

Figure 2.3 shows that we are dealing with a non-inverting converter. The Manley and Rowe relations become:

$$\frac{P_e}{f_e} + \frac{P_{(+)}}{f_{(+)}} = 0 \qquad (2.13)$$

$$\frac{P_p}{f_p} + \frac{P_{(+)}}{f_{(+)}} = 0 \qquad (2.14)$$

† For uniformity of notation, we shall use $|f_{(-)}|$ also when $f_{(-)} > 0$.

Equation (2.13) gives us the conversion power gain.†

$$G^e_{(+)} = -\frac{P_{(+)}}{P_e} = \frac{f_{(+)}}{f_e} \qquad (2.15)$$

The gain of the converter is equal to the ratio of the output and input frequencies. The difference between the output and input powers

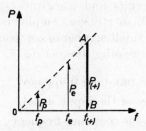

Figure 2.4. Sum-frequency converter. Case in which $f_p < f_e$

is supplied by the pump which acts as a source of power‡. As we saw in section 2.3, the converter is unconditionally stable.

A very simple representation of the Manley-Rowe equations applied to the case of a sum-frequency converter is given in Figs. 2.4 and 2.5, which correspond to the cases where $f_p < f_e$ and $f_p > f_e$

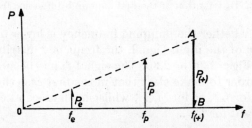

Figure 2.5. Sum-frequency converter. Case in which $f_p > f_e$

respectively. Power-frequency coordinate axes are used, the vectors representing the powers being plotted parallel to OP with an arbitrary scale on the right of the corresponding frequencies. The vectors

† It is important not to confuse the power gain discussed in this chapter with the transducer gain which will be dealt with below. The first refers to the power abstracted from the source by the input circuit of the amplifier, while the second refers to the available power of this source. It will be seen therefore that the two gains may, as functions of the parameters of the system, develop in different or even in opposite ways (see section 8).

‡ The power actually necessary for the pumping of an amplifier is much higher than this value. It is in fact determined by the inevitable losses of the non-linear element, as will be seen later.

PARAMETRIC AMPLIFIERS AND CONVERTERS 59

are directed upwards when the power entering the non-linear reactance is positive, and downwards when it is negative. The positions of the half-arrows in relation to the axes of the vectors represent the modification affecting the spectrum (inversion or non-inversion respectively).

Equations (2.13) and (2.14), which may be written:

$$P_p = -P_{(+)} \frac{f_p}{f_{(+)}} \qquad (2.16)$$

$$P_e = -P_{(+)} \frac{f_e}{f_{(+)}} \qquad (2.17)$$

show that if we decide on an output power $P_{(+)} \Leftrightarrow \overrightarrow{AB}$ (negative), then $\mathbf{P}_p$ and $\mathbf{P}_e$ lie on the straight line OA, which determines their relative magnitude. Their sign is fixed necessarily, without the need to refer to (2.16) and (2.17), by the principle of conservation of energy which requires that the sum of the projections of the vectors on OP should be zero. The sum-frequency converter is of interest as an amplifier because its gain is always higher than unity. However, once converted to the sum-frequency, the signal must be brought back to the input frequency or to a lower frequency (intermediate frequency). This second conversion is usually effected by means of the standard non-linear resistance diode converter, the conversion loss of which is independent of the ratio of the output and input frequencies. A total gain greater than unity can then be obtained by means of a sufficiently large frequency ratio $f_{(+)}/f_e$.

4.2. Difference-frequency converters

Equations (2.8) and (2.9) become:

$$-\frac{P_{(-)}}{f_{(-)}} + \frac{P_e}{f_e} = 0 \qquad (2.18)$$

$$-\frac{P_{(-)}}{f_{(-)}} - \frac{P_p}{f_p} = 0 \qquad (2.19)$$

and they are shown graphically in Figs. 2.6 and 2.7 which apply to the cases in which $f_p < f_e$ and $f_p > f_e$ respectively.

(1) Case in which $f_p < f_e$ ($f_{(-)} < 0$). This converter does not cause inversion of the spectrum. It will be seen from Fig. 2.6 that the input power is higher than the output power and that the difference is fed to the pump. Frequency conversion therefore introduces a negative resistance into the pump circuit. In practice, however, the circuit will remain stable, because the non-linear element contains a loss resistance whose value is much higher than that of the negative resistance introduced by the conversion of a low-level signal.

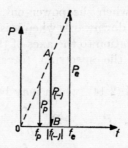

Figure 2.6. Difference-frequency converter. Case in which $f_p < f_e$

This converter is therefore unconditionally stable for a small input signal.

Relations (2.18) and (2.19) may be written

$$\frac{P_{(-)}}{|f_{(-)}|} + \frac{P_e}{f_e} = 0 \tag{2.20}$$

$$\frac{P_{(-)}}{|f_{(-)}|} - \frac{P_p}{f_p} = 0 \tag{2.21}$$

and the conversion gain is given by:

$$G^e_{(-)} = \frac{|f_{(-)}|}{f_e} \tag{2.22}$$

This converter shares with the sum-frequency converter the property of having a gain equal to the ratio of the output and input frequencies. Its gain is consequently always less than unity. A comparison of Figs. 2.4 and 2.6 shows that a difference-frequency converter with $f_p < f_e$ is the same as a sum-frequency converter in which the signal is introduced at the frequency $f_{(+)}$ and is received after conversion at the frequency f_e. This explains the similarity in their properties.

(2) Case in which $f_p > f_e$ ($f_{(-)} > 0$). This converter causes inversion

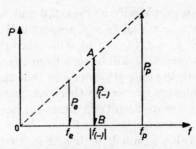

Figure 2.7. Difference-frequency converter. Case in which $f_p > f_e$

PARAMETRIC AMPLIFIERS AND CONVERTERS 61

of the spectrum. It can be seen from Fig. 2.7 that the signal generator receives energy ($P_e < 0$). The pump supplies energy to the two other circuits ($P_p > 0$). As we saw earlier the system is potentially unstable.

Relations (2.18) and (2.19) may be written:

$$\frac{P_{(-)}}{|f_{(-)}|} - \frac{P_e}{f_e} = 0 \qquad (2.23)$$

$$\frac{P_{(-)}}{|f_{(-)}|} + \frac{P_p}{f_p} = 0 \qquad (2.24)$$

The conversion gain is given by:

$$G^e_{(-)} = -\frac{|f_{(-)}|}{f_e} \qquad (2.25)$$

Its expression is negative,† which means that the real part of the input immittance of the converter is negative. It will be seen later that in these conditions, the transducer gain can be rendered as great as required, whatever the ratio between the input and output frequencies.

5. Four-frequency converters

5.1. Sum-frequency converters

In this type of converter the signal is introduced at f_e and received at $f_{(+)}$ after conversion. A load at $f_{(-)}$ ('idler') absorbs energy. In this case equations (2.8) and (2.9) may be written:

$$\frac{P_e}{f_e} + \frac{P_{(+)}}{f_{(+)}} - \frac{P_{(-)}}{f_{(-)}} = 0 \qquad (2.26)$$

$$\frac{P_p}{f_p} + \frac{P_{(+)}}{f_{(+)}} + \frac{P_{(-)}}{f_{(-)}} = 0 \qquad (2.27)$$

(1) Case in which $f_p < f_e$ ($f_{(-)} < 0$). Equations (2.26) and (2.27) become:

$$\frac{P_e}{f_e} + \frac{P_{(+)}}{f_{(+)}} + \frac{P_{(-)}}{|f_{(-)}|} = 0 \qquad (2.28)$$

$$\frac{P_p}{f_p} + \frac{P_{(+)}}{f_{(+)}} - \frac{P_{(-)}}{|f_{(-)}|} = 0 \qquad (2.29)$$

† The reader is referred back to the note in section 4.1. If the conversion gain has a negative sign, the same does not apply to the transducer gain (see section 8).

and (2.28) tells us that P_e is positive, since $P_{(+)}$ and $P_{(-)}$ are negative and the corresponding circuits do not contain any generators. We see from (2.29) that P_p is positive or negative depending on the respective magnitudes of $P_{(+)}/f_{(+)}$ and $P_{(-)}/|f_{(-)}|$. For the reasons mentioned in section 4.2 (1) the converter is unconditionally stable for a small input signal. The conversion gain is given by:

$$G^e_{(+)} = -\frac{P_{(+)}}{P_e} = \frac{f_{(+)}}{f_e} \cdot \frac{1}{1 + (P_{(-)}/P_{(+)})(f_{(+)}/|f_{(-)}|)} \quad (2.30)$$

It can be seen that this gain decreases with respect to that of the three-frequency sum-frequency converter according to a function that depends on the power absorbed at $|f_{(-)}|$.

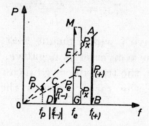

Figure 2.8. Four-frequency converter. Case in which $f_p < f_e$

The graphical representation used earlier is still applicable (Fig. 2.8). Consider an output power $P_{(+)}$ and an idler power $P_{(-)}$ ($\overrightarrow{AB}$ and $\overrightarrow{CD}$). Equations (2.28) and (2.29) tell us that:

$$P_e = -P_{(-)}\frac{f_e}{|f_{(-)}|} - P_{(+)}\frac{f_e}{f_{(+)}} \Leftrightarrow -\overrightarrow{FG} - \overrightarrow{EG} = \overrightarrow{GM} \quad (2.31)$$

$$P_p = -P_{(+)}\frac{f_p}{f_{(+)}} + P_{(-)}\frac{f_p}{|f_{(-)}|} \Leftrightarrow -\overrightarrow{KH} + \overrightarrow{JH} = \overrightarrow{JK} \quad (2.32)$$

As before, once the signs of $P_{(+)}$ and $P_{(-)}$ are fixed, those of P_e and P_p are fixed automatically. The graphical representation shows clearly the operation of the four-frequency sum-frequency converter. The input power $P_e \Leftrightarrow \overrightarrow{GM}$ is equal to the power $\overrightarrow{GE}$ corresponding to the operation as a three-frequency converter, increased by a power $P_x \Leftrightarrow \overrightarrow{GF}$ function of the idler power $P_{(-)} \Leftrightarrow \overrightarrow{CD}$. The power P_x is the input power corresponding to the operation as a three-frequency difference-frequency converter for the same power $P_{(-)}$. Similarly, the pumping power P_p is equal to the algebraic sum of the pumping powers of the three-frequency sum-frequency and difference-frequency converters. The effect of the load at $|f_{(-)}|$ is thus to decrease the gain.

PARAMETRIC AMPLIFIERS AND CONVERTERS

(2) Case in which $f_p > f_e$ ($f_{(-)} > 0$). Equations (2.8) and (2.9) may be written:

$$\frac{P_e}{f_e} + \frac{P_{(+)}}{f_{(+)}} - \frac{P_{(-)}}{|f_{(-)}|} = 0 \qquad (2.33)$$

$$\frac{P_p}{f_p} + \frac{P_{(+)}}{f_{(+)}} + \frac{P_{(-)}}{|f_{(-)}|} = 0 \qquad (2.34)$$

and the conversion gain is given by:

$$G_{(+)}^e = \frac{f_{(+)}}{f_e} \cdot \frac{1}{1 - (P_{(-)}/P_{(+)})(f_{(+)}/|f_{(-)}|)} \qquad (2.35)$$

The graphical representation is given in Fig. 2.9.

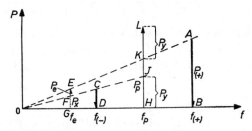

Figure 2.9. Four-frequency converter. Case in which $f_p > f_e$

We will take $P_{(+)} \Leftrightarrow \overrightarrow{AB}$ and $P_{(-)} = \overrightarrow{CD}$.

$$P_e = P_{(-)} \frac{f_e}{|f_{(-)}|} - P_{(+)} \frac{f_e}{f_{(+)}} \Leftrightarrow \overrightarrow{FG} - \overrightarrow{EG} = \overrightarrow{FE} \qquad (2.36)$$

$$P_p = -P_{(+)} \frac{f_p}{f_{(+)}} - P_{(-)} \frac{f_p}{|f_{(-)}|} \Leftrightarrow -\overrightarrow{KH} - \overrightarrow{JH} = \overrightarrow{HL} \qquad (2.37)$$

As before, the four-frequency converter is the superimposition of a sum-frequency converter and a difference-frequency converter (Fig. 2.9). The input power $P_e \Leftrightarrow \overrightarrow{FE}$ is equal to the power $\overrightarrow{GE}$ corresponding to the operation as a pure sum-frequency converter, decreased by P_x. The input power corresponds to the operation as a pure difference-frequency converter for the same power $P_{(-)}$. The input power may become negative and the system unstable for:

$$\frac{P_{(-)}}{|f_{(-)}|} < \frac{P_{(+)}}{f_{(+)}} \qquad (2.38)$$

Where $P_{(+)}$ and $P_{(-)}$ are negative, (2.38) is equal to:

$$\frac{|P_{(-)}|}{|f_{(-)}|} > \frac{|P_{(+)}|}{f_{(+)}} \tag{2.39}$$

The effect of the load at $|f_{(-)}|$ is to increase the gain.

5.2. Difference-frequency converter

In this converter the signal is introduced at f_e and received at $|f_{(-)}|$ after conversion. A load at $f_{(+)}$ ('idler') absorbs energy. Equations (2.26) and (2.27) are still applicable.

(1) Case in which $f_p < f_e$ ($f_{(-)} < 0$). Figure 2.8 can still be used. The converter is unconditionally stable and the gain is given by:

$$G^e_{(-)} = \frac{|f_{(-)}|}{f_e} \cdot \frac{1}{1 + (P_{(+)}/P_{(-)})(|f_{(-)}|/|f_{(+)}|)} \tag{2.40}$$

and we see that the effect of the load at $f_{(+)}$ is to decrease the gain.

(2) Case in which $f_p > f_e$ ($f_{(-)} > 0$). Figure 2.9 can still be used. The converter is potentially unstable in the condition (2.38). Its gain is given by:

$$G^e_{(-)} = -\frac{|f_{(-)}|}{f_e} \cdot \frac{1}{1 - (P_{(+)}/P_{(-)})(|f_{(-)}|/|f_{(+)}|)} \tag{2.41}$$

The effect of the load at $f_{(+)}$ is again to decrease the gain.

6. Negative-resistance amplifiers

We have seen that in the case of the three-frequency difference-frequency converter for $f_p > f_e$ and in that of the four-frequency converter for $f_p > f_e$ and $|P_{(-)}|/|f_{(-)}| > |P_{(+)}|/|f_{(+)}|$, the signal generator at f_e receives energy. An amplifier may therefore be constructed by placing the working load in the source circuit whilst at the same time maintaining a passive load or 'idler' in the branch corresponding to $|f_{(-)}|$.

In the case of the four-frequency circuit, the effect of the load at $f_{(+)}$ is to impair the performance of the amplifier. In fact, equation (2.33) put in the form

$$P_e = -f_e \left[\frac{|P_{(-)}|}{|f_{(-)}|} - \frac{|P_{(+)}|}{f_{(+)}} \right] \tag{2.42}$$

shows that P_e, which (2.38) causes to be negative, decreases in absolute value as $|P_{(+)}|$ increases.

Table 2.1. Properties of three- and four-frequency parametric amplifiers

Class	Type	f_p	$f_{(-)}$	Spectrum of output signal	Stability	Effect of idler load
Three-frequency converter	Sum-frequency	$\gtrless f_e$		Non-inverted	Unconditionally stable	—
	Difference-frequency	$< f_e$	< 0	Non-inverted	Unconditionally stable	—
		$> f_e$	> 0	Inverted	Potentially unstable	—
Four-frequency converter	Sum-frequency	$< f_e$	< 0	Non-inverted	Unconditionally stable if $\dfrac{\|P_{(-)}\|}{\|f_{(-)}\|} > \dfrac{\|P_{(+)}\|}{f_{(+)}}$	$G\nearrow$
		$> f_e$	> 0	Non-inverted	Potentially unstable if $\dfrac{\|P_{(-)}\|}{\|f_{(-)}\|} > \dfrac{\|P_{(+)}\|}{f_{(+)}}$	$G\nearrow$
	Difference-frequency	$< f_e$	< 0	Non-inverted	Unconditionally stable	$G\nearrow$
		$> f_e$	> 0	Inverted	Potentially unstable if $\dfrac{\|P_{(-)}\|}{\|f_{(-)}\|} > \dfrac{\|P_{(+)}\|}{f_{(+)}}$	$G\nearrow$
Amplifier	Three frequencies	$> f_e$	> 0	Non-inverted	Potentially unstable	$G\nearrow$
	Four frequencies	$> f_e$	> 0	Non-inverted	Potentially unstable if $\dfrac{\|P_{(-)}\|}{\|f_{(-)}\|} > \dfrac{\|P_{(+)}\|}{f_{(+)}}$	$\begin{cases} f_{(-)}, G\nearrow \\ f_{(+)}, G\nearrow \end{cases}$

7. Classification of three- and four-frequency parametric systems

The properties of the amplifiers and converters discussed so far are summarized in Table 2.1.

The three-frequency sum-frequency converter is unconditionally stable and non-inverting. The same applies to three-frequency difference-frequency converters and to all four-frequency converters provided $f_p < f_e$. The power gain of these converters cannot exceed the value of the ratio of the output and input frequencies. The necessary condition for having potential instability is that the frequency $f_{(-)}$ should be used (either as an output or an idler frequency), and that it should be positive ($f_p > f_e$). This essential condition is all that is required for three-frequency converters. With four-frequency converters a further condition is required for potential instability, namely:

$$\frac{|P_{(-)}|}{|f_{(-)}|} > \frac{|P_{(+)}|}{f_{(+)}} \qquad (2.43)$$

The transducer gain of the converters which are potentially unstable may, in principle, be made as high as required.

8. Relation between the power gain and the transducer gain

Let us consider the circuit in Fig. 2.10. It represents schematically an amplifier fed by a source of internal impedance $Z_e = R_e + jX_e$ and of e.m.f. E_e. The load consists of a passive impedance $Z_L = R_L + jX_L$ and $Z_i = R_i + jX_i$ is the input impedance of the amplifier.

Let G_L^e and $(G_L^e)_{\text{Tr}}$ represent the power gain and the transducer gain respectively of the amplifier. We may write:

$$G_L^e = \frac{R_L|I_L|^2}{\dfrac{|E_e|^2 R_i}{|Z_e + Z_i|^2}} \qquad (2.44)$$

$$(G_L^e)_{\text{Tr}} = \frac{R_L|I_L|^2}{|E_e|^2/4R_e} \qquad (2.45)$$

Hence:

$$(G_L^e)_{\text{Tr}} = \frac{4R_e R_L}{|Z_e + Z_i|^2} G_L^e \qquad (2.46)$$

Equation (2.44) tells us that G_L^e and R_i are necessarily of the same sign. It follows that the product $R_i G_L^e$ and consequently $(G_L^e)_{\text{Tr}}$ are

PARAMETRIC AMPLIFIERS AND CONVERTERS 67

always positive. When R_i is negative, $(G_L^e)_{\mathrm{Tr}}$ may be made as large as desired for $Z_i \approx -Z_e$ and $|R_i| \leqslant R_e$. The concept of power gain is essential in order to demonstrate the transfer of power that

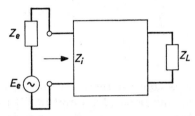

Figure 2.10. Block diagram of a two-port amplifier

can be effected by non-linear reactances. In practice, however, we are of course interested in the value of $(G_L^e)_{\mathrm{Tr}}$ and not that of G_L^e. It is for this reason that we will use the transducer gain in the subsequent chapters. In order to simplify the presentation we will in future omit the subscript 'Tr'.

9. Four-frequency double-conversion converters

With inverters the incoming signal is at a frequency f_e and at the output the frequency is $f_{(2-)}$ defined by:

$$f_{(2-)} = 2f_p - f_e \qquad (2.47)$$

A load or idler absorbs power at the frequency $f_{(-)} = f_p - f_e$. Figure 2.11 shows the spectrums of these converters. It will be

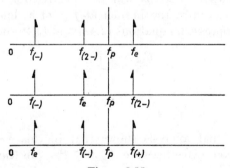

Figure 2.11

noticed (see the bottom of the figure) that this converter is in fact a four-frequency converter for $f_p > f_e$ as seen earlier, where the signal will enter at $f_{(+)}$ and leaves at $f_{(-)}$ or conversely, and where the idler will be at f_e. It is for this reason that we refer to this circuit amongst those in which harmonic combinations of f_p and f_e occur. The Manley and Rowe equations are written:

$$\frac{P_e}{f_e} - \frac{P_{(-)}}{f_{(-)}} - \frac{P_{(2-)}}{f_{(2-)}} = 0 \qquad (2.48)$$

$$\frac{P_p}{f_p} + \frac{P_{(-)}}{f_{(-)}} + \frac{2P_{(2-)}}{f_{(2-)}} = 0 \qquad (2.49)$$

As these equations concern the same circuit as that in section 5, they should enable equations (2.26) and (2.27) to be recalled if we substitute the indices $(-)$, e and $(2-)$ for indices e, $(-)$ and $(+)$ respectively. We then have:

$$\frac{P_{(-)}}{f_{(-)}} - \frac{P_e}{f_e} - \frac{P_{(+)}}{f_{(+)}} = 0 \qquad (2.50)$$

$$\frac{P_p}{f_p} + \frac{P_e}{f_e} + \frac{2P_{(+)}}{f_{(+)}} = 0 \qquad (2.51)$$

Equation (2.50) is identical with (2.26) and by adding together (2.50) and (2.51) we revert to (2.27). With non-inverting converters the signal enters at frequency f_e and leaves at frequency $f_{(2+)} = 2f_p + f_e$. A load absorbs power at frequency $f_{(+)}$. This type of device is derived from the four-frequency converter for $f_p < f_e$.

Discussion of this type of circuit gives us an opportunity to make an important observation: the Manley and Rowe equations give the relations between the powers and the frequencies in a non-linear reactance, but they do not in any way guarantee that a converter governed by these equations will be effective for the transfer of power. If, for example, the branch at $f_{(-)}$ of a double-conversion converter is suppressed, equations (2.48) and (2.49) become:

$$\frac{P_{(2-)}}{P_e} = \frac{f_{(2-)}}{f_e} \qquad (2.52)$$

$$\frac{P_p}{P_{(2-)}} = -\frac{2f_p}{f_{(2-)}} \qquad (2.53)$$

Now small-signal analysis shows that in this case no transfer of power can take place between P_e and $P_{(2-)}$ in the absence of S_2 (second harmonic of the elastance of the non-linear pumped reactance). As pumping can take place without difficulty in the absence of S_2 (cf. Chapter 8, current pumping of a square-root varactor), we can in this way obtain a circuit to which equations (2.52) and (2.53) apply but only allowing the trivial solution $P_e = P_{(2-)} = P_p = 0$.

In the absence of S_2, the transfer of power from f_e to $f_{(2-)}$ can take place only by a first conversion from f_e to $f_{(-)}$ followed by a second conversion from $f_{(-)}$ to $f_{(2-)}$, whence the nomenclature chosen for this type of circuit.

References

[1] Manley and Rowe, 'Some general properties of non-linear elements—Part I, General energy relations', *Proc. I.R.E.*, vol. 44, pp. 904–913, July 1956.

[2] Adams, D. K., 'An analysis of four-frequency nonlinear reactance circuits', *I.R.E. Trans. Microwave Theory Techniques*, vol. MTT6, pp. 274–283, March 1960.

[3] Duinker, S., 'General energy relations for parametric amplifying devices', *Tijdschrift van het Nederlands Radiogenootschap*, vol. 24, pp. 287–310, 1959.

3

SMALL-SIGNAL ANALYSIS
by L. LAURENT

The Manley and Rowe equations do no more than give us information on the general operating conditions of a parametric amplifier using a non-linear non-dissipative reactance. The reactances used in practice are semiconductor diodes (varactors) in which losses inevitably occur. The following discussion is intended to enable us to calculate the following quantities, taking into account the losses just mentioned:

the gains of the different types of amplifiers;
their optimum terminal impedances;
their effective noise temperature.

All these characteristics may be derived from one basic equation.

1. Small-signal theory for an ideal junction

This theory establishes the relations between the current through a non-linear capacitance and the voltage at its terminals, for signals whose amplitude is assumed to be infinitely small compared to that of the pumping voltage. The non-linear capacitance is assumed to be lossless.
 Let
$$q = f(v) \tag{3.1}$$
represent the law relating the load and the voltage at the terminals of the non-linear capacitance. Let us apply to its terminals a pumping voltage $v_p(t)$, periodic with respect to time; we then have:
$$q_p = f(v_p) \tag{3.2}$$
If we now apply in addition a small voltage, $\delta v \ll v_p$, (3.2) becomes:
$$q_p + \delta q = f(v_p + \delta v) \tag{3.3}$$

and the series development of $f(v_p + \delta v)$, restricted to the first term, gives:

$$q_p + \delta q = [f(v)]_{v=v_p} + \delta v \left[\frac{df(v)}{dv}\right]_{v=v_p} \tag{3.4}$$

hence:

$$\delta q = \delta v \left[\frac{df(v)}{dv}\right]_{v=v_p} \tag{3.5}$$

Let $C = (dq/dv) = (df(v)/dv)$ represent the differential capacitance of the junction, in contrast to its mean capacitance:

$$\bar{C} = \frac{q}{v} \tag{3.6}$$

Equation (3.5) is written:

$$dq = C(v_p)\,\delta v = C(t)\,\delta v \tag{3.7}$$

In order to find the relation between the current through the non-linear capacitance and the voltage at its terminals, we can start directly from relation (3.7), or from this relation written in the form:

$$\delta v = \frac{1}{C(t)}\,\delta q = S(t)\,\delta q \tag{3.8}$$

where $S(t)$ is the differential elastance. As $v_p(t)$ is periodic, $C(t)$ and $S(t)$ are also periodic, and we can develop them as Fourier series:

$$C(t) = \sum_{n=-\infty}^{+\infty} C_n\, e^{jn\omega_p t} \tag{3.9}$$

$$S(t) = \sum_{m=-\infty}^{+\infty} S_n\, e^{jn\omega_p t} \tag{3.10}$$

with

$$C_n = \frac{1}{2\pi} \int_0^{2\pi} C(t)\, e^{-jn\omega_p t}\, d\omega t \tag{3.11}$$

$$S_n = \frac{1}{2\pi} \int_0^{2\pi} S(t)\, e^{-jn\omega_p t}\, d\omega t \tag{3.12}$$

$$C_{-n} = C_n^* \tag{3.13}$$

$$S_{-n} = S_n^* \tag{3.14}$$

Note. It will be seen from the above that the small-signal hypothesis allows one to substitute for the varactor and its pumping circuit a capacitance or an elastance that varies periodically with time.

As $C(t)$ and $S(t)$ are biunivocal functions of $v(t)$ (no hysteresis) they are even if $v_p(t)$ is an even function of time, in which case we have:

$$C_n = C_n^* \qquad (3.15)$$

$$S_n = S_n^* \qquad (3.16)$$

Let us assume now that a small signal of frequency f_e is applied to the non-linear capacitance. Quite generally, the resulting charges and potential differences should be connected by equations (3.7) or (3.8), where $C(t)$ and $S(t)$ are periodic at ω_p; δv and δq are therefore periodic at ω_e and ω_p and they can be developed in the form:

$$\delta v = \sum_{m=-1}^{1} \sum_{n=-\infty}^{+\infty} V_{m,n}\, e^{j(m\omega_e + n\omega_p)t} \qquad (3.17)$$

$$\delta q = \sum_{m=-1}^{1} \sum_{n=-\infty}^{+\infty} Q_{m,n}\, e^{j(m\omega_e + n\omega_p)t} \qquad (3.18)$$

(The terms corresponding to the harmonic combinations of ω_e, $V_{m,n}$ and $Q_{m,n}$ for $|m| > 1$ are negligible in accordance with the small-signal hypothesis.) Equations (2.1) and (2.7) give us:

$$Q_{m,n} = \frac{I_{m,n}}{j\omega_{m,n}} \qquad (3.19)$$

and equations (3.7) and (3.8) respectively may be written:

$$\sum_{m=-1}^{+1} \sum_{n=-\infty}^{+\infty} \frac{I_{m,n}}{j\omega_{m,n}}\, e^{j(m\omega_e + n\omega_p)t}$$
$$= \left(\sum_{k=-\infty}^{+\infty} C_k\, e^{jk\omega_p t} \right) \left(\sum_{m=-1}^{+1} \sum_{n=-\infty}^{+\infty} V_{m,n}\, e^{j(m\omega_e + n\omega_p)t} \right) \qquad (3.20)$$

$$\sum_{m=-1}^{+1} \sum_{n=-\infty}^{+\infty} V_{m,n}\, e^{j(m\omega_e + n\omega_p)t}$$
$$= \left(\sum_{k=-\infty}^{+\infty} S_k\, e^{jk\omega_p t} \right) \left(\sum_{m=-1}^{+1} \sum_{n=-\infty}^{+\infty} \frac{I_{m,n}}{j\omega_{m,n}}\, e^{j(m\omega_e + n\omega_p)t} \right) \qquad (3.21)$$

Because of the properties of the Fourier series in equations (3.9) to (3.22), the values C_n, S_n, $V_{m,n}$ and $Q_{m,n}$ are half-amplitudes. However, the r.m.s. amplitudes are usually in more general use and in more convenient form for the representation of the voltages and currents, and the equations (3.20) and (3.21) are still valid in this case. In what follows, we will, unless otherwise stated, use the r.m.s. amplitudes for the voltages and currents. The C_n's and S_n's, on the other hand, will remain half-amplitudes.

SMALL-SIGNAL ANALYSIS

Equations (3.20) and (3.21) may be written in matrix form for $m = 1$:

$$\begin{bmatrix} \vdots \\ I_{(1,-2)} \\ I_{(1,-1)} \\ I_{(1,0)} \\ I_{(1,1)} \\ I_{(1,2)} \\ \vdots \end{bmatrix} = \begin{bmatrix} \vdots & \vdots & \vdots & \vdots & \vdots & \vdots & \vdots \\ \cdots & j\omega_{(1,-2)}C_0 & j\omega_{(1,-2)}C_1^* & j\omega_{(1,-2)}C_2^* & j\omega_{(1,-2)}C_3^* & j\omega_{(1,-2)}C_4^* & \cdots \\ \cdots & j\omega_{(1,-1)}C_1 & j\omega_{(1,-1)}C_0 & j\omega_{(1,-1)}C_1^* & j\omega_{(1,-1)}C_2^* & j\omega_{(1,-1)}C_3^* & \cdots \\ \cdots & j\omega_{(1,0)}C_2 & j\omega_{(1,0)}C_1 & j\omega_{(1,0)}C_0 & j\omega_{(1,0)}C_1^* & j\omega_{(1,0)}C_2^* & \cdots \\ \cdots & j\omega_{(1,1)}C_3 & j\omega_{(1,1)}C_2 & j\omega_{(1,1)}C_1 & j\omega_{(1,1)}C_0 & j\omega_{(1,1)}C_1^* & \cdots \\ \cdots & j\omega_{(1,2)}C_4 & j\omega_{(1,2)}C_3 & j\omega_{(1,2)}C_2 & j\omega_{(1,2)}C_1 & j\omega_{(1,2)}C_0 & \cdots \\ \vdots & \vdots & \vdots & \vdots & \vdots & \vdots & \vdots \end{bmatrix} \begin{bmatrix} \vdots \\ V_{(1,-2)} \\ V_{(1,-1)} \\ V_{(1,0)} \\ V_{(1,1)} \\ V_{(1,2)} \\ \vdots \end{bmatrix} \quad (3.22)$$

and

$$\begin{bmatrix} \vdots \\ V_{(1,-2)} \\ V_{(1,-1)} \\ V_{(1,0)} \\ V_{(1,1)} \\ V_{(1,2)} \\ \vdots \end{bmatrix} = \begin{bmatrix} \vdots & \vdots & \vdots & \vdots & \vdots & \vdots & \vdots \\ \cdots & \dfrac{S_0}{j\omega_{(1,-2)}} & \dfrac{S_1^*}{j\omega_{(1,-1)}} & \dfrac{S_2^*}{j\omega_{(1,0)}} & \dfrac{S_3^*}{j\omega_{(1,1)}} & \dfrac{S_4^*}{j\omega_{(1,2)}} & \cdots \\ \cdots & \dfrac{S_1}{j\omega_{(1,-2)}} & \dfrac{S_0}{j\omega_{(1,-1)}} & \dfrac{S_1^*}{j\omega_{(1,0)}} & \dfrac{S_2^*}{j\omega_{(1,1)}} & \dfrac{S_3^*}{j\omega_{(1,2)}} & \cdots \\ \cdots & \dfrac{S_2}{j\omega_{(1,-2)}} & \dfrac{S_1}{j\omega_{(1,-1)}} & \dfrac{S_0}{j\omega_{(1,0)}} & \dfrac{S_1^*}{j\omega_{(1,1)}} & \dfrac{S_2^*}{j\omega_{(1,2)}} & \cdots \\ \cdots & \dfrac{S_3}{j\omega_{(1,-2)}} & \dfrac{S_2}{j\omega_{(1,-1)}} & \dfrac{S_1}{j\omega_{(1,0)}} & \dfrac{S_0}{j\omega_{(1,1)}} & \dfrac{S_1^*}{j\omega_{(1,2)}} & \cdots \\ \cdots & \dfrac{S_4}{j\omega_{(1,-2)}} & \dfrac{S_3}{j\omega_{(1,-1)}} & \dfrac{S_2}{j\omega_{(1,0)}} & \dfrac{S_1}{j\omega_{(1,1)}} & \dfrac{S_0}{j\omega_{(1,2)}} & \cdots \\ \vdots & \vdots & \vdots & \vdots & \vdots & \vdots & \vdots \end{bmatrix} \begin{bmatrix} \vdots \\ I_{(1,-2)} \\ I_{(1,-1)} \\ I_{(1,0)} \\ I_{(1,1)} \\ I_{(1,2)} \\ \vdots \end{bmatrix} \quad (3.23)$$

For $m = -1$, (3.22) and (3.23) are rewritten in the conjugate form, which is without interest, and for $m = 0$, we obtain the relation between the pumping currents and voltages, which will be dealt with elsewhere. It should be noted that in equations (3.22) and (3.23) the frequencies are algebraic quantities. In order to render the formulae to be established valid whatever the relative values of f_p and f_e, we will no longer use the absolute value of the frequency $f_{(-)}$, as we did in Chapter 2, for the purpose of the classification of parametric amplifiers and converters. The effect of the sign of $f_{(-)}$ will be examined in each particular case.

The systems of equations (3.22) and (3.23) consist of an infinite number of equations, and their solution would require a knowledge of the behaviour of the circuit up to infinite frequencies. In practice, we are concerned with only three or four frequencies. In these conditions, the equations can be simplified by the elimination of the unwanted components. Depending on the configuration of the parametric device, this elimination will affect either the voltage or current components.

(1) In the series configuration (Fig. 2.2) filters are used which produce short circuits across the varactor at unwanted frequencies, without affecting the signals at wanted frequencies.

We then have $V_{(i,k)} = 0$ in (3.22) and (3.23), for the indices i and k corresponding to unused frequencies. In this case, only equation (3.22) is simplified. When we retain only $f_{(1,0)}, f_{(1,-1)}$ and $f_{(1,1)}$, (3.22) gives us:

$$\begin{bmatrix} I_{(1,-1)} \\ I_{(1,0)} \\ I_{(1,1)} \end{bmatrix} = \begin{bmatrix} j\omega_{(1,-1)}C_0 & j\omega_{(1,-1)}C_1^* & j\omega_{(1,-1)}C_2^* \\ j\omega_{(1,0)}C_1 & j\omega_{(1,0)}C_0 & j\omega_{(1,0)}C_1^* \\ j\omega_{(1,1)}C_2 & j\omega_{(1,1)}C_1 & j\omega_{(1,1)}C_0 \end{bmatrix} \begin{bmatrix} V_{(1,-1)} \\ V_{(1,0)} \\ V_{(1,1)} \end{bmatrix} \quad (3.24)$$

Using the same notation as in Chapter 2:

$$\left.\begin{aligned} \omega_{(1,0)} &= \omega_e \\ \omega_{(1,1)} &= \omega_p + \omega_e = \omega_{(+)} \\ \omega_{(1,-1)} &= -\omega_p + \omega_e = -\omega_{(-)} \end{aligned}\right\} \quad (3.25)$$

and putting:

$$
\left.\begin{aligned}
I_{(1,-1)} &= I^*_{(-1,1)} = I^*_{(-)} \\
I_{(1,0)} &= I_e \\
I_{(1,1)} &= I_{(+)} \\
V_{(1,-1)} &= V^*_{(-1,1)} = V^*_{(-)} \\
V_{(1,0)} &= V_e \\
V_{(1,1)} &= V_{(+)}
\end{aligned}\right\} \qquad (3.26)
$$

(3.24) then becomes:

$$
\begin{bmatrix} I^*_{(-)} \\ I_e \\ I_{(+)} \end{bmatrix} = \begin{bmatrix} -j\omega_{(-)}C_0 & -j\omega_{(-)}C^*_1 & -j\omega_{(-)}C^*_2 \\ j\omega_e C_1 & j\omega_e C_0 & j\omega_e C^*_1 \\ j\omega_{(+)}C_2 & j\omega_{(+)}C_1 & j\omega_{(+)}C_0 \end{bmatrix} \begin{bmatrix} V^*_{(-)} \\ V_e \\ V_{(+)} \end{bmatrix} \qquad (3.27)
$$

(2) In the parallel configuration (Fig. 2.1), filters are used which cause open circuits at the terminals of the varactor at the unwanted frequencies, without affecting the signals at the wanted frequencies. This time the $I_{(i,k)}$'s are zero and (3.23) is simplified. Keeping to the same frequencies as in the previous case, (3.23) gives us:

$$
\begin{bmatrix} V^*_{(-)} \\ V_e \\ V_{(+)} \end{bmatrix} = \begin{bmatrix} -\dfrac{S_0}{j\omega_{(-)}} & \dfrac{S^*_1}{j\omega_e} & \dfrac{S^*_2}{j\omega_{(+)}} \\ -\dfrac{S_1}{j\omega_{(-)}} & \dfrac{S_0}{j\omega_e} & \dfrac{S^*_1}{j\omega_{(+)}} \\ -\dfrac{S_1}{j\omega_{(-)}} & \dfrac{S_1}{j\omega_e} & \dfrac{S^*_0}{j\omega_{(+)}} \end{bmatrix} \begin{bmatrix} I^*_{(-)} \\ I_e \\ I_{(+)} \end{bmatrix} \qquad (3.28)
$$

It will be seen that there is a simpler method of calculation for each of these two types of configuration: analysis by series development of the capacitance applies to the series circuit, and by elastance development to the parallel circuit.

2. Introduction of terminal immittances in the equations

The terms of the principal diagonals of the matrices in equations (3.27) and (3.28) represent the susceptance and reactance respectively of the constant term of the development, C_0 or S_0, at the frequency considered. Assuming that there are filters separating the different branches from each other, all we need to do is to add directly the terminal immittances of these branches to the corresponding terms of the principal diagonals and to replace currents I and voltages V of the first members of relations (3.27) and (3.28) respectively, by sources of current J, and electromotive forces E, in order to obtain the relations between the voltages and currents of the complete amplifier. Figures 3.1 and 3.2 show the equivalent circuits for the two configurations.

Series configuration (Fig. 3.1)

$$\begin{bmatrix} 0 \\ J_e \\ 0 \end{bmatrix} = \begin{bmatrix} Y^*_{(-)} & -j\omega_{(-)}C^*_1 & -j\omega_{(-)}C^*_2 \\ j\omega_e C_1 & Y_e & j\omega_e C^*_1 \\ j\omega_{(+)}C_2 & j\omega_{(+)}C_1 & Y_{(+)} \end{bmatrix} \begin{bmatrix} V^*_{(-)} \\ V_e \\ V_{(+)} \end{bmatrix} \quad (3.29)$$

with:

$$\left. \begin{array}{l} Y_{(-)} = G_{(-)} + jB_{(-)} + j\omega_{(-)}C_0 \\ Y_e = G_e + jB_e + j\omega_e C_0 \\ Y_{(+)} = G_{(+)} + jB_{(+)} + j\omega_{(+)}C_0 \end{array} \right\} \quad (3.30)$$

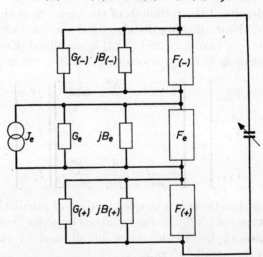

Figure 3.1. Series configuration for small signals

Parallel configuration (Fig. 3.2)

$$\begin{bmatrix} 0 \\ E_e \\ 0 \end{bmatrix} = \begin{bmatrix} Z^*_{(-)} & -j\dfrac{S^*_1}{\omega_e} & -j\dfrac{S^*_2}{\omega_{(+)}} \\ j\dfrac{S_1}{\omega_{(-)}} & Z_e & -j\dfrac{S^*_1}{\omega_{(+)}} \\ j\dfrac{S_2}{\omega_{(+)}} & -j\dfrac{S_1}{\omega_e} & Z_{(+)} \end{bmatrix} \begin{bmatrix} I^*_{(-)} \\ I_e \\ I_{(+)} \end{bmatrix} \quad (3.31)$$

with

$$Z_{(-)} = R_{(-)} + jX_{(-)} + \frac{S_0}{j\omega_{(-)}}$$
$$Z_e = R_e + jX_e + \frac{S_0}{j\omega_e}$$
$$Z_{(+)} = R_{(+)} + jX_{(+)} + \frac{S_0}{j\omega_{(+)}}$$
(3.32)

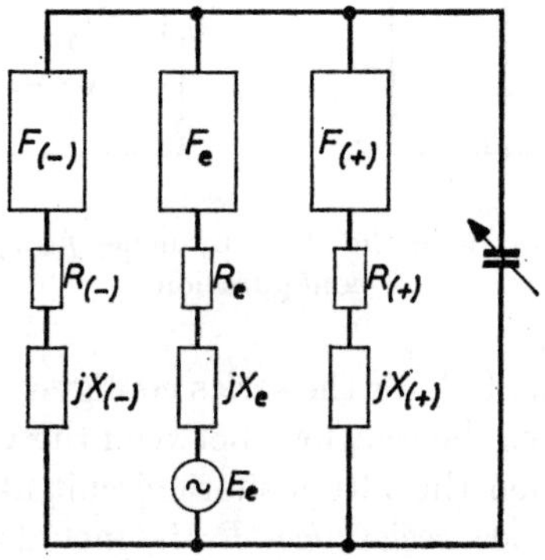

Figure 3.2. Parallel configuration for small signals

3. Introduction of the varactor losses in the equations

As will be seen in Chapter 8, the effect of the losses in the diode is to introduce a constant resistance R_s in series with the non-linear capacitance. Assuming the presence of filters separating the different branches, it is easy to introduce R_s in the equations corresponding to the parallel configuration. In fact, the voltage drop at the terminals of the resistance R_s common to the three branches influences the circuit of each of them only by its component at the frequency corresponding to that branch. We can thus transfer R_s in each of the three branches and add it to the terminal resistances. The same may be done for S_0 (Fig. 3.3). Equation (3.31) then becomes:

$$\begin{bmatrix} 0 \\ E_e \\ 0 \end{bmatrix} = \begin{bmatrix} R_s + Z^*_{(-)} & -\dfrac{S_1^*}{j\omega_e} & -j\dfrac{S_2^*}{\omega_{(+)}} \\ j\dfrac{S_1}{\omega_{(-)}} & R_s + Z_e & -j\dfrac{S_1^*}{\omega_{(+)}} \\ j\dfrac{S_2}{\omega_{(-)}} & -j\dfrac{S_1}{\omega_e} & R_s + Z_{(+)} \end{bmatrix} \begin{bmatrix} I^*_{(-)} \\ I_e \\ I_{(+)} \end{bmatrix}$$
(3.33)

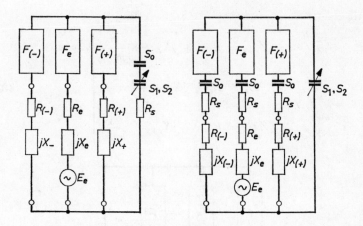

Figure 3.3. Transfer of R_s in the three branches $f_{(-)}$, f_e and $f_{(+)}$ for parallel configuration

The introduction of R_s for the series configuration is more difficult; equation (3.27) gives the relations between the voltages and currents of the junction when the filters short circuit all the unwanted voltages. When a series resistance R_s is included in the varactor, this condition cannot be realized. On the other hand, the parallel representation of the varactor, which is the only method of introducing the losses directly into (3.27), is not very convenient. The conductance G corresponding to R_s depends on the voltage applied to the junction and should also be developed as a Fourier series. An approximate method consists in limiting the matrix of (3.22) to nine elements, ignoring the high-order components and then inverting this matrix. We then obtain a relation in the form:

$$[V] = [Z][I] \tag{3.34}$$

into which we can introduce R_s. This calculation will not be made here. The comparison of the results obtained for the two configurations shows that the differences in performance are slight, with some small advantage in favour of the 'open-circuit' configuration.

It should be noted that the two configurations considered can only be obtained in practice in an approximate manner as it is, in fact, impossible to obtain either an open circuit, or a short circuit for all the unwanted frequencies. We must be content to design the circuit for the pumping frequency and for the two or three retained main frequencies. This is already a complex problem: each filter must not only transmit its corresponding signal but also reject without losses the signals at the other main frequencies. The circuit impedances at

SMALL-SIGNAL ANALYSIS

the unwanted frequencies are then usually arbitrary reactances and their behaviour is intermediate between that of open circuits and short circuits. We can therefore without hesitation choose the method of calculation by development of the elastance.

4. Introduction of reduced variables

The fundamental relations (3.32) and (3.33) will serve as a basis for a more detailed study of the different types of converters and amplifiers which were mentioned in Chapter 2. A first simplification can be made by choosing the time origin so that S_1 is real and positive; we then have

$$S_1 = S_1^* \tag{3.35}$$

The second simplification will be effected by the introduction of reduced variables (shown in lower-case letters). It will be seen during the process of calculation that the various circuit impedances figure in the formulae for the gains and the noise factors only in the form of their ratio to the series resistance R_s of the varactor. We will therefore make:

$$\left. \begin{array}{l} r_i = \dfrac{R_i}{R_s} \\[6pt] x_i = \dfrac{X_i}{R_s} \\[6pt] z_i = \dfrac{Z_i}{R_s} \end{array} \right\} \tag{3.36}$$

We will also make:

$$\left. \begin{array}{l} \omega_c = 2\pi f_c = \dfrac{S_0}{R_s} \\[6pt] \omega_c' = 2\pi f_c' = \dfrac{S_1}{R_s} \\[6pt] \omega_c'' = 2\pi f_c'' = \dfrac{|S_2|}{R_s} \end{array} \right\} \tag{3.37}$$

The quantities z_i, r_i, x_i are numbers without dimension whilst the quantities f_c, f_c' and f_c'' possess the dimension of a frequency. The symbol f_c is called the static cut-off frequency of the pumped varactor. Thus must not be confused with the static cut-off frequency of the non-pumped varactor corresponding to a biasing voltage V defined by:

$$f_c(V) = \frac{1}{2\pi R_s C(V)} = \frac{S(V)}{2\pi R_s} \qquad (3.38)$$

(this latter characteristic is the one generally quoted by the manufacturers). Similarly, f_c' is called the dynamic cut-off frequency corresponding to the first harmonic of the elastance. It will be seen that the performance of parametric systems improves as the value of f_c' increases and that f_c'', the dynamic cut-off frequency corresponding to the second harmonic of the elastance, appears only in four-frequency circuits. Relations (3.32) and (3.33) become:

$$\left.\begin{aligned} Z_{(-)} &= R_s\left[r_{(-)} + jx_{(-)} - j\frac{\omega_c}{\omega_{(-)}}\right] \\ Z_e &= R_s\left[r_e + jx_e - j\frac{\omega_c}{\omega_e}\right] \\ Z_{(+)} &= R_s\left[r_{(+)} + jx_{(+)} - j\frac{\omega_c}{\omega_{(+)}}\right] \end{aligned}\right\} \qquad (3.39)$$

$$\begin{bmatrix} 0 \\ E_e \\ 0 \end{bmatrix} = R_s \begin{bmatrix} 1 + z_{(-)}^* & -j\dfrac{\omega_c'}{\omega_e} & -j\dfrac{\omega_c'' e^{-j\phi}}{\omega_{(+)}} \\ j\dfrac{\omega_c'}{\omega_{(-)}} & 1 + z_e & -j\dfrac{\omega_c'}{\omega_{(+)}} \\ j\dfrac{\omega_c'' e^{j\phi}}{\omega_{(-)}} & -j\dfrac{\omega_c'}{\omega_e} & 1 + z_{(+)} \end{bmatrix} \begin{bmatrix} I_{(-)}^* \\ I_e \\ I_{(+)} \end{bmatrix} \qquad (3.40)$$

The complete equation (3.40) applies to four-frequency systems. These devices will be discussed in Chapter 6.

5. Special case of three-frequency systems

5.1. Sum-frequency converter

The filters allow only the currents I_e and $I_{(+)}$ to pass. We therefore have $I_{(-)} = 0$, and (3.40) reduces to:

$$\begin{bmatrix} E_e \\ 0 \end{bmatrix} = R_s \begin{bmatrix} 1 + z_e & -j\dfrac{\omega_c'}{\omega_{(+)}} \\ -j\dfrac{\omega_c'}{\omega_e} & 1 + z_{(+)} \end{bmatrix} \begin{bmatrix} I_e \\ I_{(+)} \end{bmatrix} \qquad (3.41)$$

5.2. Difference-frequency converter

The filters allow only currents I_e and $I_{(-)}$ to pass. We therefore have $I_{(+)} = 0$, and (3.40) reduces to:

$$\begin{bmatrix} 0 \\ E_e \end{bmatrix} = R_s \begin{bmatrix} 1 + z^*_{(-)} & -j\dfrac{\omega'_c}{\omega_e} \\ j\dfrac{\omega'_c}{\omega_{(-)}} & 1 + z_e \end{bmatrix} \begin{bmatrix} I^*_{(-)} \\ I_e \end{bmatrix} \qquad (3.42)$$

(a) Case in which $\omega_p < \omega_e$. As $\omega_{(-)}$ is negative, the frequency $\omega_{|-|}$ is defined by:

$$\omega_{|-|} = -\omega_{(-)} \qquad (3.43)$$

and we will make:

$$\begin{aligned} z_{|-|} &= z^*_{(-)} \\ I_{|-|} &= I^*_{(-)} \end{aligned} \qquad (3.44)$$

(3.42) can then be written:

$$\begin{bmatrix} E_e \\ 0 \end{bmatrix} = R_s \begin{bmatrix} 1 + z_e & -j\dfrac{\omega'_e}{\omega_{|-|}} \\ -j\dfrac{\omega'_c}{\omega_e} & 1 + z_{|-|} \end{bmatrix} \begin{bmatrix} I_e \\ I_{(-)} \end{bmatrix} \qquad (3.45)$$

(3.45) will be identical with (3.41) if the suffix $|-|$ is replaced by $(+)$. We would therefore expect that the gains, noise temperatures and the other characteristics would be exactly the same for the sum-frequency converter, and for the difference-frequency converter, with $f_p < f_e$ as pointed out in the previous chapter. It is for this reason that they will be studied together in detail in Chapter 4. As neither inverts the spectrum of the output signal with respect to the input signal, this chapter will be entitled: Non-inverting three-frequency converters.

(b) Case in which $\omega_p > \omega_e$. Equation (3.42) is applicable without change, as all the frequencies appearing in it are positive. We will call the converter at $f_{(-)} > 0$: the three-frequency inverting converter.

5.3. Reflection amplifier

As we saw in Chapter 2, this device is derived from the difference-frequency converter with $f_p > f_e$; it is therefore also governed by equation (3.42). For this reason the detailed studies of the inverting converter and the reflection amplifier will be made together in Chapter 5.

5.4. Degenerate amplifiers

When the pumping frequency is nearly twice the input frequency, the inverting converter and the reflection amplifier exhibit special properties resulting from the impossibility of producing filters selective enough to separate frequencies f_e and $f_{(-)}$. These devices are called degenerate amplifiers; they will be discussed in Chapter 7.

References

[1] Manley and Rowe, 'Some general properties of non-linear elements—Part II, Small-signal theory', *Proc. I.R.E.*, vol. 46, pp. 850–860, May 1958.
[2] Leenov, 'Gain and noise figure of a variable-capacitance up-converter', *Bell System Technical Journal*, vol. 37, pp. 989–1008, July 1958.
[3] Kurokawa, K. and Uenohara, M., 'Minimum noise figure of the variable-capacitance amplifier', *Bell System Technical Journal*, vol. 40, pp. 695–722, May 1961.
[4] Blackwell and Kotzebue, *Semiconductor-Diode Parametric Amplifiers*, Prentice-Hall, Inc., Englewood Cliffs, N.J., 1961.
[5] Laurent, L., 'Les amplificateurs paramétriques (1ere partie)', *Revue M.B.L.E.*, vol. VI (No. 2), June 1963.

4

THREE-FREQUENCY NON-INVERTING CONVERTERS

by L. LAURENT

1. Sum-frequency converter

1.1. Equivalent circuit

The frequency $f_{(-)}$ is not used and the equivalent circuit of Fig. 3.3 reduces to that in Fig. 4.1.

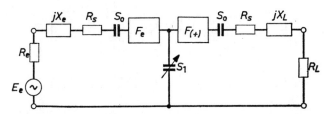

Figure 4.1. Equivalent circuit of a sum-frequency converter

Reverting to equation (3.41).

$$\begin{bmatrix} E_e \\ 0 \end{bmatrix} = R_s \begin{bmatrix} 1 + z_e & -j\dfrac{\omega'_c}{\omega_{(+)}} \\ -j\dfrac{\omega'_c}{\omega_e} & 1 + z_{(+)} \end{bmatrix} \begin{bmatrix} I_e \\ I_{(+)} \end{bmatrix} \quad (4.1)$$

Equations (3.39) may be written:

$$z_e = \frac{R_e}{R_s} + j\left(\frac{X_e}{R_s} - \frac{\omega_c}{\omega_e}\right) \quad (4.2)$$

$$z_{(+)} = \frac{R_L}{R_s} + j\left(\frac{X_L}{R_s} - \frac{\omega_c}{\omega_{(+)}}\right) \quad (4.3)$$

where R_e is the internal resistance of the generator, E_e is the electromotive force of the generator, R_L is the resistance of the load, and jX_e and jX_L are matching reactances. It will be seen that the terms in S_2 no longer appear in the equation. The properties of the sum-frequency converter therefore depend only on the first and second terms of the development of the elastance S_0 and S_1. If we introduce the auxiliary variables:

$$n_{(+)} = \frac{\omega_{(+)}}{\omega_e} \tag{4.4}$$

$$q = \frac{\omega'_c}{\omega_e} \tag{4.5}$$

(4.1) gives us

$$E_e = R_s I_e \left(1 + z_e + \frac{q^2/n_{(+)}}{1 + z_{(+)}} \right) \tag{4.6}$$

$$I_{(+)} = \frac{E_e}{R_s} \left[\frac{jq}{q^2/n_{(+)} + (1 + z_e)(1 + z_{(+)})} \right] \tag{4.7}$$

hence the input and output equivalent circuits shown in Fig. 4.2.

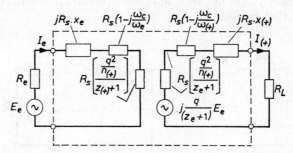

Figure 4.2. Equivalent circuits corresponding to equations (4.6) and (4.7)

We see that the generator of the input circuit, at frequency f_e, gives rise to an electromotive force of frequency $f_{(+)}$ in the output circuit and that the internal impedance of the generator at $f_{(+)}$ depends on the impedance of the circuit at f_e.

1.2. Transducer gain

The power dissipated in the load R_L is equal to:

$$P_{(+)} = R_L |I_{(+)}|^2 \tag{4.8}$$

The available power of the generator at frequency f_e is equal to

$$P_e = \frac{|E_e|^2}{4R_e} \tag{4.9}$$

The expression for the transducer gain is then:

$$G^e_{(+)} = \frac{P_{(+)}}{P_e} = 4R_e R_L \frac{|I_{(+)}|^2}{|E_e|^2} \quad (4.10)$$

which, by virtue of (4.6) and (4.7), may be written:

$$G^e_{(+)} = \frac{4r_e r_L q^2}{|q^2/n_{(+)} + (1+z_e)(1+z_{(+)})|^2} \quad (4.11)$$

1.3. Instability factor of the gain

As will be seen in Chapter 8, ω'_c depends on the pumping conditions and any drift in the pumping source may cause it to vary. If the input and output frequencies are given, q depends only on ω'_c. Since q appeared in the second power in formula (4.11) we will define the instability factor of the gain by:†

$$S^e_{(+)} = \frac{\partial G^e_{(+)}/G^e_{(+)}}{\partial(q^2)/q^2} = \frac{\partial G^e_{(+)}/G^e_{(+)}}{\partial(\omega'^2_c)/\omega'^2_c} \quad (4.12)$$

We find, in the condition of resonance:

$$S^e_{(+)} = 1 - \frac{2q^2/n_{(+)}}{q^2/n_{(+)} + (1+r_e)(1+r_L)} \quad (4.13)$$

1.4. Effective noise temperature

If we replace the input generator of internal resistance R_e by a simple resistance of value R_e cooled to 0°K, the only source of noise is R_s. The noise power fed to the load R_L is then coming from:

(1) a generator of e.m.f. $e_e = \sqrt{4kT_v R_s \Delta f}$, which accounts for the noise generated by R_s in the band Δf centered on f_e, and which produces, after conversion to a frequency $f_{(+)}$, an output power of ΔN_e. (T_v represents the temperature of the varactor.)
(2) a generator of an e.m.f. $e_{(+)} = \sqrt{4kT_v R_s \Delta f}$, which accounts for the noise generated in R_s in the band Δf centered on $f_{(+)}$, and which produces an output power $\Delta N_{(+)}$.

† The definition of the instability factor with respect to q^2 is particularly advantageous in the study of negative-resistance amplifiers, as we shall see in Chapter 5. In order to be consistent we have used this definition for all the types of converters and amplifiers considered.

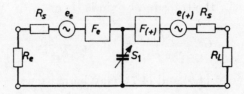

Figure 4.3. Noise generators of the sum-frequency converter

The equivalent circuit used for the calculation of the noise temperature is shown in Fig. 4.3.

In order to calculate ΔN_e, all that is needed is to replace E_e by e_e in the circuit shown in Fig. 4.2. Then, if we take account of (4.11) it is easy to obtain:

$$\Delta N_e = \frac{kT_v \Delta f \, G^e_{(+)}}{r_e} \qquad (4.14)$$

Similarly, to calculate $\Delta N_{(+)}$ we need only make $E_e = 0$ in the circuit of Fig. 4.2, and add the generator $e_{(+)}$ in the circuit $f_{(+)}$. We then find:

$$\Delta N_{(+)} = 4r_{(+)} kT_v \Delta f \left| \frac{1 + z_e}{q^2/n_{(+)} + (1 + z_e)(1 + z_{(+)})} \right|^2 \qquad (4.15)$$

Assuming that the input and output circuits are tuned ($z_e = r_e$ and $z_{(+)} = r_L$), the expression for the effective noise temperature is:

$$T^e_{(+)} = \frac{\Delta N_e + \Delta N_{(+)}}{k \, \Delta f G^e_{(+)}} = \frac{1}{r_e} \left[1 + \left(\frac{1 + r_e}{q} \right)^2 \right] T_v \qquad (4.16)$$

It can be seen that one can in principle make $T^e_{(+)}$ as low as required by a sufficient degree of cooling of the varactor.

1.5. Optimization of the gain

Examination of (4.11) shows immediately that at $n_{(+)}$, q, r_e and r_L given, one can obtain an optimum value of the gain by making z_e and $z_{(+)}$ real, which is equivalent to tuning the input and output circuits by means of the reactances x_e and x_L. We then have, with the aid of (4.39), (4.40), (4.41) and (4.42):

$$X_E = R_s \frac{\omega_c}{\omega_e} \qquad (4.17)$$

$$X_L = R_s \frac{\omega_c}{\omega_{(+)}} \qquad (4.18)$$

We will assume in what follows that this condition is satisfied.

(1) *Optimization of the gain with respect to r_e and r_L.* It is clear that in the conditions of optimization the resistance of the load r_L should be matched with the output impedance of the amplifier. The equivalent circuit of Fig. 4.2 shows that in this case we should have:

$$r_L = 1 + \frac{q^2/n_{(+)}}{1 + r_e} \tag{4.19}$$

Substituting this value in the expression obtained for the gain in the resonance condition we find:

$$G^e_{(+)} = \frac{r_e}{(1 + r_e)(q^2/n_{(+)} + 1 + r_e)} \tag{4.20}$$

If we equate to zero the differential of (4.20) with respect to r_e:

$$\frac{d\,G^e_{(+)}}{d\,r_e} = -(1 + r_e)^2(q^2/n_{(+)} + 1 + r_e)^2[r_e^2 - (1 + q^2/n_{(+)})] = 0 \tag{4.21}$$

Hence,

$$r_e = \sqrt{1 + q^2/n_{(+)}} \tag{4.22}$$

This gives us the optimum value of r_L if we substitute (4.22) in (4.19). We then have:

$$r_L = \sqrt{1 + q^2/n_{(+)}} = r_e \tag{4.23}$$

and the maximum gain corresponding to these conditions is equal to:

$$G^e_{(+)\max} = n_{(+)} \frac{\sqrt{1 + q^2/n_{(+)}} - 1}{\sqrt{1 + q^2/n_{(+)}} + 1} \tag{4.24}$$

(4.24) may be written in the form:

$$\frac{G^e_{(+)\max}}{n_{(+)}} = \frac{\sqrt{1 + q^2/n_{(+)}} - 1}{\sqrt{1 + q^2/n_{(+)}} + 1} \tag{4.25}$$

In the case of an ideal varactor, $q = \infty$ and (4.24) becomes:

$$[G^e_{(+)\max}]_{q\to\infty} = n_{(+)} \tag{4.26}$$

The second member of (4.25) then gives the deterioration factor of the gain with respect to that under ideal conditions.

(2) *Optimization of the gain with respect to r_e, r_L and $n_{(+)}$.* This optimization leads to an infinite value of $n_{(+)}$. The corresponding gain limit is equal to:

$$[G^e_{(+)\max}]_{n_{(+)}\to\infty} = q^2/4 \tag{4.27}$$

The factors limiting the permissible value of $n_{(+)}$ will be discussed later (section 4.1.9). Tables 4.1 and 4.2 and Figs. 4.4 and 4.5 show

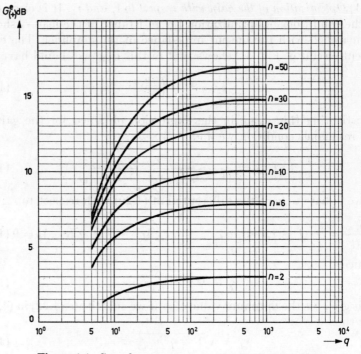

Figure 4.4. Sum-frequency converter. Maximum gain

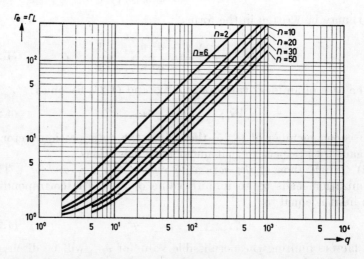

Figure 4.5. Sum-frequency converter, r_e and r_L for maximum gain

Table 4.1. Maximum gain (in dB) of the sum-frequency converter

	$n_{(+)}$							
q	2	4	6	8	10	12	14	16
2	−2·71	−1·63	−1·18	−0·93	−0·76	−0·65	−0·56	−0·50
5	0·50	2·63	3·68	4·35	4·82	5·17	5·45	5·68
7	1·27	3·57	4·80	5·61	6·20	6·65	7·02	7·31
10	1·79	4·29	5·67	6·61	7·30	7·84	8·28	8·65
15	2·19	4·87	6·37	7·40	8·18	8·80	9·32	9·75
20	2·40	5·15	6·72	7·81	8·63	9·29	9·85	10·32
50	2·76	5·67	7·36	8·54	9·45	10·19	10·81	11·35
70	2·83	5·77	7·48	8·68	9·61	10·36	11·00	11·55
100	2·89	5·85	7·57	8·79	9·73	10·49	11·14	11·69
150	2·93	5·90	7·64	8·87	9·82	10·59	11·24	11·81
200	2·95	5·93	7·68	8·91	9·86	10·64	11·30	11·87
500	2·99	5·99	7·74	8·98	9·95	10·73	11·40	11·97
700	2·99	6·00	7·75	9·00	9·96	10·75	11·41	11·99
1000	3·00	6·00	7·76	9·01	9·97	10·76	11·43	12·01

	$n_{(+)}$							
q	18	20	25	30	35	40	45	50
2	−0·45	−0·40	−0·33	−0·28	−0·24	−0·21	−0·19	−0·17
5	5·86	6·02	6·32	6·54	6·71	6·84	6·95	7·03
7	7·57	7·78	8·21	8·53	8·77	8·97	9·14	9·27
10	8·97	9·24	9·80	10·22	10·56	10·84	11·07	11·27
15	10·13	10·46	11·14	11·67	12·10	12·46	12·77	13·03
20	10·72	11·08	11·83	12·42	12·91	13·32	13·67	13·98
50	11·82	12·23	13·11	13·82	14·42	14·92	15·37	15·77
70	12·03	12·46	13·36	14·09	14·71	15·24	15·70	16·11
100	12·18	12·62	13·55	14·30	14·93	15·47	15·95	16·38
150	12·31	12·75	13·69	14·45	15·10	15·65	16·14	16·58
200	12·37	12·82	13·76	14·53	15·18	15·75	16·24	16·68
500	12·48	12·93	13·89	14·68	15·34	15·91	16·42	16·87
700	12·50	12·95	13·92	14·70	15·37	15·94	16·45	16·90
1000	12·52	12·97	13·94	14·72	15·39	15·97	16·47	16·93

$G^e_{(+)\text{max}}$ and the coresponding terminal resistances as functions of q for different values of $n_{(+)}$, and we see that from a certain value of q it is practically useless to select a better varactor, as the gain increases only very slowly with the value of q.

It should be noted that formula (4.25) enables us to plot a single reduced curve: that of $G^e_{(+)\text{max}}/n_{(+)}$ as a function of $q^2/n_{(+)}$. However, the network of Fig. 4.4 is considered preferable, as it is more suitable for immediate use.

Table 4.2. Sum-frequency converter. Reduced terminating impedances for maximum gain

q	\multicolumn{8}{c}{$n_{(+)}$}
	2	4	6	8	10	12	14	16
2	1·7	1·4	1·3	1·2	1·2	1·2	1·1	1·1
5	3·7	2·7	2·3	2·0	1·0	1·8	1·7	1·6
7	5·0	3·6	3·0	2·7	2·4	2·3	2·1	2·0
10	7·1	5·1	4·2	3·7	3·3	3·1	2·9	2·7
15	10·7	7·6	6·2	5·4	4·8	4·4	4·1	3·9
20	14·2	10·0	8·2	7·1	6·4	5·9	5·4	5·1
50	35·4	25·0	20·4	17·7	15·8	14·5	13·4	12·5
70	49·5	35·0	28·6	24·8	22·2	20·2	18·7	17·5
100	70·7	50·0	40·8	35·4	31·6	28·9	26·7	25·0
150	106·1	75·0	61·2	53·0	47·4	43·3	40·1	37·5
200	141·4	100·0	81·7	70·7	63·3	57·7	53·5	50·0
500	353·6	250·0	204·1	176·8	158·1	144·3	133·6	125·0
700	495·0	350·0	285·8	247·5	221·4	202·1	187·1	175·0
1000	707·1	500·0	408·2	353·6	316·2	288·7	267·3	250·0

q	\multicolumn{8}{c}{$n_{(+)}$}
	18	20	25	30	35	40	45	50
2	1·1	1·1	1·1	1·1	1·1	1·0	1·0	1·0
5	1·5	1·5	1·4	1·4	1·3	1·3	1·2	1·2
7	1·9	1·9	1·7	1·6	1·5	1·5	1·4	1·4
10	2·6	2·4	2·2	2·1	2·0	1·9	1·8	1·7
15	3·7	3·5	3·2	2·9	2·7	2·6	2·4	2·3
20	4·8	4·6	4·1	3·8	3·5	3·3	3·1	3·0
50	11·8	11·2	10·0	9·2	8·5	8·0	7·5	7·1
70	16·5	15·7	14·0	12·8	11·9	11·1	10·5	9·9
100	23·6	22·4	20·0	18·3	16·9	15·8	14·9	14·2
150	35·4	33·6	30·0	27·4	25·4	23·7	22·4	21·2
200	47·2	44·7	40·0	36·5	33·8	31·6	29·8	28·3
500	117·9	111·8	100·0	91·3	84·5	79·1	74·5	70·7
700	165·0	156·5	140·0	127·8	118·3	110·7	104·4	99·0
1000	235·7	223·6	200·0	182·6	169·0	158·1	149·1	141·4

(3) *Limiting values of the optimum terminal resistances.* As shown in formula (4.23), the optimum reduced resistances r_e and r_L increase with q (for high values of q, proportionally to q). The non-reduced resistances R_e and R_L follow the same law if the improvement in q is obtained by an increase in $|S_1|$ for a constant R_s. On the other hand,

if we make q tend to infinity by a decrease in R_s for a constant $|S_1|$, R_e and R_L tend to a finite limiting value given by

$$R_{e\,\text{lim}} = R_{L\,\text{lim}} = \frac{|S_1|}{\omega_e \sqrt{n_{(+)}}} \quad (4.28)$$

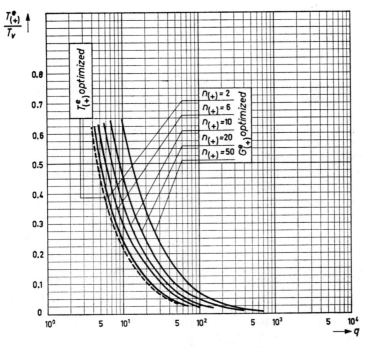

Figure 4.6. Sum-frequency converter. $T^e_{(+)}$ as a function of q for optimization of the gain and of the noise temperature

(4) *Noise temperature corresponding to maximum gain.* If we substitute (4.23) in (4.16) we will obtain the expression for the noise temperature corresponding to the maximum gain:

$$[T^e_{(+)}]_{G_{\max}} = \frac{T_v}{\sqrt{1+q^2/n_{(+)}}} \left[1 + \frac{1}{n_{(+)}} \cdot \frac{\sqrt{1+q^2/n_{(+)}}-1}{\sqrt{1+q^2/n_{(+)}}+1}\right] \quad (4.29)$$

Table 4.3 and Fig. 4.6 show $[T^e_{(+)}/T_v]_{G_{\max}}$ as a function of q for different values of $n_{(+)}$.

1.6. Optimization of the intrinsic noise temperature

Relation (4.16) shows that $T^e_{(+)}$ depends only on r_e. All that is necessary, therefore, is to optimize with respect to this parameter.

Table 4.3. Sum-frequency converter. $T^e_{(+)}/T_v$ corresponding to maximum gain

q	\multicolumn{7}{c}{$n_{(+)}$}						
	2	4	6	8	10	12	14	16
2	1·65	1·74	1·79	1·83	1·85	1·87	1·89	1·90
5	0·51	0·57	0·63	0·67	0·71	0·74	0·77	0·79
7	0·35	0·40	0·44	0·48	0·51	0·54	0·57	0·59
10	0·23	0·27	0·30	0·33	0·36	0·38	0·40	0·42
15	0·15	0·18	0·20	0·22	0·24	0·25	0·27	0·28
20	0·11	0·13	0·15	0·16	0·18	0·19	0·20	0·21
50	0·04	0·05	0·06	0·06	0·07	0·08	0·08	0·09
70	0·03	0·04	0·04	0·05	0·05	0·05	0·06	0·06
100	0·02	0·03	0·03	0·03	0·03	0·04	0·04	0·04
150	0·01	0·02	0·02	0·02	0·02	0·03	0·03	0·03
200	0·01	0·01	0·01	0·02	0·02	0·02	0·02	0·02
500	0·00	0·01	0·01	0·01	0·01	0·01	0·01	0·01
700	0·00	0·00	0·00	0·00	0·00	0·01	0·01	0·01
1000	0·00	0·00	0·00	0·00	0·00	0·00	0·00	0·00

q	\multicolumn{7}{c}{$n_{(+)}$}						
	18	20	25	30	35	40	45	50
2	1·91	1·91	1·93	1·94	1·95	1·95	1·96	1·96
5	0·81	0·83	0·87	0·90	0·93	0·95	0·96	0·98
7	0·61	0·63	0·67	0·70	0·73	0·76	0·78	0·79
10	0·44	0·46	0·49	0·53	0·55	0·58	0·60	0·62
15	0·30	0·31	0·34	0·37	0·39	0·41	0·43	0·45
20	0·23	0·24	0·26	0·28	0·30	0·32	0·33	0·35
50	0·09	0·09	0·10	0·11	0·12	0·13	0·14	0·14
70	0·06	0·07	0·07	0·08	0·09	0·09	0·10	0·10
100	0·04	0·05	0·05	0·06	0·06	0·06	0·07	0·07
150	0·03	0·03	0·03	0·04	0·04	0·04	0·05	0·05
200	0·02	0·02	0·03	0·03	0·03	0·03	0·03	0·04
500	0·01	0·01	0·01	0·01	0·01	0·01	0·01	0·01
700	0·01	0·01	0·01	0·01	0·01	0·01	0·01	0·01
1000	0·00	0·00	0·01	0·01	0·01	0·01	0·01	0·01

Equating to zero the differential of (4.16) with respect to r_e gives:

$$r_e = \sqrt{1 + q^2} \qquad (4.30)$$

If we substitute this expression in (4.16) we find:

$$T^e_{(+)\,\min} = \frac{2T_v}{q^2}(1 + \sqrt{1 + q^2}) \qquad (4.31)$$

Table 4.4 and Fig. 4.6 give $T^e_{(+)\min}/T_v$ as a function of q. For a high-quality varactor (high q) we have approximately

$$T^e_{(+)\min} \approx \frac{2T_v}{q} \tag{4.32}$$

As parameters r_L and $n_{(+)}$ are without influence on the effective noise temperature they can be chosen so as to obtain the maximum gain. r_L will then be matched with the output impedance of the amplifier, giving:

$$r_L = 1 + \frac{q^2/n_{(+)}}{n_{(+)}(1+r_e)} = 1 + \frac{q^2}{n_{(+)}(1+\sqrt{1+q^2})} \tag{4.33}$$

Table 4.4. Sum-frequency converter. Optimization of the noise temperature

q	r_e	$T_{(+)}/T_v$
2	2·2	1·6180
5	5·1	0·4879
7	7·1	0·3294
10	10·0	0·2210
15	15·0	0·1425
20	20·0	0·1051
50	50·0	0·0408
70	70·0	0·0290
100	100·0	0·0202
150	150·0	0·0134
200	200·0	0·0101
500	500·0	0·0040
700	700·0	0·0029
1000	1000·0	0·0020

Tables 4.4 and 4.5 and Figs. 4.6 and 4.7 give the values of r_e and r_L corresponding to equations (4.30) and (4.33). If we introduce (4.30) and (4.32) into (4.24) we obtain the expression for the maximum gain in conditions of minimum noise temperature:

$$[G^e_{(+)}]_{T_{\min}} = \frac{n_{(+)}}{\left[1+\frac{n_{(+)}}{q^2}(1+\sqrt{1+q^2})\right]\left[1+\frac{1}{\sqrt{1+q^2}}\right]} \tag{4.34}$$

The optimum value of $n_{(+)}$ is infinite and its choice will be determined by the practical considerations discussed in section 9. Table 4.6 and Fig. 4.8 give $[G^e_{(+)}]_{T_{\min}}$ as a function of q for different values of $n_{(+)}$.

Table 4.5. Sum-frequency converter. Reduced load resistance optimizing the noise temperature

	$n_{(+)}$							
q	2	4	6	8	10	12	14	16
2	1·6	1·3	1·2	1·2	1·1	1·1	1·1	1·1
5	3·0	2·0	1·7	1·5	1·4	1·3	1·3	1·3
7	4·0	2·5	2·0	1·8	1·6	1·5	1·4	1·4
10	5·5	3·3	2·5	2·1	1·9	1·8	1·6	1·6
15	8·0	4·5	3·3	2·8	2·4	2·2	2·0	1·9
20	10·5	5·8	4·2	3·4	2·9	2·6	2·4	2·2
50	25·5	13·3	9·2	7·1	5·9	5·1	4·5	4·1
70	35·5	18·3	12·5	9·6	7·9	6·8	5·9	5·3
100	50·5	25·8	17·5	13·4	10·9	9·3	8·1	7·2
150	75·5	38·3	25·8	19·6	15·9	13·4	11·6	10·3
200	100·5	50·8	34·2	25·9	20·9	17·6	15·2	13·4
500	250·5	125·8	84·21	63·4	50·9	42·6	36·6	32·2
700	350·5	175·8	117·5	88·4	70·9	59·3	50·9	44·7
1000	500·5	250·8	167·5	125·9	100·9	84·3	72·4	63·4

	$n_{(+)}$							
q	18	20	25	30	35	40	45	50
2	1·1	1·1	1·0	1·0	1·0	1·0	1·0	1·0
5	1·2	1·2	1·2	1·1	1·1	1·1	1·1	1·1
7	1·3	1·3	1·2	1·2	1·2	1·2	1·1	1·1
10	1·5	1·5	1·4	1·3	1·3	1·2	1·2	1·2
15	1·8	1·7	1·6	1·5	1·4	1·4	1·3	1·3
20	2·1	2·0	1·8	1·6	1·5	1·5	1·4	1·4
50	3·7	3·5	3·0	2·6	2·4	2·2	2·1	2·0
70	4·8	4·5	3·8	3·3	3·0	2·7	2·5	2·4
100	6·5	6·0	5·0	4·3	3·8	3·5	3·2	3·0
150	9·3	8·5	7·0	6·0	5·3	4·7	4·3	4·0
200	12·1	11·0	9·0	7·6	6·7	6·0	5·4	5·0
500	28·7	26·0	21·0	17·6	15·3	13·5	12·1	11·0
700	39·8	36·0	29·0	24·3	21·0	18·5	16·5	15·0
1000	56·5	51·0	41·0	34·3	29·5	26·0	23·2	21·0

1.7. Optimization of the total noise temperature

If T_2 is the noise temperature in the amplification chain after the converter, the expression for the total noise temperature $T^e_{(+)T}$ is

$$T^{(e)}_{(+)T} = T^e_{(+)} + \frac{T_2}{G^e_{(+)}} \qquad (4.35)$$

THREE-FREQUENCY NON-INVERTING CONVERTERS 95

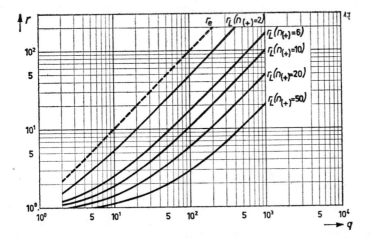

Figure 4.7. Sum-frequency converter. Reduced terminating resistance giving minimum noise temperature

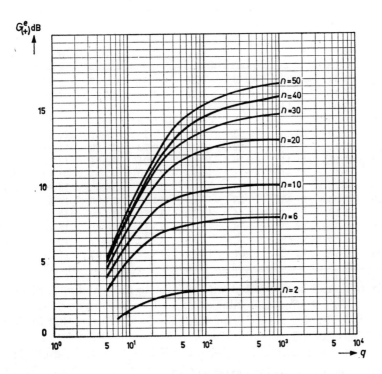

Figure 4.8. Sum-frequency converter. Gain corresponding to minimum noise temperature

Table 4.6. Sum-frequency converter. Gain corresponding to minimum noise temperature (in dB)

q	\multicolumn{8}{c}{$n_{(+)}$}							
	2	4	6	8	10	12	14	16
2	−2·77	−1·85	−1·50	−1·31	−1·19	−1·11	−1·05	−1·01
5	0·51	2·29	3·09	3·55	3·86	4·07	4·23	4·36
7	1·20	3·25	4·22	4·81	5·20	5·48	5·69	5·86
10	1·73	4·02	5·16	5·87	6·36	6·71	6·99	7·21
15	2·15	4·65	5·96	6·79	7·38	7·83	8·18	8·46
20	2·36	4·98	6·38	7·29	7·95	8·46	8·85	9·18
50	2·75	5·59	7·19	8·29	9·11	9·75	10·28	10·73
70	2·82	5·71	7·36	8·49	9·35	10·03	10·60	11·07
100	2·88	5·81	7·48	8·65	9·54	10·25	10·84	11·35
150	2·92	5·88	7·58	8·77	9·69	10·43	11·04	11·57
200	2·95	5·91	7·63	8·84	9·77	10·52	11·14	11·68
500	2·98	5·98	7·72	8·95	9·91	10·68	11·33	11·90
700	2·99	5·99	7·74	8·98	9·93	10·71	11·37	11·94
1000	3·00	6·00	7·75	8·99	9·95	10·74	11·40	11·97

q	\multicolumn{8}{c}{$n_{(+)}$}							
	18	20	25	30	35	40	45	50
2	−0·97	−0·95	−0·89	−0·86	−0·84	−0·82	−0·80	−0·79
5	4·46	4·54	4·69	4·79	4·87	4·93	4·97	5·01
7	6·00	6·11	6·31	6·46	6·56	6·64	6·71	6·76
10	7·39	7·53	7·81	8·01	8·16	8·27	8·36	8·43
15	8·69	8·88	9·26	9·53	9·73	9·89	10·01	10·12
20	9·45	9·68	10·12	10·45	10·70	10·89	11·05	11·18
50	11·11	11·44	12·10	12·61	13·01	13·34	13·62	13·85
70	11·48	11·84	12·58	13·14	13·60	13·97	14·29	14·56
100	11·78	12·17	12·96	13·58	14·08	14·50	14·86	15·17
150	12·03	12·43	13·28	13·95	14·50	14·96	15·36	15·70
200	12·16	12·57	13·44	14·14	14·72	15·20	15·63	15·99
500	12·39	12·83	13·76	14·51	15·14	15·68	16·15	16·57
700	12·44	12·88	13·82	14·58	15·22	15·77	16·25	16·68
1000	12·47	12·92	13·87	14·64	15·29	15·85	16·34	16·77

This value can be optimized with respect to r_e, r_L and $n_{(+)}$. Since r_L and $n_{(+)}$ do not affect the value of $T^e_{(+)}$ all we have to do is to choose them so as to give a maximum value for $G^e_{(+)}$. We saw that in these conditions, for a fixed value of $n_{(+)}$, r_L is given by the relation (4.19). By expanding (4.35) and applying condition (4.19) to it, we find:

$$T^e_{(+)T} = \frac{T_v}{r_e}\left[1 + \left(\frac{1+r_e}{q}\right)^2\right] + \frac{T_e}{r_e}(1+r_e)\left(1 + r_e + \frac{q^2}{n_{(+)}}\right) \quad (4.36)$$

THREE-FREQUENCY NON-INVERTING CONVERTERS 97

and by optimizing (4.36) with respect to r_e, we obtain:

$$r_e = \sqrt{1 + q^2 \frac{T_v + T_2/n_{(+)}}{T_v + T_2}} \qquad (4.37)$$

The expression for the total optimized noise temperature is then:

$$T^e_{(+)T_{\min}} = \frac{2(T_v + T_2)}{q^2} + \frac{T_2}{n_{(+)}}$$

$$+ 2\sqrt{\frac{T_v + T_2/n_{(+)}}{q^2}(T_v + T_2) + \left(\frac{T_v + T_2}{q^2}\right)^2} \qquad (4.38)$$

We see that, as before, it is impossible to optimize with respect to $n_{(+)}$, as we could obtain an infinite value. In practice, $n_{(+)}$ would above all be determined by the considerations discussed in section 4.1.9.

1.8. Optimization of the stability

The condition for optimum stability ($S^e_{(+)} = 0$) is obtained for:

$$(1 + r_e)(1 + r_L) = q^2/n_{(+)} \qquad (4.39)$$

In the three cases of optimization we have studied, r_L is matched with the output impedance of the amplifier.

Equation (4.19) may be written in the form:

$$(1 + r_e)(r_L - 1) = q^2/n_{(+)} \qquad (4.40)$$

a relation which is very close to (4.39) if $r_L \gg 1$. Examination of the formula (4.23) shows that this last condition will be satisfied, if

$$q^2 \gg n_{(+)} \qquad (4.41)$$

that is, if:

$$\omega'_c \gg \sqrt{\omega_e \omega_{(+)}} \qquad (4.42)$$

which would usually be the case for high-quality varactor. We may conclude that a sum-frequency converter giving a maximum gain is very stable.

1.9. Choice of $n_{(+)}$

The above theory does not allow us to choose an optimum value for $n_{(+)}$, as it is theoretically infinite. The value of $n_{(+)}$ will then be fixed by other considerations:

(1) When the frequency becomes too high, the representation of the varactor by a resistance of constant value in series with a

non-linear capacitance ceases to be accurate. A conductance appears in parallel with the junction.

(2) A high value of $n_{(+)}$ leads us to very close values of ω_p, $\omega_{(+)}$ ω_t and consequently to difficulties in the production of the filters.

(3) The converter should usually be followed by a diode mixer to convert the sum-frequency either back to the input frequency f_e or to a lower intermediate frequency. The noise temperature of this mixer will deteriorate as the input frequency increases. We must then select a value of n which will give a minimum value for (4.37) with $T_2 = f(n)$.

(4) The question of economics must also be considered. For a given varactor, the optimum pumping power increases as the square of the frequency. On the other hand, the price of klystrons or multipliers increases fairly steeply with frequency and the output power. The choice of n therefore depends essentially on the material available.

(5) We must not forget that the power of the pump is dissipated in the varactor and increases T_v. Optimization, taking into account this phenomenon, leads to a formula which is difficult to use. It is easier to trace the curve of $T^e_+ = f(n_{(+)}, T_v)$, calculating T_v from the pump power (see Chapter 8) and the thermal dissipation of the varactor for different values of the pump coefficients.

1.10. Bandwidth

Let us consider the general formula (4.11) for the gain:

the numerator is not very selective;
the denominator is formed by the square of the modulus of the sum of two terms, one of which, $q^2/n_{(+)}$ which is also not very selective, 'dilutes' the variation of the term $(1 + z_e)(1 + z_L)$ which contains the tuning reactances.

If the term $(1 + z_e)(1 + z_L)$ is not too selective, as in a well-designed device, we might expect a wide bandwidth to result. As a consequence, we must not ignore in our calculation the variation of $n_{(+)}$ and q as we would for a difference-frequency converter or for a negative-resistance amplifier. The formula obtained, taking these terms into account, would be too complicated to be of practical use and the most convenient procedure for calculating the bandwidth is to trace the gain versus frequency from (4.11).

1.11. Conclusions

The sum-frequency converter is characterized by a very stable gain, a wide bandwidth and a low intrinsic noise temperature. Its gain is always lower than $n_{(+)}$ and it is therefore limited by the fact that, for practical reasons, one cannot choose a very high value of $n_{(+)}$. The contribution made by the amplification chain after the converter to the total noise temperature is thus considerable and reduces the advantage of the low intrinsic noise temperature.

2. Difference-frequency non-inverting converters

We saw in section 5 of Chapter 3 that this device obeys equations (3.45) which are the same as the equations of the frequency-sum converter in which the subscript $(+)$ has been replaced by the subscript $(-)$. When this substitution is made, the whole discussion in the last section applies to the non-inverting difference-frequency converter. It follows at once that the intrinsic noise temperature is the same as that of the sum-frequency converter and that the transducer gain is always lower than the ratio of the output and input frequencies, this ratio being less than unity. The total noise temperature will therefore always be higher than the noise temperature of the second stage of amplification. This last property does not in itself constitute a negative criterion for the use of the converter. Indeed, standard mixers which convert a high-frequency signal into one of lower frequency (intermediate frequency) also have a conversion loss. However, since the loss is not tied to the frequency ratio, the performance of these mixers is better than that of the converter when the ratio of the input and output frequencies is substantially greater than one (as is usually the case).

The difference-frequency non-inverting converter is therefore of very little practical interest.

Note. If we place a sum-frequency converter with gain G_1 after a non-inverting difference-frequency converter with gain G_2, the total gain G_T is given by:

$$G_T = G_1 G_2 \qquad (4.43)$$

with

$$\left. \begin{array}{l} G_1 < \omega_{(+)}/\omega_e \\ G_2 < \omega_e/\omega_{(+)} \end{array} \right\} \qquad (4.44)$$

We then have:

$$G_T < 1 \qquad (4.45)$$

It is impossible to amplify at constant frequency with a combination of non-inverting three-frequency converters.

References

[1] Leenov, 'Gain and noise figure of a variable-capacitance up-converter', *Bell System Technical Journal*, vol. 37, pp. 989–1008, July 1958.
[2] Kurokawa, K. and Uenohara, M., 'Minimum noise figure of the variable-capacitance amplifier', *Bell System Technical Journal*, vol. 40, pp. 695–722, May 1961.
[3] Blackwell and Kotzebue, *Semiconductor-Diode Parametric Amplifier*, Prentice-Hall, Inc., Englewood Cliffs, N.J., 1961.
[4] Laurent, L., 'Les amplificateurs parametriques (1ere partie)', *Revue M.B.L.E.*, vol. VI (No. 2), June 1963.

5

THE THREE-FREQUENCY INVERTING CONVERTER AND REFLECTION AMPLIFIER

by L. LAURENT

1. Basic equations

1.1. Equivalent circuit

As we saw earlier, $f_p > f_e$ and the frequency $f_{(+)}$ is not used. The equivalent circuit in Fig. 5.1 is derived from that in Fig. 3.3.

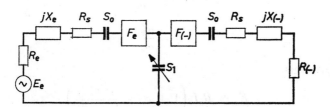

Figure 5.1. Equivalent circuit diagram of the inverting converter and of the reflection amplifier

Referring back to equation (3.42)

$$\begin{bmatrix} E_e \\ 0 \end{bmatrix} = R_s \begin{bmatrix} 1 + z_e & j\dfrac{\omega'_c}{\omega_{(-)}} \\ -j\dfrac{\omega'_c}{\omega_e} & 1 + z_{(-)} \end{bmatrix} \begin{bmatrix} I_e \\ I^*_{(-)} \end{bmatrix} \quad (5.1)$$

Equations (3.39) may be written:

$$z_e = \frac{R_e}{R_s} + j\left(\frac{X_e}{R_s} - \frac{\omega_c}{\omega_e}\right) \tag{5.2}$$

$$z_{(-)} = \frac{R_{(-)}}{R_s} + j\left(\frac{X_{(-)}}{R_s} - \frac{\omega_c}{\omega_{(-)}}\right) \tag{5.3}$$

where, R_e is the internal resistance of the generator, E_e is the e.m.f. of the generator, $R_{(-)}$ is the external resistance of the difference-frequency circuit, and jX_e and $jX_{(-)}$ are the matching reactances. It can be seen that, as for the sum-frequency converter, the terms in S_2 have disappeared from the equation. The properties of the inverting converter and of the reflection amplifier depend therefore only on the first and second terms of the development of the elastance, S_0 and S_1.

Equation (5.1) is of exactly the same form as equation (4.1). However, we are concerned with the conjugate values of the currents, voltage and impedances corresponding to $f_{(-)}$, and the term located at the intersection of the first row and the second column has changed sign. As we will see, this change of sign causes the real parts of the circuit impedances at f_e and $f_{(-)}$ to be negative; the essential properties of the devices studied in this chapter result from this. Introducing, as before, the auxiliary variables

$$n_{(-)} = \frac{\omega_{(-)}}{\omega_e} \tag{5.4}$$

$$q = \frac{\omega'_c}{\omega_e} \tag{5.5}$$

(5.1) gives us

$$E_e = R_s I_e \left(1 + z_e - \frac{q^2/n_{(-)}}{1 + z^*_{(-)}}\right) \tag{5.6}$$

$$I^*_{(-)} = \frac{E_e}{R_s} \left[\frac{jq}{(1 + z_e)(1 + z^*_{(-)}) - q^2/n_{(-)}}\right] \tag{5.7}$$

hence the equivalent circuits shown in Fig. 5.2 in which all the terms of equation (5.7) have been conjugated. This enables us to see the similarity of the equivalent circuits corresponding to the two frequencies: the load at frequency $f_{(-)}$ reflects a reduced impedance

$$-q^2/[n_{(-)}(1 + z^*_{(-)})]$$

into the circuit at frequency f_e; similarly the impedance Z_e at frequency f_e reflects a reduced impedance $-q^2/[n_{(-)}(1 + z^*_e)]$ into the

INVERTING CONVERTER AND REFLECTION AMPLIFIER

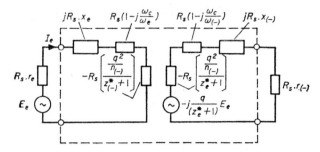

Figure 5.2. Equivalent circuit corresponding to equations (5.6) and (5.7)

circuit at frequency $f_{(-)}$. We see that the real part of these impedances may become negative.

The appearance of a negative resistance in the circuit at f_e gives to the circuit in Fig. 5.1 two possibilities of amplification:

(1) input at f_e and output at $f_{(-)}$: inverting converter,
(2) input and output at f_e: negative-resistance amplifier.

In practice inverting converters and negative-resistance amplifiers are always used with non-reciprocal elements which improve their performance. We will therefore start by recalling the essential properties of these devices (circulators and isolators).

1.2. Circulators and isolators

(1) *The scattering matrix*

If we wish to study a passive linear multiport without being concerned with the characteristic impedances of the lines terminating it, the usual relations which make use of admittance or impedance matrices are unnecessarily complicated. It is then preferable to use scattering matrices which give the relations between the normalized (r.m.s.) amplitudes of the incident waves and outgoing waves for all the ports. Let us consider a linear multiport at which q lines terminate. These lines may be of any impedance and of any type (waveguides, coaxial lines, etc.). We define a reference plane in each line sufficiently removed from the junction for any parasitic propagation modes to have disappeared. This multiport is represented by another multiport with p ports ($p \geq q$), in which each line counts as one port if it propagates only one normal mode, if not for as many ports as it propagates modes (Fig. 5.3).

Each reference plane is traversed by two waves which we will characterize by their voltage, one incident of r.m.s. value V_a, the other outgoing of r.m.s. value V_b. If we normalize these amplitudes with respect to the characteristic impedances of the corresponding lines, and if we put:

$$\left.\begin{aligned} a_k &= \frac{V_{ak}}{\sqrt{Z_{0k}}} \\ b_k &= \frac{V_{bk}}{\sqrt{Z_{0k}}} \end{aligned}\right\} \quad (5.8)$$

we can disregard the characteristic impedances Z_{0k}, and the incoming and outgoing powers are given by $a_k\, a_k^*$ and $b_k\, b_k^*$. The general relation between the normalized amplitude of the incident and outgoing waves is then given by:

$$b_i = \sum_{k=1}^{p} S_{ik} a_k \quad (5.9)$$

or

$$[b] = [S][a]$$

$[S]$ is the scattering matrix.

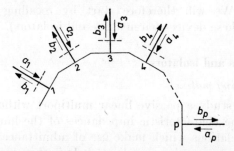

Figure 5.3. Multiport with p ports. Incident and outgoing waves

We see at once that S_{kk} is the reflection coefficient of the port k, while S_{ik} is the transmission coefficient from port k to port i. If the b_i of each port is zero when the other ports are terminated by a matched load (that is when the a_k for $k \neq i$ are zero) the multiport is said to be matched.

(2) *Chief properties of the scattering matrix*

(a) *Conservation of energy.* For a non-dissipative multiport, we can show that $[S]$ should be unitary or:

$$\sum_k S_{ik}^* S_{jk} = \delta_{ij} \quad (5.10)$$

where δ_{ij} is the Kronecker symbol ($\delta_{ij} = 0$ for $i \neq j$ and $\delta_{ij} = 1$ for $i = j$).

(b) *Reciprocity*. When the multiport is reciprocal, the matrix is symmetrical:

$$S_{ik} = S_{ki}$$

When the multiport is non-reciprocal, this relation is not satisfied. This happens with isolators and circulators in which the non-reciprocity is obtained by means of ferrites in which the tensor of the magnetic permeability is asymmetric.

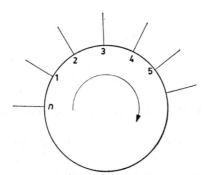

Figure 5.4. Circulator with n ports

(3) *Circulators*

An ideal circulator is a non-dissipative and non-reciprocal matched multiport so that an incoming signal through port 1 is transmitted only to port 2, a signal entering by port 2 to port 3, ..., a signal entering by port n to port 1 (Fig. 5.4). The circulator therefore absorbs no energy; it guides it. The distributive matrix of a circulator of this type is given by

$$[S] = \begin{bmatrix} 0 & 0 & 0 & 0 & \ldots & 1 \\ 1 & 0 & 0 & 0 & \ldots & 0 \\ 0 & 1 & 0 & 0 & \ldots & 0 \\ 0 & 0 & 1 & 0 & \ldots & 0 \\ \vdots & \vdots & \vdots & \vdots & & \vdots \\ 0 & 0 & 0 & 0 & & 0 \end{bmatrix} \quad (5.11)$$

It is theoretically possible to design a circulator with any number of ports, and it can be shown that such a circulator may be made up of combinations of three-port circulators. Although this method is not the only one, it will be convenient for our purposes to discuss circulators with four or five ports as formed by combinations of two or three three-port circulators respectively.

Three-port circulator. This circulator is the most common one, either alone or as a unit for the construction of circulators with a larger number of ports, because of the ease with which it can be produced. Figure 5.5 shows an example of a circulator made with 'striplines'.

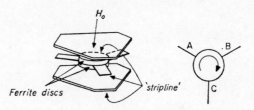

Figure 5.5. Example of design of a three-port circulator

It can be shown by means of the scattering matrix that the three-port circulator is the only T-junction which is non-dissipative and matched. Its scattering matrix is given by:

$$[S] = \begin{vmatrix} 0 & 0 & 1 \\ 1 & 0 & 0 \\ 0 & 1 & 0 \end{vmatrix} \tag{5.12}$$

Application of a three-port circulator to a reflection amplifier. Let us assume that ports A and C are matched and that port B is terminated by a load with a reflection coefficient ρ_B

(5.12) gives us:

$$\left. \begin{aligned} b_C &= a_B \\ b_B &= a_A \end{aligned} \right\} \tag{5.13}$$

Now,

$$a_B = b_B \rho_B \tag{5.14}$$

thus the signal leaving at C is equal to

$$b_C = \rho_B a_A \tag{5.15}$$

If ρ_B is greater than unity (negative resistance), we have an amplifier.

Turning from the distributive matrix to voltages and currents at the terminals of the circulator. The currents and voltages at the terminals

INVERTING CONVERTER AND REFLECTION AMPLIFIER 107

of the circulator are linked to the amplitudes of the incident and outgoing waves by

$$\left.\begin{aligned} I_k &= I_{ak} - I_{bk} = \frac{1}{\sqrt{Z_{0k}}}(a_k - b_k) \\ V_k &= V_{ak} + V_{bk} = \sqrt{Z_{0k}}(a_k + b_k) \end{aligned}\right\} \quad (5.16)$$

If we eliminate the a_k and b_k between (5.15) and (5.12), we find, for $Z_{0A} = Z_{0B} = Z_{0C} = Z_0$ (which is practically always true)

$$\left.\begin{aligned} V_A &= V_C + Z_0(I_A + I_C) \\ V_B &= V_A + Z_0(I_A + I_B) \\ V_C &= V_B + Z_0(I_B + I_C) \end{aligned}\right\} \quad (5.17)$$

These relations will be used in Chapter 7 for the study of degenerate amplifiers.

(4) *The isolator*

An ideal isolator is a two-port which transmits energy without loss in one direction and absorbs it in the other. The scattering matrix of a two-port is given by:

$$[S] = \begin{vmatrix} S_{11} & S_{12} \\ S_{21} & S_{22} \end{vmatrix} \quad (5.18)$$

It can easily be shown that a non-dissipative two-port transmitting energy in one direction, for example $1 \to 2$ and not in the other, cannot be designed: in fact, if the two-port is non-dissipative, (5.9) tells us:

$$|S_{12}|^2 + |S_{22}|^2 = 1 \quad (5.19)$$

$$S_{11}^* S_{12} + S_{21}^* S_{22} = 0 \quad (5.20)$$

If the two-port transmits no energy in the direction $2 \to 1$

$$S_{12} = 0 \quad (5.21)$$

hence, according to (5.19)

$$|S_{22}|^2 = 1 \quad (5.22)$$

and according to (5.20)

$$S_{21} = 0 \quad (5.23)$$

it does not transmit either in the direction $1 \to 2$. The isolator is therefore necessarily dissipative; in the 'cut-off' direction the energy is not reflected but absorbed. The scattering matrix of an ideal matched isolator is given by:

$$[S] = \begin{vmatrix} 0 & 0 \\ 1 & 0 \end{vmatrix} \qquad (5.24)$$

for the direction of transmission $1 \rightarrow 2$.

Representation of the isolator. If, for example, we terminate port C of a three-port circulator by a matched load and if we are concerned only with the waves leaving or entering by ports A and B, (5.11) gives:

$$\left. \begin{array}{l} b_A = 0 \\ b_B = a_A \end{array} \right\} \qquad (5.25)$$

which, put in the form of a scattering matrix, gives (5.24). It is

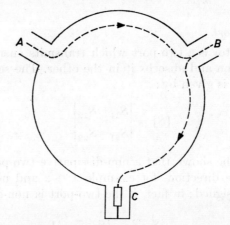

Figure 5.6. Equivalent circuit of an isolator

therefore allowable to represent the isolator by a circulator with a matched load, even if other types of construction are used in practice. The equivalent circuit of an isolator is given in Fig. 5.6. The signal is transmitted without loss from A to B; if port B is not matched, the reflected signal is absorbed by the load C. Port A is then totally 'isolated' from reflection at port B.

A consequence of the isolator being necessarily dissipative is that it is impossible to prevent port A receiving the thermal-agitation noise emitted by the load C. We will see later when studying the inverting converter using an output isolator the importance of this source of noise.

1.3. Transducer gain of a negative-resistance amplifier without circulator

Let us consider a generator with an e.m.f. E_G and internal impedance Z_g feeding into a load Z_L (Fig. 5.7a). The maximum available power of the generator is equal to:

$$P_1 = \frac{E_G^2}{4R_G} \tag{5.26}$$

making

$$Z_G = R_G + jX_G \tag{5.27}$$

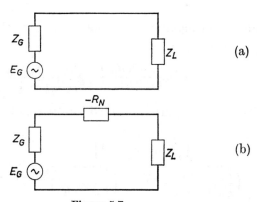

Figure 5.7

(a) Generator feeding into a load
(b) Conversion of the circuit into an amplifier by the insertion of a negative resistance

A negative resistance $(-R_N)$ is put in series with the load (Fig. 5.7b). The current through the load is equal to:

$$I = \frac{E_G}{Z_G + Z_L - R_N} \tag{5.28}$$

and the power fed to the load:

$$P_2 = \frac{E_G^2 R_L}{|Z_G + Z_L - R_N|^2} \tag{5.29}$$

making:

$$Z_L = R_L + jX_L \tag{5.30}$$

The transducer gain of the amplifier formed in this way will then be equal to:

$$G = \frac{P_2}{P_1} = \frac{4R_G R_L}{|Z_G + Z_L - R_N|^2} \tag{5.31}$$

This gain may be made as high as desired by decreasing the denominator; unfortunately this will increase its sensitivity to a variation of one of the terms of this denominator. The amplifier may even oscillate if the real part of the total circuit impedance becomes negative.

Let us indeed consider the effect on the gain of the variation of the negative resistance only and define the instability factor by:

$$S = \frac{dG/G}{dR_N/R_N} \tag{5.32}$$

we find:

$$S = 2\frac{R_N(R_G + R_L - R_N)}{|Z_G + Z_L - R_N|^2} \tag{5.33}$$

The condition for maximum gain for a given stability by combining (5.31) and (5.33) gives:

$$G = \frac{2(R_G/R_N)(R_L/R_N)}{(R_G/R_N) + (R_L/R_N) - 1} S \tag{5.34}$$

As R_G/R_N and R_L/R_N are connected by the condition

$$\frac{2\left(\dfrac{R_G}{R_N} + \dfrac{R_L}{R_N} - 1\right)}{\left(\dfrac{R_G}{R_N} + \dfrac{R_L}{R_N} - 1\right)^2 + \left(\dfrac{X_G}{R_N} + \dfrac{X_L}{R_N}\right)^2} = S = \text{const.} \tag{5.35}$$

The condition of maximum of $[(R_G/R_N)(R_L/R_N)]/[(R_G/R_N) + (R_L/R_N) - 1]$ according to (5.35) is obtained by the Lagrange method of indeterminate multipliers and is expressed by:

$$X_G + X_L = 0 \tag{5.36}$$

$$R_G = R_L \tag{5.37}$$

We then have:

$$G_{\max} = \frac{4(R_G/R_N)^2}{[2(R_G/R_N) - 1]^2} \tag{5.38}$$

$$S = \frac{2}{[2(R_G/R_N) - 1]} \tag{5.39}$$

hence

$$G_{\max} = \left(1 + \frac{S}{2}\right)^2 \tag{5.40}$$

We see that the amplifier becomes the more unstable as its gain increases.

1.4. Transducer gain of a negative-resistance amplifier with circulator

Instead of connecting the load and the negative resistance directly in series with the generator, let us assume that they are connected through a circulator of characteristic impedance Z_0 (Fig. 5.8), and let:

$$R_G = R_L = Z_0 \tag{5.41}$$

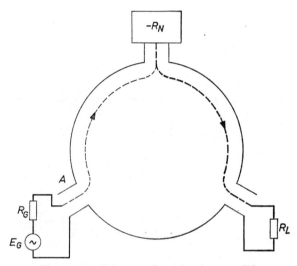

Figure 5.8. Schema of a reflection amplifier

Ports A and C are matched, while the terminating impedance of port B has a coefficient of reflection:

$$\rho_B = \frac{-R_N - Z_0}{-R_N + Z_0} \tag{5.42}$$

The useful power of the generator is given as before by (5.26); the power reflected from port B towards port C is equal to:

$$P_2 = |\rho_B|^2 P_1 \tag{5.43}$$

hence:

$$G_T = |\rho_B|^2 = \left|\frac{Z_0 + R_N}{Z_0 - R_N}\right|^2 \tag{5.44}$$

and in accordance with definition (5.32):

$$S = \frac{4 Z_0 R_N}{(Z_0 + R_N)(Z_0 - R_N)} \tag{5.45}$$

Formula (5.44) justifies the expression 'reflection amplifier' which will be used in what follows to designate negative-resistance amplifiers incorporating a circulator.

1.5. Comparison between amplifiers with and without circulator

For the sake of simplicity we will work on the assumption that the gain is high. For the amplifier without a circulator we must, by virtue of (5.38), take:

$$R_G \approx \frac{R_N}{2} \tag{5.46}$$

which from (5.40), gives:

$$G_T \approx \frac{S^2}{4} \tag{5.47}$$

For the amplifier without a circulator we must, by virtue of (5.44), take:

$$Z_0 \approx R_N \tag{5.48}$$

which gives:

$$G_T \approx \left(\frac{2R_N}{Z_0 - R_N}\right)^2$$

and (5.49)

$$S \approx \frac{2R_N}{Z_0 - R_N}$$

whence

$$G_T \approx S^2 \tag{5.50}$$

Conclusion. When the gain is high, for a given negative resistance R_N and a given instability coefficient S, the use of a circulator increases the gain four-fold. Further, if the load R_L is matched with the port C, the circulator has the advantage of 'isolating' the amplifier from any fluctuations in R_G. In fact, if the generator is not matched, there will be a loss of gain due to the reflection at input A of the circulator. This mismatching affects G_T, which must be multiplied by $1 - |\rho_A|^2$ but it leaves the denominator of (5.44) unchanged: the amplifier cannot then oscillate ($Z_0 \leq R_N$) due to a change in the impedance of the source.

1.6. Use of circulators with four and five ports

In order to 'isolate' a negative-resistance amplifier from any fluctuations of either R_G or R_L, it is necessary to use a circulator with four ports fitted with a matched termination, so that port C of the input circulator is always matched (Fig. 5.9) (this combination is equivalent to a circulator with three ports followed by an isolator). This system has also the advantage that it prevents the noise generated by the load reaching port A, from which it might be partially reflected towards port B (and amplified), if the matching at A is not perfect. The noise of the matched termination of the isolator is substituted for this noise; the equivalent noise temperature of this termination may be lower than that of the load.

We have been dealing so far with perfect circulators. If the amplifier gain is high and of the same order of magnitude as the isolation in the reverse direction, the latter can no longer be considered as infinite. In fact, the signal entering by port A is strongly amplified at B. The power returned towards port A due to the imperfect isolation

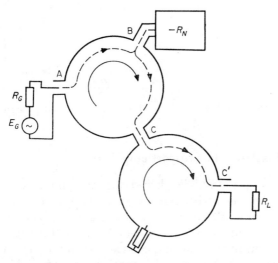

Figure 5.9. Circulator with four ports used in a reflection amplifier

is no longer negligible, compared to the input signal: a change in the impedance of the source may then have an appreciable effect on the gain. In order to avoid this trouble, the practice has been to use a circulator with five ports fitted with two terminations, equivalent to a circulator with three ports between two isolators (Fig. 5.10).

The isolation between ports B and A', when doubled, enables a higher gain to be obtained without instability, and therefore allows a decrease in the noise contribution made by the following amplification stage to the total noise temperature. Unfortunately the attenuation in the direction A'B due to the losses in the circulators is also doubled, which raises the intrinsic noise temperature of the combination formed by the amplifier and the circulators, and reduces the advantage of the system.

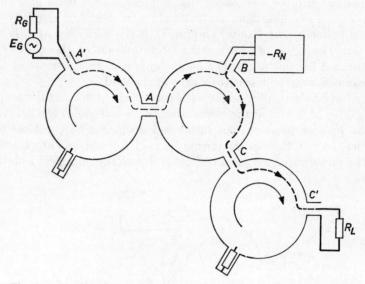

Figure 5.10. Circulator with five ports used in a reflection amplifier

1.7. Use of isolators in two-port amplifiers

When a two-port amplifier has a negative input or output impedance, as happens with the inverting converter, it is possible to 'isolate' it from any changes in the impedances of the generator and load by placing it between two isolators. In these conditions the amplifier is terminated by impedances equal to the characteristic impedances of the isolators, which are therefore invariable (Fig. 5.11).

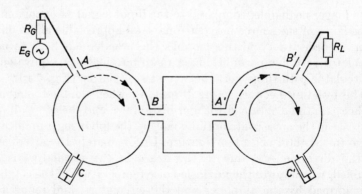

Figure 5.11. Two-port amplifier connected between two isolators

1.8. The six transducer gains of difference-frequency systems

The great usefulness of circulators and isolators in circuits in which the real part of the impedances is negative was illustrated in the previous paragraphs. In the continuation of the study of inverting converters and reflection amplifiers, we will confine our attention to the case in which the two circuits f_e and $f_{(-)}$ are provided with these devices. The equivalent diagram in Fig. 5.1 then becomes that in Fig. 5.12. Since isolators and circulators have generally normalized characteristic impedances, we must introduce the concept of ideal transformers in order to obtain impedances R_e and $R_{(-)}$ from Z_0 and Z_0'. Our calculations will be facilitated if we:

(1) reject the transformers outside the isolators or circulators and replace them by imaginary isolators or circulators with characteristic impedances equal respectively to $Z_{0e} = n_1^2 Z_0$ and $Z_{0(-)} = n_2^2 Z_0'$;
(2) incorporate the transformers in the generator and the loads which are replaced by a generator and by loads defined by the relations:

$$\left. \begin{array}{ll} E_e = n_1 E_e' & R_{(-)} = n_2^2 R_{(-)}' \\ R_e = n_1^2 R_e' & R_{L4} = n_2^2 R_{L4}' \\ R_{L3} = n_1^2 R_{L3}' & \end{array} \right\} \quad (5.51)$$

The transformation ratios thus disappear from the calculations. As a result we get a new equivalent circuit as shown in Fig. 5.13. By means of this circuit we can calculate the six transducer gains which, as we will see later, determine the performance of difference-frequency systems. We will designate these gains with a superscript to indicate the input port for the signal and a subscript for the output port.

(1) *Gain due to reflection* G_3^1 (Fig. 5.14a)

Port 2 of the first circulator looks into an impedance

$$Z_2 = R_s \left[1 + j\left(x_e - \frac{\omega_c}{\omega_e}\right) - \frac{q^2}{n_{(-)}(z_{(-)}^* + 1)} \right] \quad (5.52)$$

(cf. Fig. 5.2). On the other hand, if we assume that ports 1 and 3 are matched ($Z_{0e} = R_{L3} = R_e$), we have:

$$\rho = \frac{Z_2 - R_e}{Z_2 + R_e} \quad (5.53)$$

116   PARAMETRIC AMPLIFIERS

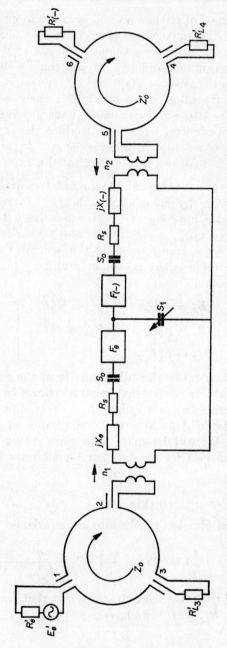

Figure 5.12. Equivalent diagram of difference-frequency systems provided with isolators or circulators

INVERTING CONVERTER AND REFLECTION AMPLIFIER 117

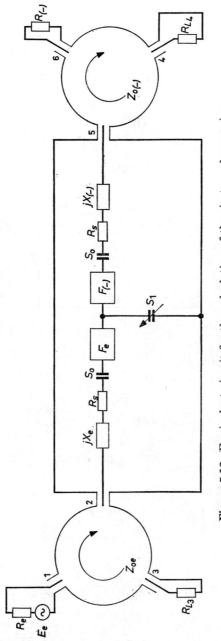

Figure 5.13. Equivalent circuit for the calculation of the six transducer gains

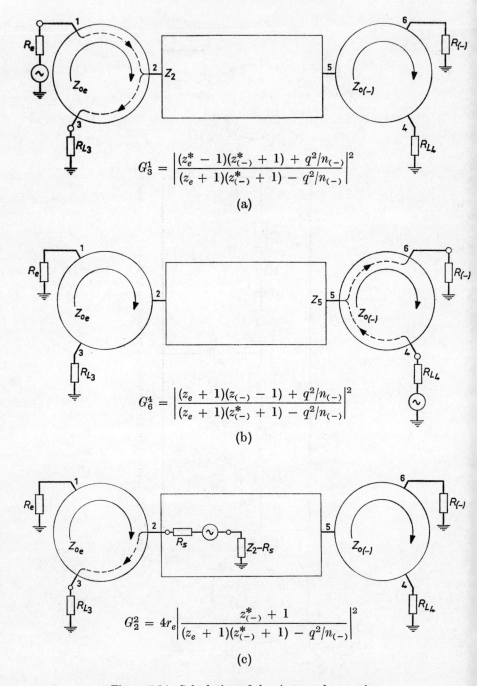

Figure 5.14. Calculation of the six transducer gains

INVERTING CONVERTER AND REFLECTION AMPLIFIER

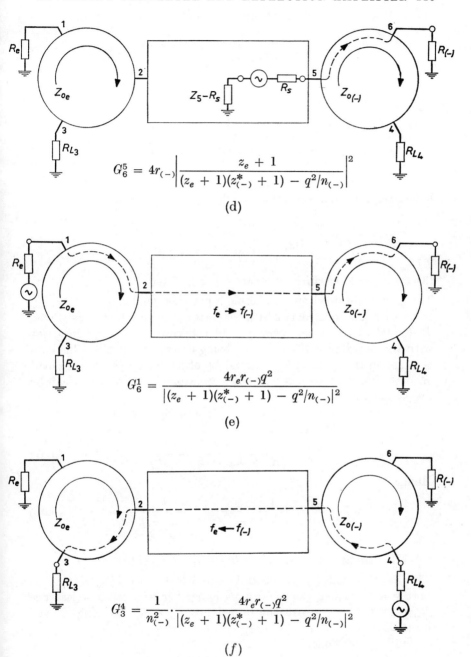

$$G_6^5 = 4r_{(-)}\left|\frac{z_e + 1}{(z_e + 1)(z_{(-)}^* + 1) - q^2/n_{(-)}}\right|^2$$

(d)

$$G_6^1 = \frac{4r_e r_{(-)} q^2}{|(z_e + 1)(z_{(-)}^* + 1) - q^2/n_{(-)}|^2}$$

(e)

$$G_3^4 = \frac{1}{n_{(-)}^2} \cdot \frac{4r_e r_{(-)} q^2}{|(z_e + 1)(z_{(-)}^* + 1) - q^2/n_{(-)}|^2}$$

(f)

Figure 5.14. (cont)

hence, if we take (5.2), (5.44) and (5.52) into account, we find:

$$G_3^1 = |\rho|^2 = \left|\frac{(z_e^* - 1)(z_{(-)}^* + 1) + q^2/n_{(-)}}{(z_e + 1)(z_{(-)}^* + 1) - q^2/n_{(-)}}\right|^2 \quad (5.54)$$

(2) *Gain due to reflection* G_6^4 (Fig. 5.14b)

Port 5 of the second circulator looks into an impedance:

$$Z_5 = R_s\left[1 + j\left(x_{(-)} - \frac{\omega_c}{\omega_{(-)}}\right) - \frac{q^2/n_{(-)}}{x_e^* + 1}\right] \quad (5.55)$$

hence, by a calculation similar to the above, we obtain:

$$G_6^4 = \left|\frac{(z_e + 1)(z_{(-)} - 1) + q^2/n_{(-)}}{(z_e + 1)(z_{(-)}^* + 1) - q^2/n_{(-)}}\right|^2 \quad (5.56)$$

(3) *Gain due to negative resistance* G_3^2 (Fig. 5.14c)

The gain is obtained by connecting a matched load $R_{L3} = R_e = Z_{0e}$ to port 3 and feeding port 2 by a generator with an internal resistance R_s (equal to the series resistance of the varactor) placed in series with the impedance $Z_2 - R_s$, Z_2 being defined by (5.52). This gain is equal to the gain G_2^2 that would be obtained by placing the load R_e directly across port 2 (without circulator), since no power can be transmitted to port 1.

If we apply formula (5.31) with $-R_N = Z_2 - R_s$, we find:

$$G_2^2 = \frac{4R_s R_e}{|R_s + R_e + (Z_2 - R_s)|^2} \quad (5.57)$$

Hence

$$G_2^2 = G_3^2 = 4r_e\left|\frac{z_{(-)}^* + 1}{(z_e + 1)(z_{(-)}^* + 1) - q^2/n_{(-)}}\right|^2 \quad (5.58)$$

(4) *The gain due to negative resistance* G_6^5 (Fig. 5.14d)

This is obtained by connecting a matched load $R_{L6} = R_{(-)}$ to port 6 and feeding port 5 from a generator of internal resistance R_s, placed in series with the impedance $Z_5 - R_s$, Z_5 being defined by (5.55).

Similarly we find:

$$G_5^5 = G_6^5 = 4r_{(-)}\left|\frac{z_e + 1}{(z_e + 1)(z_{(-)}^* + 1) - q^2/n_{(-)}}\right|^2 \quad (5.59)$$

INVERTING CONVERTER AND REFLECTION AMPLIFIER

(5) *Conversion gain* G_6^1 (Fig. 5.14e)

By a similar calculation to that used to obtain (4.11), we find:

$$G_6^1 = G_5^2 = \frac{4r_e r_{(-)} q^2}{|(z_e + 1)(z_{(-)}^* + 1) - q^2/n_{(-)}|^2} \qquad (5.60)$$

(6) *Conversion gain* G_3^4 (Fig. 5.14f)

It will easily be found, by permuting in (5.60) r_e, z_e, ω_e and $r_{(-)}$, $z_{(-)}^*$, $\omega_{(-)}$ respectively:

$$G_3^4 = G_3^5 = \frac{1}{n_{(-)}^2} \cdot \frac{4r_e r_{(-)} q^2}{|(z_e + 1)(z_{(-)}^* + 1) - q^2/n_{(-)}|^2} \qquad (5.61)$$

2. The inverting converter

2.1. Transducer gain

The equivalent diagram can be derived directly from Fig. 5.13 if we consider $R_{(-)}$ as the working load. In what follows we will therefore write R_L instead of $R_{(-)}$. The circulators are provided with matched loads R_{13} and R_{14}, thus converting them into isolators. We will assume that R_e and R_L are matched with the characteristic impedance of these isolators.

The transducer gain is given directly by formula (5.60) in which $R_{(-)}$ is replaced by R_L:

$$G_{(-)}^e = G_6^1 = \frac{4r_e r_L q^2}{|(1 + z_e)(1 + z_{(-)}^*) - q^2/n_{(-)}|^2} \qquad (5.62)$$

It can be seen that:

(1) For given r_e and r_L, this gain is maximum for real values of z_e and $z_{(-)}$ (tuned input and output circuits).
(2) The expression for this gain is analogous to that of the gain of the sum-frequency converter, but the term $q^2/n_{(-)}$ in the denominator has a negative sign. This allows us to bring the gain $G_{(-)}^e$ to as high a value as desired, for given values of $n_{(-)}$ and q by adjusting r_e or r_L. We will assume in what follows that the resonance condition has been achieved.

We then have:

$$G_{(-)}^e = \frac{4r_e r_L q^2}{[(1 + r_e)(1 + r_L) - q^2/n_{(-)}]^2} \qquad (5.63)$$

2.2. Instability coefficient

As the converter is isolated from changes in the impedances of the generator and of the load thanks to the use of isolators, the chief cause of instability is the change in the term q^2 which depends on the pumping conditions. We will then define as for the non-inverting converter, the instability factor by the relation:

$$S^e_{(-)} = \frac{\mathrm{d}G^e_{(-)}/G^e_{(-)}}{\mathrm{d}q^2/q^2} \qquad (5.64)$$

(5.63) gives us:

$$S^e_{(-)} = 1 + \frac{2q^2/n_{(-)}}{(1+r_e)(1+r_L) - q^2/n_{(-)}} \qquad (5.65)$$

We see that the condition for high gain (denominator of (5.63) near to zero) involves a high instability factor.

2.3. Effective noise temperature

We will assume that the input generator is replaced by a resistance R_e cooled to 0°K. Figure 5.15 shows the internal noise generators of the converter. It should be remembered that we have assumed that the input and output circuits are tuned. The noise power supplied to the load $R_L = r_L R_s$ consists of three terms.

(1) The noise power produced at frequency f_e by R_s and transmitted to the load after conversion: ΔN_e. The e.m.f. of the noise generator is equal to:

$$e_e = \sqrt{4kT_v R_s \Delta f} \qquad (5.66)$$

where T_v is the temperature of the varactor. This e.m.f. may be considered to be that of a generator of internal impedance R_e connected in the circuit f_e in place of port 2 of the first circulator. The available power of this generator is:

$$\frac{kT_v R_s \Delta f}{R_e} = \frac{1}{r_e} kT_v \Delta f \qquad (5.67)$$

All that is needed to find ΔN_e is to multiply this available power by the gain G_5^2:

$$\Delta N_e = \frac{1}{r_e} kT_v G_5^2 \Delta f \qquad (5.68)$$

INVERTING CONVERTER AND REFLECTION AMPLIFIER 123

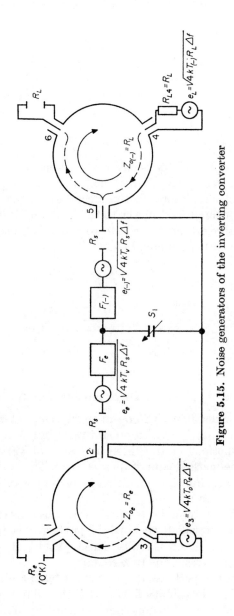

Figure 5.15. Noise generators of the inverting converter

(2) The noise power generated at $f_{(-)}$ by R_s is transmitted to the load after amplification by the negative resistance: $\Delta N_{(-)}$. The available power of the noise generator of internal resistance R_s is equal to $kT_v \Delta f$, whence:

$$\Delta N_{(-)} = kT_v G_6^5 \Delta f \tag{5.69}$$

(3) The noise power generated at $f_{(-)}$ by the termination $R_{L4} = R_L$ of port 4 is transmitted to the load after amplification by reflection ΔN_L. The usable power of the generator is equal to $kT_{(-)} \Delta f$ where $T_{(-)}$ is the temperature of the termination. We then have:

$$\Delta N_L = kT_{(-)} G_6^4 \Delta f \tag{5.70}$$

We assumed in the previous section that the terminations of ports 1 and 4 were perfectly matched with the characteristic impedances of their circulators. In these conditions, the noise sources of ports 3 and 6 cannot feed power to the load, and it is only the three terms listed above that contribute to the noise temperature of the converter, which is then equal to:

$$T_{(-)}^e = \left(\frac{1}{r_e} + \frac{G_6^5}{G_6^1}\right) T_v + \frac{G_6^4}{G_6^1} T_{(-)} \tag{5.71}$$

If we replace the G_k^j's by their value, we get:

$$T_{(-)}^e = \frac{1}{r_e}\left[1 + \left(\frac{1+r_e}{q}\right)^2\right] T_v + \frac{[(1+r_e)(r_L-1) + q^2/n_{(-)}]^2}{4r_e r_L q^2} T_{(-)} \tag{5.72}$$

2.4. Part played by the instability factor in optimizations

In the chapter dealing with non-inverting converters, the gain, the effective and total noise temperatures and the stability were studied as functions of the three independent variables r_e, r_L and $n_{(+)}$. The same method cannot be used here. The optimization of the gain without any limiting condition is meaningless, since by adjusting r_e and r_L, it can become infinite. Similarly, the optimization of the intrinsic and total noise temperatures causes us to choose operating conditions in which the instability factor is infinite. Finally, the optimization of the stability causes $(1 + r_e)(1 + r_L)$ to tend to infinity, which corresponds to zero gain.

In practice, the performance is limited by the maximum permissible value of the instability factor. It is therefore logical to optimize $G_{(-)}^e$, $T_{(-)}^e$ and the total noise temperature by fixing the value of $S_{(-)}^e$. The three parameters r_e, r_L and $n_{(-)}$ then reduce to two independent parameters.

INVERTING CONVERTER AND REFLECTION AMPLIFIER

2.5. Optimization of the gain for a given instability factor

2.5.1. OPTIMIZATION WITH RESPECT TO THE TERMINATING IMPEDANCES

If we eliminate the variable r_L by means of condition $S^e_{(-)} = \text{const.}$, (5.65) gives us:

$$r_L = \frac{1}{1+r_e} \cdot \frac{q^2}{n_{(-)}} \cdot \frac{S^e_{(-)}+1}{S^e_{(-)}-1} - 1 \qquad (5.73)$$

Substitute this value in (5.63) and we obtain:

$$G^e_{(-)} = \frac{r_e}{1+r_e} n_{(-)}(S^{e2}_{(-)} - 1) - r_e \left[\frac{n_{(-)}}{q}(S^e_{(-)} - 1)\right]^2 \qquad (5.74)$$

If we optimize with respect to r_e, we obtain:

$$\frac{dG^e_{(-)}}{dr_e} = \frac{1}{(1+r_e)^2} n_{(-)}(S^{e2}_{(-)} - 1) - \left[\frac{n_{(-)}}{q}(S^e_{(-)} - 1)\right]^2 = 0 \qquad (5.75)$$

hence:

$$r_{e\,\text{opt}} = \sqrt{\frac{q^2}{n_{(-)}} \cdot \frac{S^e_{(-)}+1}{S^e_{(-)}-1}} - 1 \qquad (5.76)$$

and, from (5.73):

$$r_{L\,\text{opt}} = \sqrt{\frac{q^2}{n_{(-)}} \cdot \frac{S^e_{(-)}+1}{S^e_{(-)}-1}} - 1 = r_{e\,\text{opt}} \qquad (5.77)$$

The gain $G^e_{(-)\,\text{max}}$ optimized with respect to the terminating impedances is then equal to:

$$G^e_{(-)\,\text{max}} = n_{(-)}(S^e_{(-)} - 1)^2 \left[\sqrt{\frac{S^e_{(-)}+1}{S^e_{(-)}-1}} - \sqrt{\frac{n_{(-)}}{q^2}}\right]^2 \qquad (5.78)$$

The corresponding noise temperature may be written:

$$[T^e_{(-)}]_{G\,\text{max}} = \frac{(n_{(-)}+1)S^e_{(-)} - (n_{(-)}-1)}{q\sqrt{n_{(-)}(S^{e2}_{(-)}-1)} - n_{(-)}(S^e_{(-)}-1)} T_v$$

$$+ \frac{1}{n_{(-)}} \cdot \frac{S^e_{(-)}+1}{S^e_{(-)}-1} \left[\frac{\sqrt{\frac{q^2}{n_{(-)}} \cdot \frac{2S^e_{(-)}}{S^e_{(-)}+1}} - 1}{\sqrt{\frac{q^2}{n_{(-)}} \cdot \frac{S^e_{(-)}+1}{S^e_{(-)}-1}} - 1}\right]^2 \qquad (5.79)$$

In normal working conditions $S^e_{(-)}$ is large compared to unity and the quality of the varactor is high enough for $q \gg 1$. Relations (5.78) and (5.79) may then be written:

$$G^e_{(-)\,\mathrm{max}} \approx n_{(-)} S^{e\,2}_{(-)} \left(1 - \frac{\sqrt{n_{(-)}}}{q}\right)^2 \tag{5.80}$$

$$[T^e_{(-)}]_{G\,\mathrm{max}} \approx \frac{n_{(-)} + 1}{q\sqrt{n_{(-)}} - n_{(-)}} T_v + \frac{1}{n_{(-)}} \left[\frac{q\sqrt{2/n_{(-)}} - 1}{q(1/\sqrt{n_{(-)}}) - 1}\right]^2 T_{(-)} \tag{5.81}$$

2.5.2. OPTIMIZATION WITH RESPECT TO r_e, r_L AND $n_{(-)}$

If we equate to zero the differential of (5.78) with respect to $n_{(-)}$ we find:

$$n_{(-)\,\mathrm{opt}} = \frac{q^2}{4} \cdot \frac{S^e_{(-)} + 1}{S^e_{(-)} - 1} \tag{5.82}$$

The gain $G^e_{(-)}$ optimized with respect to all the parameters is then equal to:

$$G^e_{(-)\,\mathrm{opt}} = \frac{q^2}{16} (S^e_{(-)} + 1)^2 \tag{5.83}$$

The corresponding noise temperature may then be written:

$$[T^e_{(-)}]_{G\,\mathrm{opt}} = \left(1 + \frac{4}{q^2}\right) T_v + \frac{4}{q^2} \left[\frac{2\sqrt{2S^e_{(-)}(S^e_{(-)} - 1)}}{S^e_{(-)} + 1} - 1\right]^2 T_{(-)} \tag{5.84}$$

In ordinary conditions $S^e_{(-)}$ and q^2 are large compared to unity, and relations (5.83) and (5.84) may be written:

$$G^e_{(-)\,\mathrm{opt}} \approx \left(\frac{q S^e_{(-)}}{4}\right)^2 \tag{5.85}$$

$$[T^e_{(-)}]_{G\,\mathrm{opt}} \approx T_v + \frac{4(2\sqrt{2} - 1)^2}{q^2} T_{(-)} \tag{5.86}$$

We see from (5.86) that the optimization of the gain gives a noise temperature higher than T_v. It is not likely therefore that this case will be of much practical interest.

2.6. Optimization of the intrinsic noise temperature for a given instability factor

If we introduce the condition $S^e_{(-)} = $ const., (5.73) in (5.72) and if we attempt to optimize $T^e_{(-)}$, the calculations become far too complicated, and it is preferable to use an approximate method. First, we

INVERTING CONVERTER AND REFLECTION AMPLIFIER   127

will show that the reduced optimum resistance r_e is usually large compared to unity, and this allows us to take $(1 + r_e)$ as r_e in formulae (5.72) and (5.73). Further, we limit ourselves to the case in which $T_{(-)} = T_v$ (the varactor and the isolator of circuit $f_{(-)}$ are at the same temperature). The calculations will then be greatly simplified.

2.6.1. ORDER OF MAGNITUDE OF THE OPTIMUM VALUE OF r_e

If we eliminate variable r_L by means of the $S^e_{(-)} = $ const. condition by substituting (5.73) in (5.72) we obtain:

$$T^e_{(-)} = F_1(r_e) + \left(1 + \frac{1}{r_e}\right) F_2(r_e) \qquad (5.87)$$

with:

$$F_1(r_e) = \left[\left(1 + \frac{1}{q^2}\right)\frac{1}{r_e} + \frac{2}{q^2} + \frac{1}{q^2} r_e\right] T_v$$

$$F_2(r_e) = \left\{ \frac{1}{q^2} \left[\frac{q^2}{n_{(-)}} \cdot \frac{S^e_{(-)} + 1}{S^e_{(-)} - 1} - (1 + r_e)\right] - \frac{2}{n_{(-)}} \cdot \frac{1}{S^e_{(-)} - 1} \right. \\
\left. + \frac{\frac{q^2}{n^2_{(-)}} \cdot \frac{1}{(S^e_{(-)} - 1)^2}}{\left[\frac{q^2}{n_{(-)}} \cdot \frac{S^e_{(-)} + 1}{S^e_{(-)} - 1} - (1 - r_e)\right]} \right\} \qquad (5.88)$$

$F_1(r_e)$ is, of course, always positive; while $F_2(r_e)$ is a perfect square:

$$F_2(r_e) \geqslant 0 \qquad (5.89)$$

Let r'_e and r'' represent respectively the values of r which cause functions $F_1(r_e)$ and $F_2(r_e)$ to be a minimum. We find:

$$r'_e = \sqrt{1 + q^2}$$

$$r''_e = \frac{q^2}{n_{(-)}} \cdot \frac{S^e_{(-)}}{S^e_{(-)} - 1} - 1 \qquad (5.90)$$

On the other hand, the differential of (5.87) is expressed by:

$$\frac{dT^e_{(-)}}{dr_e} = \frac{dF_1}{dr_e} - \frac{1}{r_e^2} F_2 + \left(1 + \frac{1}{r_e}\right) \frac{dF_2}{dr_e} \qquad (5.91)$$

If r_e is smaller than the smaller of the two values r'_e and r''_e, dF_1/dr_e and dF_2/dr_e are negative, and if we take account of (5.89), we have:

$$\left[\frac{dT^e_{(-)}}{dr_e}\right]_{r_e \leqslant r'_e, r''_e} \leqslant 0 \qquad (5.92)$$

The root $r_{e\,\text{opt}}$ of the equation $dT^e_{(-)}/dr_e = 0$ which makes $T^e_{(-)}$ a minimum is then higher than the smaller of the two values r'_e and r''_e. If we assume that q and $q^2/n_{(-)}$ are large compared to unity (as is usual), we may write in accordance with (5.90):

$$\left.\begin{array}{r} r'_e \gg 1 \\ r''_e \gg 1 \end{array}\right\} \tag{5.93}$$

whence:

$$r_{e\,\text{opt}} \gg 1 \tag{5.94}$$

2.6.2. OPTIMIZATION OF $T^e_{(-)}$ WITH RESPECT TO THE TERMINATING IMPEDANCES

The use of equation (5.87) in which 1 is ignored by comparison with r_e again involves very long calculations. It is preferable to revert to equation (5.72) and to remove r_e from it by means of condition $S^e_{(-)} = \text{const.}$, (5.65) gives us:

$$(1 + r_e) = \frac{1}{1 + r_L} \cdot \frac{q^2}{n_{(-)}} \cdot \frac{S^e_{(-)} + 1}{S^e_{(-)} - 1} \tag{5.95}$$

If we introduce (5.95) in (5.72) and take r_e as equivalent to $(1 + r_e)$, we obtain:

$$T^e_{(-)} = T_v\left[\frac{1}{n_{(-)}} \cdot \frac{1}{S^{e2}_{(-)} - 1} \cdot \frac{1}{r_L} + \frac{n_{(-)}}{q^2} \cdot \frac{S^e_{(-)} - 1}{S^e_{(-)} + 1} \right.$$
$$\left. + \frac{1}{n_{(-)}} \frac{S^{e2}_{(-)}}{S^{e2}_{(-)} - 1} + \frac{n_{(-)}}{q^2} \frac{S^e_{(-)} - 1}{S^e_{(-)} + 1} r_L \right] \tag{5.96}$$

The optimum value of r_L is then:

$$r_{L\,\text{opt}} = \frac{q}{n_{(-)}(S^e_{(-)} - 1)} \tag{5.97}$$

whence, in accordance with (5.95), we find:

$$r_{e\,\text{opt}} = \frac{q^2(S^e_{(-)} + 1)}{q + n_{(-)}(S^e_{(-)} - 1)} - 1 \tag{5.98}$$

and if we introduce these values in (5.72), we obtain the expression for the minimum noise temperature for a given stability:

$$T^e_{e\,\text{min}} =$$
$$\frac{\begin{array}{c}n_{(-)}(S^e_{(-)} - 1)[q + n_{(-)}(S^e_{(-)} - 1)]^2 \\ + n_{(-)}q^2(S^e_{(-)} - 1)(S^e_{(-)} + 1)^2 + q[S^e_{(-)}(q - n_{(-)}) + n_{(-)}]^2\end{array}}{n_{(-)}(S^e_{(-)} + 1)[q + n_{(-)}(S^e_{(-)} - 1)][S^e_{(-)}(q^2 - n_{(-)}) + q^2 - q - n_{(-)}]}$$
$$\tag{5.99}$$

INVERTING CONVERTER AND REFLECTION AMPLIFIER

At high gain, $T^e_{(-)\,\min}$ has the limiting value:

$$[T^e_{(-)\,\min}]_{S^e_{(-)}\to\infty} = \frac{(q^2/n^2_{(-)}) + 1}{(q^2/n_{(-)}) - 1} T_v \qquad (5.100)$$

This value is a lower limit: from the noise point of view, it is advantageous to use the inverting converter at high gain.

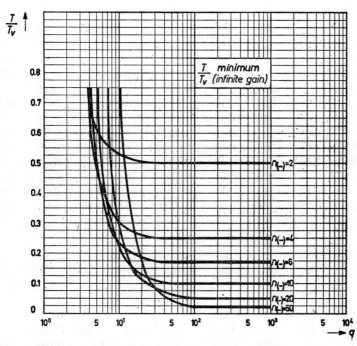

Figure 5.16. Inverting converter and reflection amplifier minimum T/T_v (infinite gain)

The terminal resistances are in these conditions:

$$\left.\begin{aligned} r_e &= q^2/n_{(-)} - 1 \\ r_L &= 0 \end{aligned}\right\} \qquad (5.101)$$

Figure 5.16 and Table 5.1 give $[T^e_{(-)\,\min}/T_v]S \to \infty$ for different values of $n_{(-)}$ and of q. For a perfect varactor ($q = \infty$).

$$T^e_{(-)\,\min} = \frac{T_v}{n_{(-)}} \qquad (5.102)$$

We see that the non-cooled inverting converter is of interest as a low-noise amplifier only if the output frequency is appreciably higher than the input frequency.

Table 5.1. Inverting converter and reflection amplifier minimum T/T (infinite gain)

	$n_{(-)}$							
q	2	4	6	8	10	12	14	16
2	2·00							
5	0·63	0·49	0·54	0·65	0·83	1·08	1·44	1·95
7	0·56	0·36	0·33	0·34	0·38	0·43	0·50	0·58
10	0·53	0·30	0·24	0·22	0·22	0·23	0·25	0·26
15	0·54	0·27	0·20	0·17	0·15	0·14	0·14	0·14
20	0·51	0·26	0·18	0·15	0·13	0·12	0·11	0·11
50	0·50	0·25	0·17	0·13	0·10	0·09	0·08	0·07
70	0·50	0·25	0·17	0·13	0·10	0·09	0·07	0·07
100	0·50	0·25	0·17	0·13	0·10	0·08	0·07	0·06
150	0·50	0·25	0·17	0·13	0·10	0·08	0·07	0·06
200	0·50	0·25	0·17	0·13	0·10	0·08	0·07	0·06
500	0·50	0·25	0·17	0·13	0·10	0·08	0·07	0·06
700	0·50	0·25	0·17	0·13	0·10	0·08	0·07	0·06
1000	0·50	0·25	0·17	0·13	0·10	0·08	0·07	0·06

	$n_{(-)}$							
q	18	20	25	30	35	40	45	50
2								
5	2·77	4·25						
7	0·67	0·77	1·12	1·66	2·60	4·58	11·52	
10	0·29	0·31	0·39	0·48	0·58	0·71	0·86	1·04
15	0·15	0·15	0·17	0·19	0·22	0·25	0·28	0·31
20	0·11	0·11	0·11	0·12	0·13	0·14	0·15	0·17
50	0·06	0·06	0·05	0·05	0·04	0·04	0·04	0·04
70	0·06	0·05	0·05	0·04	0·04	0·03	0·03	0·03
100	0·06	0·05	0·04	0·04	0·03	0·03	0·03	0·03
150	0·06	0·05	0·04	0·03	0·03	0·03	0·02	0·02
200	0·06	0·05	0·04	0·03	0·03	0·03	0·02	0·02
500	0·06	0·05	0·04	0·03	0·03	0·03	0·02	0·02
700	0·06	0·05	0·04	0·03	0·03	0·03	0·02	0·02
1000	0·06	0·05	0·04	0·03	0·03	0·03	0·02	0·02

INVERTING CONVERTER AND REFLECTION AMPLIFIER

2.6.3. OPTIMIZATION OF $T^e_{(-)}$ WITH RESPECT TO $n_{(-)}$, r_e AND r_L

If we now optimize $T^e_{(-)}$ with respect to $n_{(-)}$, equation (5.99) shows that the optimum value of $n_{(-)}$ is:

$$n_{(-)\,\mathrm{opt}} \approx q \frac{S^e_{(-)}}{S^e_{(-)} - 1} \tag{5.103}$$

The temperature $T^e_{(-)\,\mathrm{opt}}$ optimized with respect to all the variables is then equal to:

$$T^e_{(-)\,\mathrm{opt}} = \frac{2}{q} \frac{S^e_{(-)}}{S^e_{(-)} + 1} T_v \tag{5.104}$$

and the corresponding gain:

$$[G^e_{(-)}]_{T_{\mathrm{opt}}} = q S^e_{(-)} \tag{5.105}$$

and the corresponding terminal impedances:

$$r_{e\,\mathrm{opt}} = q \tag{5.106}$$

$$r_{L\,\mathrm{opt}} = \frac{1}{S^e_{(-)}} \tag{5.107}$$

For high gain ($S^e_{(-)}$ high), relations (5.103) and (5.104) may be written:

$$n_{(-)\,\mathrm{opt}} \approx q \tag{5.108}$$

$$T^e_{(-)\,\mathrm{opt}} \approx \frac{2T_v}{q} \tag{5.109}$$

which is the same as formula (4.32): for a given varactor, of sufficiently high quality, the minimum noise temperatures of the sum-frequency converter and of the inverting converter are equal.

2.7. Optimization of the total noise temperature for a given instability factor

2.7.1. OPTIMIZATION WITH RESPECT TO THE TERMINATING IMPEDANCES

If T_2 is the noise temperature of the amplification chain after the converter, the expression for the total noise temperature is:

$$T^e_{(-)T} = T^e_{(-)} + \frac{T_2}{G^e_{(-)}} \tag{5.110}$$

This value must be optimized with respect to r_e, r_L and $n_{(-)}$ taking into account condition $S^e_{(-)} = $ const. It is clear that the optimum value of r_e will lie between the values corresponding to the maximum gain and the minimum intrinsic noise respectively, and that it will become more nearly equal to the second as the values of $G^e_{(-)}$ and therefore $S^e_{(-)}$ increase.

In normal conditions (q and $S^e_{(-)}$ high), taking (5.63), (5.72) and (5.95) and assuming as before that $T_{(-)} = T_v$, then (5.110) becomes:

$$T^e_{(-)t} = \left[\frac{1}{n_{(-)}} \cdot \frac{1}{S^{e2}_{(-)} - 1} \cdot \frac{1}{r_L} + \frac{n_{(-)}}{q} \frac{S^e_{(-)} - 1}{S^e_{(-)} + 1} \right.$$

$$+ \frac{1}{n_{(-)}} \cdot \frac{S^{e2}_{(-)}}{S^{e2}_{(-)} - 1} + \frac{n_{(-)}}{q^2} \cdot \frac{S^e_{(-)} - 1}{S^e_{(-)} + 1} r_L \Bigg] T_v$$

$$+ \left[\left(1 + \frac{1}{r_L} \right) \frac{1}{n_{(-)}} \cdot \frac{1}{S^{e2}_{(-)} - 1} \right] T_2 \quad (5.111)$$

which gives us:

$$r_{L\text{ opt}} = \frac{q}{n_{(-)}} \cdot \frac{1}{S^e_{(-)} - 1} \sqrt{1 + \frac{T_2}{T_v}} \quad (5.112)$$

$$r_{e\text{ opt}} = \frac{q(S^e_{(-)} + 1)}{\sqrt{1 + T_2/T_v} + (n_{(-)}/q)(S^e_{(-)} - 1)} \quad (5.113)$$

The total noise temperature $T^e_{(-)t\text{ min}}$, optimized with respect to the impedances is then equal to:

$$T^e_{(-)t\text{ min}} = \frac{2\sqrt{1 + T_2/T_v}}{q(S^e_{(-)} + 1)} T_v$$

$$+ \frac{n_{(-)}}{q^2} \cdot \frac{S^e_{(-)} - 1}{S^e_{(-)} + 1} T_v + \frac{1}{n_{(-)}} \cdot \frac{S^{e2}_{(-)} T_v + T_2}{S^{e2}_{(-)} - 1} \quad (5.114)$$

2.7.2. OPTIMIZATION WITH RESPECT TO $n_{(-)}$

The optimum value of $n_{(-)}$ can be obtained immediately from (5.114)

$$n_{(-)\text{ opt}} = \frac{q\sqrt{S^{e2}_{(-)} + T_2/T_v}}{S^e_{(-)} - 1} \quad (5.115)$$

whence the total noise temperature $T^e_{(-)\text{ opt}}$, optimized with respect to all the variables

$$T^e_{(-)t\text{ opt}} = \frac{2}{q} \cdot \frac{1}{S^e_{(-)} + 1} \left[\sqrt{1 + \frac{T_2}{T_v}} + \sqrt{S^{e2}_{(-)} + \frac{T_2}{T_v}} \right] T_v \quad (5.116)$$

2.8. High-gain bandwidth

Let us consider once again the general formula for the gain:

$$G^e_{(-)} = \frac{4r_e r_L q^2}{|(z_e + 1)(z^*_{(-)} + 1) - q^2/n_{(-)}|^2} \qquad (5.62)$$

It will be assumed that the selectivity of the term $(z_e + 1)(z^*_{(-)} + 1)$, a function of the tuning reactances, is predominant. We can then take q and $n_{(-)}$ as constants in the band-pass interval. It should be noted on the other hand that, unlike the case of the sum-frequency converter, the denominator consists of the square of the modulus of the difference of the two terms. We can then see that for a high gain a slight variation in the reactances will result in a considerable variation in the gain: the bandwidth is narrow and a first-order approximation can be used for the calculation.

Let:

$$\omega_e = \omega_{e_0} + \delta\omega \qquad (5.117)$$

in which ω_{e_0} is the nominal input frequency. As the bandwidth is narrow we may write $(z_e + 1)$ and $(z_{(-)} + 1)$ in the form:

$$z_e + 1 = (r_e + 1)\left(1 + 2jQ_e \frac{\delta\omega}{\omega_{e_0}}\right) \qquad (5.118)$$

$$z^*_{(-)} + 1 = (r_L + 1)\left(1 - 2j\frac{Q_{(-)}}{n_{(-)}} \frac{\delta\omega}{\omega_{e_0}}\right) \qquad (5.119)$$

assuming that the input and output branches have simple tuning, that is to say that they are equivalent to an LRC series circuit in the frequency interval defined by the bandwidth. Q_e and $Q_{(-)}$ are the 'cold' (for $S_1 = 0$) quality factors of the input and output circuits, and are easily calculable. The value of $2\delta\omega$ corresponding to the bandwidth at 3 dB is then the root of the equation:

$$\left\{(r_e + 1)(r_L + 1)\left[1 - \frac{Q_e Q_{(-)}}{n_{(-)}}\left(\frac{2\delta\omega}{\omega_{e_0}}\right)^2\right] - \frac{q^2}{n_{(-)}}\right\}^2$$

$$+ \left[(r_e + 1)(r_L + 1)\left(Q_e + \frac{Q_{(-)}}{n_{(-)}}\right)\left(\frac{2\delta\omega}{\omega_{e_0}}\right)\right]^2$$

$$= 2[(r_e + 1)(r_L + 1) - q^2/n_{(-)}]^2 \qquad (5.120)$$

using (5.95) and ignoring

$$\left[\frac{Q_e Q_{(-)}}{n_{(-)}}\left(\frac{2\delta\omega}{\omega_{e_0}}\right)\right]^2 \quad \text{compared to} \quad \left[1 - 2\frac{Q_e Q_{(-)}}{n_{(-)}}\left(\frac{2\delta\omega}{\omega_{e_0}}\right)^2\right]$$

we find the relative bandwidth:

$$B^e_{(-)} = \frac{2\delta\omega}{\omega_{e_0}} =$$

$$= \frac{2}{S^e_{(-)} + 1} \frac{1}{\sqrt{(Q_e + Q_{(-)}/n_{(-)}) - (4/S^e_{(-)} + 1)(Q_e Q_{(-)}/n_{(-)})}} \quad (5.121)$$

We see that $B^e_{(-)}$ decreases and approaches zero when $S^e_{(-)}$ and consequently the gain tends to infinity.

As $S^e_{(-)}$ is high:

$$B^e_{(-)} \approx \frac{2}{S^e_{(-)}(Q_e + Q_{(-)}/n_{(-)})} \quad (5.122)$$

We will not calculate the optimization because Q_e and $Q_{(-)}$ depend on the levels of the input and output impedance in a non-predictable manner, the circuits becoming in general more selective as the transformation ratio increases.

3. The reflection amplifier

3.1. Transducer gain

As we saw in the first part of this chapter, the circuit at $\omega_{(-)}$ reflects a negative resistance into the input circuit. It is therefore possible to amplify directly at ω_e without frequency conversion. We also saw the advantages of using an input circulator for a negative-resistance amplifier. So we will limit our discussion to the circulator- or 'reflection'-amplifier, which is the only one used in practice. Figure 5.17 shows the equivalent schema of the reflection amplifier. It is derived directly from Fig. 5.13 by taking R_{L3} as the working load and the circulator of the circuit at $\omega_{(-)}$ terminated by its loads $R_{(-)}$ and R_{L4} as the idler load.

The formulae for Fig. 5.14 are therefore still applicable and the transducer gain of the reflection amplifier is given by:

$$G^e_e = G^1_3 = \left| \frac{(z^*_e - 1)(z^*_{(-)} + 1) + q^2/n_{(-)}}{(z_e + 1)(z^*_{(-)} + 1) - q^2/n_{(-)}} \right|^2 \quad (5.123)$$

We see that:

(1) For given values of q, $n_{(-)}$, r_e and r_L, the gain is a maximum when z_e and $z_{(-)}$ are real (resonance condition of the input and output circuits).

INVERTING CONVERTER AND REFLECTION AMPLIFIER 135

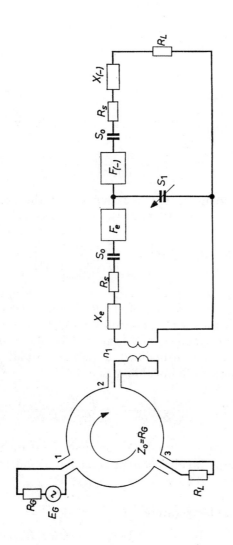

Figure 5.17. Equivalent schema of the reflection amplifier

(2) If $q^2/n_{(-)}$ has a high enough value (greater than unity when in the resonance condition) this gain may be made arbitrarily high by adjusting r_e and r_L.

We will assume in what follows that the resonance condition has been realized; (5.123) may then be written:

$$G_e^e = \left[\frac{(r_e - 1)(r_{(-)} + 1) + q^2/n_{(-)}}{(r_e + 1)(r_{(-)} + 1) - q^2/n_{(-)}}\right]^2 \qquad (5.124)$$

3.2. Instability factor

As for all the circuits discussed so far, the instability factor will be defined by:

$$S_e^e = \frac{dG_e^e/G_e^e}{dq^2/q^2} \qquad (5.125)$$

The equivalent circuit in Fig. 5.2 shows that the negative input resistance of the amplifier when in the resonance condition is equal to:

$$-|R_N| = R_s\left[1 - \frac{q^2}{n_{(-)}(r_{(-)} + 1)}\right] \qquad (5.126)$$

If the term $q^2/[n_{(-)}(r_{(-)} + 1)]$ is large compared to unity, which is usually the case as we will see later, R_N is proportional to q^2 and (5.125) is in accordance with the definition (5.32) used in the general study of negative-resistance amplification. The choice of the definition (5.125) was therefore justified. Equation (5.136) gives us the general expression for the instability factor:

$$S_e^e = \frac{G_e^e - 1}{\sqrt{G_e^e}} \frac{q^2}{q^2 - n_{(-)}(r_{(-)} + 1)} \qquad (5.127)$$

It will be seen that the condition for high gain (denominator of (5.124) near to zero), involves a high value for the instability factor.

3.3. Effective noise temperature

The input generator is replaced by a resistance R_e cooled to 0°K. Figure 5.18 shows the internal noise generators of the amplifier. It should be remembered that we have assumed that the input and idler circuits are in the resonance condition. The noise power fed to the load R_L is the sum of the following three terms:

(1) The noise power generated at frequency f_e by R_s and transmitted to the load after negative-resistance amplification ΔN_e. The available power of the noise generator is equal to $kT_v \Delta f$, hence:

$$\Delta N_e = kT_v G_3^2 \Delta f \qquad (5.128)$$

(2) The noise power generated at frequency $f_{(-)}$ by R_s and transmitted to the load after conversion $\Delta N_{(-)}$. By reasoning similar to that in section 2.3.1 we find:

$$\Delta N_{(-)} = \frac{1}{r_{(-)}} kT_v G_3^5 \Delta f \qquad (5.129)$$

(3) The noise power generated at frequency $f_{(-)}$ by $R_{(-)}$ and transmitted to the load after conversion ΔN_i. We find:

$$\Delta N_i = kT_{(-)} G_3^5 \Delta f \qquad (5.130)$$

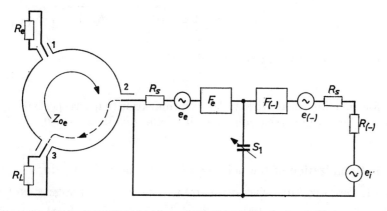

Figure 5.18. Noise generators of the reflection amplifier

We assumed in the last section that the termination R_e was perfectly matched with Z_{0e}. The noise of the load R_L will not be amplified, and it is only the above three terms that contribute to the noise temperature of the amplifier, which is equal to:

$$T_e^e = \left(\frac{G_3^2}{G_3^1} + \frac{1}{r_{(-)}} \frac{G_3^5}{G_3^1}\right) T_v + \frac{G_3^5}{G_3^1} T_{(-)} \qquad (5.131)$$

If we replace the G_k^i's by their values we obtain:

$$T = 4r_e \frac{(r_{(-)} + 1)^2 T_v + q^2/n_{(-)}^2 (T_v + r_{(-)} T_{(-)})}{[(r_e - 1)(r_{(-)} + 1) + q^2/n_{(-)}]^2} \qquad (5.132)$$

3.4. Optimization of the gain for a given instability factor

For the reasons discussed in section 2.4, the optimizations will be carried out for a given stability. The three parameters r_e, r_e and $n_{(-)}$ can then be reduced to two independent parameters. If we express the gain as a function of the instability factor, (5.127) gives us:

$$G_e^e = \left\{ \frac{S_e^{e2}}{2} \left[1 - \frac{n_{(-)}}{q^2} (r_{(-)} + 1) \right] + \sqrt{\frac{S_e^{e2}}{4} \left[1 - \frac{n_{(-)}}{q^2} (r_{(-)} + 1) \right]^2 + 1} \right\}^2 \quad (5.133)$$

It will be seen at once that, for given values of $n_{(-)}$, q and S_e^e, the gain is a maximum for:

$$r_{(-)} = 0 \quad (5.134)$$

that is when the idler load consists only of the resistance R_s of the varactor. We then have:

$$G_{e\,\max}^e = \left\{ \frac{S_e^{e2}}{2} \left[1 - \frac{n_{(-)}}{q^2} \right] + \sqrt{\frac{S_e^{e2}}{4} \left[1 - \frac{n_{(-)}}{q^2} \right] + 1} \right\}^2 \quad (5.135)$$

which can usually be reduced, when q^2/n and S_e^e are large, to:

$$G_{\max} \approx S_e^{e2} \quad (5.136)$$

a formula identical with (5.50). Optimization with respect to $n_{(-)}$ would be without meaning, since it would give us $n_{(-)} = 0$.

3.5. Optimization of the noise temperature for a given gain

3.5.1. OPTIMIZATION WITH RESPECT TO THE TERMINAL RESISTANCES

We will commence with the optimization for a given gain, the computations being simpler. If we combine relations (5.124) and (5.132) we obtain in the usual case when $T_v = T_{(-)}$:

$$T_e^e = \left(1 + \frac{1}{\sqrt{G_e^e}} \right)^2 \left[\frac{1}{r_e} \left(1 + \frac{1}{n_{(-)}} \right) + \frac{1}{n_{(-)}} \frac{\sqrt{G_e^e} - 1}{\sqrt{G_e^e} + 1} \right] T_v \quad (5.137)$$

It will be seen at once that we obtain the minimum noise temperature for a given gain when r_e is a maximum, that is for condition (5.134), the idler load then consists only of R_s. We then have:

$$\left. \begin{array}{l} r_{(-)\,\mathrm{opt}} = 0 \\[4pt] r_{e\,\mathrm{opt}} = \left(\dfrac{q^2}{n_{(-)}} - 1 \right) \dfrac{\sqrt{G_e^e} + 1}{\sqrt{G_e^e} - 1} \end{array} \right\} \quad (5.138)$$

INVERTING CONVERTER AND REFLECTION AMPLIFIER

and

$$T^e_{e\,\min} = \left(1 - \frac{1}{G^e_e}\right) \frac{q^2/n^2_{(-)} + 1}{q^2/n_{(-)} - 1} T_v \qquad (5.139)$$

When the gain is high, we have:

$$[T^e_{e\,\min}]_{G \to \infty} \approx \frac{q^2/n^2_{(-)} + 1}{q^2/n_{(-)} - 1} T_v \qquad (5.140)$$

a formula identical with (5.100). Table 5.1 and the group of curves in Fig. 5.16 are therefore still applicable.

The value of $T^e_{e\,\min}$ corresponding to a finite gain always lies between that given by (5.140) and zero, where the zero value corresponds to unity gain. For a perfect varactor (q infinite) the limiting value is:

$$[T^e_{e\,\min}]_{q \to \infty} = \frac{T_v}{n_{(-)}} \qquad (5.141)$$

identical with that given by (5.101).

3.5.2. OPTIMIZATION WITH RESPECT TO $n_{(-)}$

If in (5.139) we equate to zero the differential of $T^e_{e\,\min}$ with respect to $n_{(-)}$ we find:

$$n_{(-)\,\mathrm{opt}} = \sqrt{q^2 + 1} - 1 \qquad (5.142)$$

a value identical with that obtained for the high-gain inverting converter. The noise temperature optimized with respect to all the parameters is equal to:

$$T^e_{e\,\mathrm{opt}} = 2\left(1 - \frac{1}{G^e_e}\right) \frac{T_v}{\sqrt{q^2 + 1} - 1} \qquad (5.143)$$

For a high-quality varactor (high q), we have:

$$n_{(-)\,\mathrm{opt}} \approx q \qquad (5.144)$$

hence, for a high gain:

$$T^e_{e\,\mathrm{opt}} \approx 2\frac{T_v}{q} \qquad (5.145)$$

as for the two converters.

3.6. Optimization of the noise temperature for a given instability factor

We saw that in the usual conditions ($q^2/n_{(-)}$ high and high gain), the gain and the stability are connected by relation (5.136). In these conditions the optimization for a given gain and for a given instability factor are then equivalent and we have:

$$\left. \begin{array}{l} r_{(-)\,\text{opt}} \approx 0 \\ r_{e\,\text{opt}} \approx \left(\dfrac{q^2}{n_{(-)}} - 1\right) \cdot \dfrac{S_e^e + 1}{S_e^e - 1} \end{array} \right\} \quad (5.146)$$

$$T_{e\,\text{min}}^e \approx \left(1 - \frac{1}{S_e^{e2}}\right) \frac{q^2/n_{(-)}^2 + 1}{q^2/n_{(-)} - 1} T_v \quad (5.147)$$

$$T_{e\,\text{opt}}^e \approx 2\left(1 - \frac{1}{S_e^{e2}}\right) \frac{T_v}{\sqrt{q^2 + 1} - 1} \quad (5.148)$$

and for a high q:

$$T_{e\,\text{opt}}^e \approx 2\left(1 - \frac{1}{S_e^{e2}}\right) \frac{T_v}{q} \quad (5.149)$$

The limiting values given by (5.141) (5.142) and (5.144) will of course be unchanged.

3.7. Minimization of the total noise temperature

If we take (5.139) into account, the minimum total noise temperature is equal to:

$$T_{e\,\text{min}\,T}^e = \frac{q^2/n_{(-)}^2 + 1}{q^2/n_{(-)} - 1} T_v + \frac{1}{G_e^e}\left(T_2 - \frac{q^2/n_{(-)}^2 + 1}{q^2/n_{(-)} - 1} T_v\right) \quad (5.150)$$

This formula shows that if $T_2 > [T_{e\,\text{min}}^e]_{G \to \infty}$, which is of course always true in practice, the noise temperature decreases as the gain increases. It is therefore always desirable to work at high gain, the limiting value of which is fixed by the permissible instability factor.

3.8. High-gain bandwidth

Let us again consider formula (5.135) which gives the general expression for the transducer gain as:

$$G_e^e = \left| \frac{(z_e^* - 1)(z_{(-)}^* + 1) + q^2/n_{(-)}}{(z_e + 1)(z_{(-)}^* + 1) - q^2/n_{(-)}} \right|^2 \quad (5.123)$$

As before we will assume that the selectivity of the terms containing the tuning reactances is predominant and that we may ignore the variation of $q^2/n_{(-)}$ in the bandwidth. Further, it is clear that for a high gain, the relative variation of the denominator, as a function of

INVERTING CONVERTER AND REFLECTION AMPLIFIER

the frequency, is much larger than that of the numerator. We can then expect that the bandwidth will be narrow and we will ignore the variation of the numerator. We will also assume as before that the circuits have simple tuning.

Equations (5.117), (5.118) and (5.119) are still applicable and the value of $2\delta\omega$ corresponding to the bandwidth at 3 dB is the root of the equation:

$$\left\{(r_e + 1)(r_{(-)} + 1)\left[1 - \frac{Q_e Q_{(-)}}{n_{(-)}}\left(\frac{2\delta\omega}{\omega_{e_0}}\right)^2\right] - \frac{q^2}{n_{(-)}}\right\}^2$$
$$+ \left[(r_e + 1)(r_{(-)} + 1)\left(Q_e + \frac{Q_{(-)}}{n_{(-)}}\right)\frac{2\delta\omega}{\omega_{e_0}}\right]^2$$
$$= 2[(r_e + 1)(r_{(-)} + 1) - q^2/n_{(-)}]^2 \quad (5.151)$$

As we have assumed that the gain is high, taking $(r_e - 1)$ as equal to $(r_e + 1)$, we obtain from (5.124) and (5.127):

$$(r_e + 1)(r_{(-)} + 1) \approx \frac{q^2}{n}\frac{S_e^e - 1}{S_e^e + 1} \quad (5.152)$$

Introducing this relation into (5.121) we obtain, if we limit ourselves to second-order terms in $2\delta\omega/\omega_{e_0}$, the relative bandwidth:

$$B_e^e = \frac{2\delta\omega}{\omega_{e_0}} = \frac{2}{S_e^e + 1}\frac{1}{\sqrt{(Q_e + Q_{(-)}/n_{(-)})^2 - (4/S_e^e + 1)(Q_e Q_{(-)}/n_{(-)})}} \quad (5.153)$$

or, as S_e^e is large:

$$B_e^e \approx \frac{2}{S_e^e(Q_e + Q_{(-)}/n_{(-)})} \quad (5.154)$$

These equations are similar to equations (5.121) and (5.122).

4. Conclusion

The inverting converter and the reflection amplifier have the common property that they can be adjusted to an arbitrarily high gain, limited only by the permissible instability factor, and their limiting noise temperatures are identical. In practice, however, the reflection amplifier is the one much more commonly used; its advantages are that:

(1) it requires only one non-reciprocal element instead of the two required by the converter;
(2) the output frequency is equal to the input frequency; it is easier to design the frequency converter which follows, as it operates at a lower frequency

(3) the phase of the output signal is independent of that of the pump (this property may be of importance in interferometry for example);

(4) the output circuit of the converter raises a difficult problem for the transformation of the impedance, as the value of $r_{L\,opt}$ is very low—this problem is aggravated by the fact that the output is usually connected to a waveguide (see Chapter 11).

References

[1] Lax, B. and Button, K., *Microwave Ferrites and Ferrimagnetics*, Lincoln Laboratory Publications, McGraw-Hill, New York, 1962.

[2] Kurokawa, K. and Uenohara, M., *Minimum Noise Figure of the Variable Capacitance Amplifier*, Bell System Technical Journal, vol. 40, pp. 695–722, May 1961.

[3] Laurent, L., 'Les amplificateurs paramétriques (2eme partie)', *Revue M.B.L.E.*, vol. VI (No. 4), December 1963.

6

FOUR-FREQUENCY CONVERTERS AND AMPLIFIERS

by L. LAURENT

1. Introduction

In four-frequency converters and amplifiers, the four main frequencies f_p, f_e, $f_{(+)}$ and $f_{(-)}$ are used. Devices of this type are little used in practice, because of the difficulty in producing them satisfactorily. The only one which is of interest, and of which some practical examples have been mentioned in technical literature, is the sum-frequency converter [1]. The representation of four-frequency circuits by equations is much more complicated than for three-frequency circuits. It is for this reason that we have studied the two families of circuits independently, instead of considering the second as special cases derived from the first by suppression of $f_{(+)}$ or $f_{(-)}$.

In spite of the limited use of four-frequency circuits, the study of them has nevertheless a certain interest. It allows us to calculate the disturbance introduced into a three-frequency circuit when the fourth frequency is not completely 'open', which often happens due to the imperfection of the filters employed. We will not study them in as great detail as the preceding circuits and will consider only the conditions for stability and the gains.

2. Equivalent diagram and general equations for four-frequency converters

The general equivalent diagram of a four-frequency circuit is shown in Fig. 6.1. The signal enters at f_e and leaves at $f_{(+)}$, $f_{(-)}$ or f_e depending on the circumstances. The branch which contains no working load is called the 'idler' as before. When the signal is reflected at

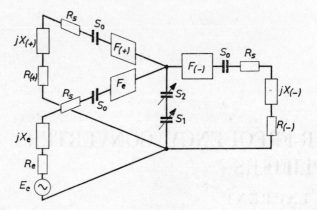

Figure 6.1

f_e (negative-resistance amplifier) there are two idlers, one at $f_{(+)}$, and the other at $f_{(-)}$. Equation (3.40) is the general equation for the relations between the voltages and currents corresponding to the three branches at ω_e, $\omega_{(+)}$ and $\omega_{(-)}$.

$$\begin{bmatrix} 0 \\ E_e \\ 0 \end{bmatrix} = R_s \begin{bmatrix} 1 + z^*_{(-)} & -j\dfrac{\omega'_c}{\omega_e} & -j\dfrac{\omega''_c e^{-j\varphi}}{\omega_{(+)}} \\ j\dfrac{\omega'_c}{\omega_{(-)}} & 1 + z_e & -j\dfrac{\omega_c}{\omega_{(+)}} \\ j\dfrac{\omega''_c e^{j\varphi}}{\omega_{(-)}} & -j\dfrac{\omega'_c}{\omega_e} & 1 + z_{(+)} \end{bmatrix} \begin{bmatrix} I^*_{(-)} \\ I_e \\ I_{(+)} \end{bmatrix} \quad (3.40)$$

3. Calculation of the total reduced impedances of the three branches

3.1. Calculation of the total reduced input impedance $\bar{z}_e$

Unlike the three-frequency circuits, the four-frequency circuits use the second harmonic of the elastance S_2. Recalling that:

$$S_2 = |S_2|\, e^{j\phi} \quad (6.1)$$

and assuming the time origin chosen for S_1 to be real and positive. The total reduced impedance in the input branch is equal to:

$$\bar{z}_e = \dfrac{E_e}{R_s I_e} \quad (6.2)$$

FOUR-FREQUENCY CONVERTERS AND AMPLIFIERS

The matrix relation (3.40) gives us:

$$\bar{z}_e = (1 + z_e) + \frac{\omega_c'}{\omega_e} \frac{(1 + z_{(+)}) \frac{\omega_{(+)}}{\omega_c'} - (1 + z_{(-)}^*) \frac{\omega_{(-)}}{\omega_c'} + 2j \frac{\omega_c''}{\omega_c'} e^{j\phi}}{- (1 + z_{(+)})(1 + z_{(-)}^*) \frac{\omega_{(+)}\omega_{(-)}}{\omega_c'^2} + \frac{\omega_c''^2}{\omega_c'^2}} \quad (6.3)$$

When $\omega_p < \omega_e$, if we use only the positive frequencies this relation may be written:

$$\bar{z}_e = (1 + z_e) + \frac{\omega_c'}{\omega_e} \frac{(1 + z_{(+)}) \frac{\omega_{(+)}}{\omega_c'} + (1 + z_{(-)}) \frac{|\omega_{(-)}|}{\omega_c'} + 2j \frac{\omega_c''}{\omega_c'} e^{j\phi}}{(1 + z_{(+)})(1 - z_{(-)}) \frac{\omega_{(+)}|\omega_{(-)}|}{\omega_c'^2} + \frac{\omega_c''^2}{\omega_c'^2}} \quad (6.4)$$

3.2. Calculation of the equivalent circuit of the branch at $\omega_{(+)}$

By elimination, (3.40) gives us the current in the branch at $\omega_{(+)}$:

$$I_{(+)} = j \frac{E_e}{R_s} \cdot \frac{\omega_c'}{\omega_e} \cdot \frac{-(1 + z_{(-)}^*) + j \frac{\omega_c''}{\omega_{(-)}} e^{j\phi}}{(1 + z_e)(1 + z_{(+)})(1 + z_{(-)}^*) + 2j \frac{\omega_c'' \omega_c' e^{j\phi}}{\omega_e \omega_{(+)} \omega_{(-)}}}$$

$$+ \frac{(1 + z_e)\omega_c''^2}{\omega_{(+)}\omega_{(-)}} + \frac{(1 + z_{(+)})\omega_c'^2}{\omega_e \omega_{(-)}} + \frac{(1 + z_{(-)}^*)\omega_c'^2}{\omega_e \omega_{(+)}} \quad (6.5)$$

Putting

$$I_{(+)} = \frac{E_{(+)}}{\bar{z}_{(+)} R_s} \quad (6.6)$$

and noting that $\bar{z}_{(+)}$ must necessarily be of the form

$$\bar{z}_{(+)} = (1 + z_{(+)}) + A$$

in which A is independent of $(1 + z_{(+)})$, it is easy to make $E_{(+)}$ and $\bar{z}_{(+)}$ appear in (6.5), and we thus obtain:

$$\bar{z}_{(+)} = (1 + z_{(+)}) + \frac{\omega_c'}{\omega_{(+)}}$$

$$\times \frac{(1 + z_e) \frac{\omega_e}{\omega_c'} \cdot \frac{\omega_c''^2}{\omega_c'^2} - (1 + z_{(-)}^*) \frac{\omega_{(-)}}{\omega_c'} + 2j \frac{\omega_c''}{\omega_c'} e^{j\phi}}{1 - (1 + z_e)(1 + z_{(-)}^*) \frac{\omega_e \omega_{(-)}}{\omega_c'^2}} \quad (6.7)$$

and:

$$E_{(+)} = jE_e \frac{-(1 + z^*_{(-)})\frac{\omega_{(-)}}{\omega'_c} + j\frac{\omega''_c}{\omega'_c} e^{j\phi}}{1 - (1 + z_e)(1 + z^*_{(-)})\frac{\omega_e\omega_{(-)}}{\omega'^2_c}} \quad (6.8)$$

When $\omega_p < \omega_e$, (6.7) and (6.8) may be written in the form:

$$\bar{z}_{(+)} = (1 + z_{(+)})$$

$$+ \frac{\omega'_e}{\omega_{(+)}} \frac{(1 + z_e)\frac{\omega_e}{\omega'_c} \cdot \frac{\omega''^2_c}{\omega'^2_c} + (1 + z_{(-)})\frac{|\omega_{(-)}|}{\omega'_c} + 2j\frac{\omega''_c}{\omega'_e} e^{j\phi}}{1 + (1 + z_e)(1 + z_{(-)})\frac{\omega_e|\omega_{(-)}|}{\omega'^2_c}} \quad (6.9)$$

and

$$E_{(+)} = jE_e \frac{(1 + z_{(-)})\frac{|\omega_{(-)}|}{\omega'_c} + j\frac{\omega''_c}{\omega'_c} e^{j\phi}}{1 + (1 + z_e)(1 + z_{(-)})\frac{\omega_e|\omega_{(-)}|}{\omega'^2_c}} \quad (6.10)$$

3.3. Calculation of the equivalent circuit of the branch at $\omega_{(-)}$

A similar calculation gives us

$$\bar{z}^*_{(-)} = \frac{E^*_{(-)}}{R_s I^*_{(-)}} = (1 + z^*_{(-)}) - \frac{\omega'_c}{\omega_{(-)}}$$

$$\times \frac{(1 + z_e)\frac{\omega_e}{\omega'_c} \cdot \frac{\omega''^2_c}{\omega'^2_c} + (1 + z_{(+)})\frac{\omega_{(+)}}{\omega'_c} + 2j\frac{\omega''_c}{\omega'_c}}{1 + (1 + z_e)(1 + z_{(+)})\frac{\omega_e\omega_{(+)}}{\omega'^2_c}} e^{j\phi} \quad (6.11)$$

$$E^*_{(-)} = jE_e \frac{(1 + z_{(+)})\frac{\omega_{(+)}}{\omega'_c} + j\frac{\omega''_c}{\omega'_c} e^{j\phi}}{1 + (1 + z_e)(1 + z_{(+)})\frac{\omega_e\omega_{(+)}}{\omega'^2_c}} \quad (6.12)$$

and for $\omega_p < \omega_e$

$$\bar{z}_{(-)} = (1 + z_{(-)}) + \frac{\omega'_c}{|\omega_{(-)}|}$$

$$\times \frac{(1 + z_e)\frac{\omega_e}{\omega'_c} \cdot \frac{\omega''^2_c}{\omega'^2_c} + (1 + z_{(+)})\frac{\omega_{(+)}}{\omega'_c} + 2j\frac{\omega''_c}{\omega'_c} e^{j\phi}}{1 + (1 + z_e)(1 + z_{(+)})\frac{\omega_e\omega_{(+)}}{\omega'^2_c}} \quad (6.13)$$

FOUR-FREQUENCY CONVERTERS AND AMPLIFIERS

$$E_{(-)} = jE_e \frac{(1+z_{(+)})\frac{\omega_{(+)}}{\omega'_c} + j\frac{\omega''_c}{\omega'_c}e^{j\phi}}{1 + (1+z_e)(1+z_{(+)})\frac{\omega_e \omega_{(+)}}{\omega'^2_c}} \tag{6.14}$$

4. Introduction of new reduced variables

These relations can be considerably simplified by the introduction of new reduced variables. Let:

$$\left. \begin{array}{l} (1+z_e)\dfrac{\omega_e}{\omega'_c} = b_e \\[6pt] (1+z_{(+)})\dfrac{\omega_{(+)}}{\omega'_c} = b_{(+)} \\[6pt] (1+z_{(-)})\dfrac{\omega_{(-)}}{\omega'_c} = b_{(-)} \end{array} \right\} \tag{6.15}$$

$$\frac{\omega''_c}{\omega'_c} = a \tag{6.16}$$

Putting as before $\omega'_c/\omega_e = q$, the equivalent circuits of the three branches will then be represented in Table 6.1.

5. Calculation of the transducer gains

The establishment of equivalent circuits for the three branches will enable us to calculate the transducer gains in the three different working conditions. Repeating the classification outlined in Chapter 2, we can distinguish:

(1) The sum-frequency converter for $f_p > f_e$.
(2) The sum-frequency converter for $f_p < f_e$.
(3) The difference-frequency converter for $f_p > f_e$.
(4) The difference-frequency converter for $f_p < f_e$.
(5) The negative-resistance amplifier.

5.1. Transducer gain of the sum-frequency converter for $f_p > f_e$

The signal enters at f_e and leaves at $f_{(+)}$; the energy is absorbed by a termination at the frequency $f_{(-)}$. This termination can, if necessary,

148 PARAMETRIC AMPLIFIERS

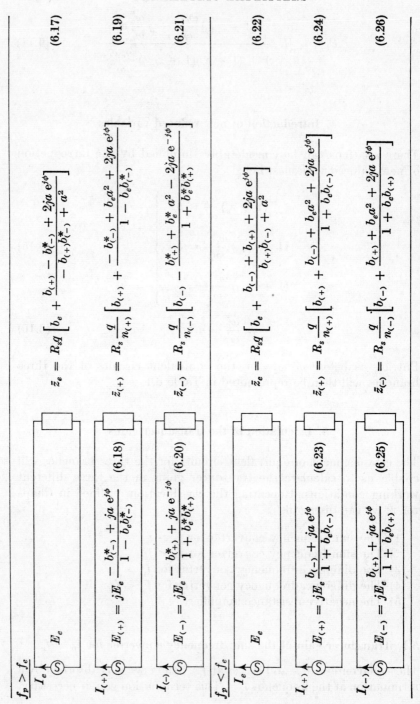

Table 6.1. Equivalent circuits of four-frequency converters

$f_p > f_e$

E_e : $\bar{z}_e = R_s q \left[b_e + \dfrac{b_{(+)} - b^*_{(-)} + 2ja\, e^{j\phi}}{-b_{(+)} b^*_{(-)} + a^2} \right]$ (6.17)

$E_{(+)} = jE_e \dfrac{-b^*_{(-)} + ja\, e^{j\phi}}{1 - b_e b^*_{(-)}}$ (6.18)

$\bar{z}_{(+)} = R_s \dfrac{q}{n_{(+)}} \left[b_{(+)} + \dfrac{-b^*_{(-)} + b_e a^2 + 2ja\, e^{j\phi}}{1 - b_e b^*_{(-)}} \right]$ (6.19)

$E_{(-)} = jE_e \dfrac{-b^*_{(+)} + ja\, e^{-j\phi}}{1 + b_e b^*_{(+)}}$ (6.20)

$\bar{z}_{(-)} = R_s \dfrac{q}{n_{(-)}} \left[b_{(-)} - \dfrac{b^*_{(+)} + b^*_e a^2 - 2ja\, e^{-j\phi}}{1 + b^*_e b_{(+)}} \right]$ (6.21)

$f_p < f_e$

E_e : $\bar{z}_e = R_s q \left[b_e + \dfrac{b_{(-)} + b_{(+)} + 2ja\, e^{j\phi}}{b_{(+)} b_{(-)} + a^2} \right]$ (6.22)

$E_{(+)} = jE_e \dfrac{b_{(-)} + ja\, e^{j\phi}}{1 + b_e b_{(-)}}$ (6.23)

$\bar{z}_{(+)} = R_s \dfrac{q}{n_{(+)}} \left[b_{(+)} + \dfrac{b_{(-)} + b_e a^2 + 2ja\, e^{j\phi}}{1 + b_e b_{(-)}} \right]$ (6.24)

$E_{(-)} = jE_e \dfrac{b_{(+)} + ja\, e^{j\phi}}{1 + b_e b_{(+)}}$ (6.25)

$\bar{z}_{(-)} = R_s \dfrac{q}{n_{(-)}} \left[b_{(-)} + \dfrac{b_{(+)} + b_e a^2 + 2ja\, e^{j\phi}}{1 + b_e b_{(+)}} \right]$ (6.26)

FOUR-FREQUENCY CONVERTERS AND AMPLIFIERS   149

be reduced to R_s. It is easy to calculate the transducer gain with the aid of equivalent circuits. We have:

$$G^e_{(+)} = 4R_e R_{(+)} \frac{|I_{(+)}|^2}{E_e^2} \qquad (6.27)$$

which gives us from (6.5) by means of (6.15) and (6.16).

$$\frac{I_{(+)}}{E_e} = j\frac{n_{(+)}}{qR_s} \cdot \frac{b^*_{(-)} - ja\,e^{j\phi}}{b_e b_{(+)} b^*_{(-)} - 2ja\,e^{j\phi} - b_e a^2 - b_{(+)} + b^*_{(-)}} \qquad (6.28)$$

$$G^e_{(+)} = 4r_e r_{(+)} \frac{n^2_{(+)}}{q^2} \left| \frac{b^*_{(-)} - ja\,e^{j\phi}}{b_e b_{(+)} b^*_{(-)} - b_{(+)} + b^*_{(-)} - 2ja\,e^{j\phi} - b_e a^2} \right|^2 \qquad (6.29)$$

5.2. Transducer gain of the sum-frequency converter for $f_p < f_e$

Similarly, we obtain:

$$\frac{I_{(+)}}{E_e} = j\frac{n_{(+)}}{qR_s} \cdot \frac{b_{(-)} + ja\,e^{j\phi}}{b_e b_{(+)} b_{(-)} + b_{(+)} + b_{(-)} + 2ja\,e^{j\phi} + b_e a^2} \qquad (6.30)$$

and

$$G^e_{(+)} = 4r_e r_{(+)} \frac{n^2_{(+)}}{q^2} \left| \frac{b_{(-)} + ja\,e^{j\phi}}{b_e b_{(+)} b_{(-)} + b_{(+)} + b_{(-)} + 2ja\,e^{j\phi} + b_e a^2} \right|^2 \qquad (6.31)$$

5.3. Transducer gain of the difference-frequency converter for $f_p > f_e$

$$\frac{I^*_{(-)}}{E_e} = j\frac{n_{(-)}}{qR_s} \cdot \frac{b_{(+)} + ja\,e^{j\phi}}{b_e b_{(+)} b^*_{(-)} - b_{(+)} + b^*_{(-)} - 2ja\,e^{j\phi} - b_e a^2} \qquad (6.32)$$

$$G^e_{(-)} = 4r_e r_{(-)} \frac{n^2_{(-)}}{q^2} \left| \frac{b_{(+)} + ja\,e^{j\phi}}{b_e b_{(+)} b^*_{(-)} - b_{(+)} + b^*_{(-)} - 2ja\,e^{j\phi} - b_e a^2} \right|^2 \qquad (6.33)$$

5.4. Transducer gain of the difference-frequency converter for $f_p < f_e$

$$\frac{I_{(-)}}{E_e} = j\frac{n_{(-)}}{qR_s} \cdot \frac{b_{(+)} + ja\,e^{j\phi}}{b_e b_{(+)} b_{(-)} + b_{(+)} + b_{(-)} + 2ja\,e^{j\phi} + b_e a^2} \qquad (6.34)$$

$$G^e_{(-)} = 4r_e r_{(-)} \frac{n^2_{(-)}}{q^2} \left| \frac{b_{(+)} + ja\,e^{j\phi}}{b_e b_{(+)} b_{(-)} + b_{(+)} + b_{(-)} + 2ja\,e^{j\phi} + b_e a^2} \right|^2 \qquad (6.35)$$

5.5. Transducer gain of the four-frequency reflection amplifier

We saw in Chapter 5 that the transducer gain of a reflection amplifier is equal to:

$$G_e^e = |\rho|^2$$

Now,

$$\rho = \frac{\bar{z}_e - 2r_e}{\bar{z}_e} \tag{6.36}$$

hence, from (6.3)

$$G_e^e = \left| \frac{b_e b_{(+)} b^*_{(-)} - b_{(+)} + b^*_{(-)} - 2ja\,e^{j\phi} - b_e a^2 \big|^2 - 2(r_e/q)(b_{(+)} b^*_{(-)} - a)^2}{b_e b_{(+)} b^*_{(-)} - b_{(+)} + b^*_{(-)} - 2ja\,e^{j\phi} - b_e a^2} \right| \tag{6.37}$$

Equations (6.17) to (6.26) and (6.29), (6.31), (6.33), (6.35) and (6.37) give the equivalent circuits and the exact transducer gains of the four-frequency converters in the general case, that is: for any given terminations, for any given value of a and for any given value of ϕ.

These results are in themselves difficult to use for a discussion, because of the number of parameters (b_e, $b_{(+)}$, $b_{(-)}$, a, ϕ). For the sake of simplicity, therefore, we will assume, as is often justified in practice, that parameter a is negligible by comparison with the other terms. This is especially true in current pumping of a square-root varactor (see Chapter 8). This assumption could have been made at the beginning of the chapter, but we thought it preferable to introduce it at the latest stage possible. The method of calculation up to and including the calculation of the gains, is identical for the two cases. We have the advantage that we have two exact formulae which could serve either for a numerical calculation, or for a more detailed discussion involving S_2.

6. Summary of the formulae when S_2 is negligible

Table 6.2. Total reduced impedance of each branch for $a = 0$

$f_p > f_e$

$$\bar{z}_e = q\left[b_e - \frac{1}{b^*_{(-)}} + \frac{1}{b_{(+)}}\right] \tag{6.38}$$

$$\bar{z}_{(+)} = \frac{q}{n_{(+)}}\left[b_{(+)} - \frac{b^*_{(-)}}{1 - b_e b^*_{(-)}}\right] \tag{6.39}$$

$$\bar{z}_{(-)} = \frac{q}{n_{(-)}}\left[b_{(-)} - \frac{b^*_{(+)}}{1 + b^*_e b^*_{(+)}}\right] \tag{6.40}$$

FOUR-FREQUENCY CONVERTERS AND AMPLIFIERS

Table 6.2—*continued*

$f_p < f_e$

$$\bar{z}_e = q\left[b_e + \frac{1}{b_{(-)}} + \frac{1}{b_{(+)}}\right] \tag{6.41}$$

$$\bar{z}_{(+)} = \frac{q}{n_{(+)}}\left[b_{(+)} + \frac{b_{(-)}}{1 + b_e b_{(-)}}\right] \tag{6.42}$$

$$\bar{z}_{(-)} = \frac{q}{n_{(-)}}\left[b_{(-)} + \frac{b_{(+)}}{1 + b_e b_{(+)}}\right] \tag{6.43}$$

Table 6.3. Transducer gains of four-frequency converters for $a = 0$

Sum-frequency converter

$f_p > f_e$

$$G^e_{(+)} = 4r_e r_{(+)} \frac{n^2_{(+)}}{q^2}\left|\frac{b^*_{(-)}}{b_e b_{(+)} b^*_{(-)} - b_{(+)} + b^*_{(-)}}\right|^2 \tag{6.44}$$

$f_p < f_e$

$$G^e_{(+)} = 4r_e r_{(+)} \frac{n^2_{(+)}}{q^2}\left|\frac{b_{(-)}}{b_e b_{(+)} b_{(-)} + b_{(+)} + b_{(-)}}\right|^2 \tag{6.45}$$

Difference-frequency converter

$f_p > f_e$

$$G^e_{(-)} = 4r_e r_{(-)} \frac{n^2_{(-)}}{q^2}\left|\frac{b_{(+)}}{b_e b_{(+)} b^*_{(-)} - b_{(+)} + b^*_{(-)}}\right|^2 \tag{6.46}$$

$f_p < f_e$

$$G^e_{(-)} = 4r_e r_{(-)} \frac{n^2_{(-)}}{q^2}\left|\frac{b_{(+)}}{b_e b_{(+)} b_{(-)} + b_{(+)} + b_{(-)}}\right|^2 \tag{6.47}$$

Reflection amplifier

$$G^e_e = \left|\frac{b_e b_{(+)} b^*_{(-)} - b_{(+)} + b^*_{(-)} - 2(r_e/q)b_{(+)}b^*_{(-)}}{b_e b_{(+)} b^*_{(-)} - b_{(+)} + b^*_{(-)}}\right|^2 \tag{6.48}$$

7. Stability conditions for the four-frequency converter when S_2 is negligible

We will now investigate the stability conditions of the four-frequency-converter. The circuit is stable when the real parts of the total impedances of the three branches are positive. Equations (6.41),

(6.42) and (6.43) show that the circuit is always stable for $f_p < f_e$ since the equations contain only positive signs. Conversely, equations (6.38), (6.39) and (6.40) are likely to lead to instability regions when $f_p > f_e$. Since we are dealing only with the real parts of the impedances, we will assume, in order to simplify the calculations, that each branch is in the resonance condition: b_e, $b_{(-)}$ and $b_{(+)}$ are then real, and we will mark them with the index (') in order to remind ourselves of this restriction. The stability condition of the branch at ω_e is obtained from (6.38):

$$b'_{(-)} > \frac{1}{b'_e + (1/b'_{(+)})} \qquad (6.49)$$

Similarly, for the branch at $\omega_{(+)}$, from (6.39)

$$\frac{(b'_{(+)}/b'_{(-)}) - (1 + b'_e b'_{(+)})}{(1/b'_{(-)}) - b'_e} > 0 \qquad (6.50)$$

Noting that (6.49) causes the numerator to be negative, this condition becomes:

$$b'_{(-)} > \frac{1}{b'_e} \qquad (6.51)$$

Equation (6.40) tells us that the stability condition of the branch at $\omega_{(-)}$ is also given by (6.49). From condition (6.51) which includes (6.49), we see that as $b'_{(-)}$ decreases the instability appears first in the branch at $\omega_{(+)}$.

The inequality (6.51) therefore constitutes the necessary and sufficient condition for the stability of the system. On the other hand, in (6.38) b'_e corresponding to the internal resistance of the generator, and $(1/b'_{(+)} - 1/b'_{(-)})$ corresponding to the input resistance of the converter, we see that this latter resistance is negative if:

$$b'_{(-)} < b'_{(+)} \qquad (6.52)$$

In this case amplification may occur at ω_e. Conversely, the introduction of the conditions for stability (6.51) in (6.39) gives

$$\frac{b_{(-)}}{1 - b'_e b'_{(-)}} < 0 \qquad (6.53)$$

At $\omega_{(+)}$ the internal impedance of the converter is always positive. This internal impedance increases as $b'_{(-)}$ decreases and passes from $+\infty$ to $-\infty$ on each side of the point $b'_{(-)} = 1/b'_e$ causing the system to be unstable. Equation (6.40) shows that the internal impedance of the branch at $\omega_{(-)}$ is always negative whatever the values of b_e, $b_{(+)}$ and $b_{(-)}$.

FOUR-FREQUENCY CONVERTERS AND AMPLIFIERS   153

8. Applications

8.1. The sum-frequency converter

When $b^*_{(-)} = \infty$, which corresponds to the absence of the idler termination, (6.44) and (6.45) become:

$$G^e_{(+)0} = \frac{4r_e r_{(+)} n^2_{(+)}/q^2}{|b_e b_{(+)} + 1|^2} \tag{6.54}$$

which is identical with (4.11) if we replace b_e and $b_{(+)}$ by their values taken from (6.15).

Let β represent the ratio of the gains with and without the idler at $\omega_{(-)}$; we then have when $f_p > f_e$, according to (6.44) and (6.54)

$$\beta = \left| \frac{b_e b_{(+)} b^*_{(-)} + b^*_{(-)}}{b_e b_{(+)} b^*_{(-)} + b^*_{(-)} - b_{(+)}} \right|^2 \tag{6.55}$$

and when $f_p < f_e$, according to (6.45) and (6.54)

$$\beta = \left| \frac{b_e b_{(+)} b_{(-)} + b_{(-)}}{b_e b_{(+)} b_{(-)} + b_{(-)} + b_{(+)}} \right|^2 \tag{6.56}$$

This again gives the property established in Chapter 2, namely, that the effect of the idler load is to increase the gain when $\omega_p > \omega_e$ and to decrease it when $\omega_p < \omega_e$. Let us now proceed to find the limits in the variation of β. We will assume a resonant condition and we will deal only with the case when $f_p > f_e$, which is the only one used in practice. Let

$$b'_{(-)} = \frac{1}{\mu b'_e} \tag{6.57}$$

The condition for stability requires

$$\mu < 1$$

If we introduce this value of $b'_{(-)}$ in (6.44), we obtain

$$G^e_{(+)} = \frac{4 r_e r_{(+)} (n^2_{(+)}/q^2)}{[b'_e b'_{(+)}(1 - \mu) + 1]^2} \tag{6.58}$$

and by substituting b'_e and $b'_{(+)}$ in this equation by their values

$$b'_e = \frac{1 + r_e}{q} \tag{6.59}$$

$$b'_{(+)} = \frac{n}{q}(1 + r_{(+)}) \tag{6.60}$$

We find

$$G^e_{(+)} = \frac{4q^2}{(1-\mu)^2} \cdot \frac{r_e r_{(+)}}{[(1+r_e)(1+r_{(+)}) + q^2/n_{(+)}(1-\mu)]^2} \quad (6.61)$$

By analogy with the case of the sum-frequency converter without the idler discussed in Chapter 4, we see at once that the gain is maximum for:

$$r_e = r_{(+)} = \sqrt{1 + \frac{q^2}{n_{(+)}(1-\mu)}} \quad (6.62)$$

If we assume that the varactor is of a high enough quality ($q^2/n_{(+)}$ large), (6.62) gives us

$$r_e r_{(+)} \approx (1+r_e)(1+r_{(+)}) \approx \frac{q^2}{n_{(+)}(1-\mu)} \quad (6.63)$$

and (6.61) becomes

$$G^e_{(+)} \approx \frac{n_{(+)}}{(1-\mu)} \quad (6.64)$$

Hence the property: the addition of an idler to the sum-frequency converter makes it possible to obtain an arbitrarily large gain. Let us now assume that we maintain the terminal impedances corresponding to the optimization of the sum-frequency converter without idler. We saw (Chapter 3) that in this case:

$$r_e = r_{(+)} = \sqrt{1 + q^2/n_{(+)}} \quad (6.65)$$

Assuming as before that $q^2/n_{(+)}$ is large, (6.61) becomes:

$$G^e_{(+)} \approx \frac{4n_{(+)}}{(2-\mu)^2} \quad (6.66)$$

and when $\mu \approx 1$

$$G^e_{(+)} \approx 4n_{(+)} \quad (6.67)$$

hence $\beta \approx 4$.

If we optimize the gain of a sum-frequency converter, maintaining the impedances of the generator and of the load, the addition of an idler results at the most in a fourfold increase of gain. However, these improvements in the gain can only be obtained at the expense of the stability and of the bandwidth, and they involve a more complicated construction and, above all, greater difficulty of adjustment.

FOUR-FREQUENCY CONVERTERS AND AMPLIFIERS

8.2. Deterioration in the performance of the reflection amplifier due to power consumption at $\omega_{(+)}$

Making $b_{(+)} = 0$, formula (6.48) becomes:

$$G_e^e = \left|\frac{b_e b_{(-)}^* - 1 - 2(r_e/q)b_{(-)}^*}{b_e b_{(-)}^* - 1}\right|^2 \quad (6.68)$$

It can be seen that if we replace b_e and $b_{(-)}^*$ by their values obtained from (6.15), (6.68) is identical with (5.63). Let us now calculate the disturbance introduced by an idler at $\omega_{(+)}$. We cannot proceed in the same way as for the sum-frequency converter, since the gain cannot directly be optimized as a function of the terminal impedances.

Since the operation of a reflection amplifier is based on the effect of negative resistance at the frequency ω_e, let us examine formula (6.38) which gives the total reduced input impedance $\tilde{z}_e$. If we replace $b_{(-)}$, $b_{(+)}$ and b_e by their values obtained from (6.15), we obtain, in the resonant condition

$$\tilde{z}_e = r_e - \left[q^2\left(\frac{1}{n_{(-)}(1 + r_{(-)})} - \frac{1}{n_{(+)}(1 + r_{(+)})}\right) - 1\right] \quad (6.69)$$

a formula to be compared with that obtained in the absence of $r_{(+)}$:

$$\tilde{z}_e = r_e - \left[q^2 \frac{1}{n_{(-)}(1 + r_{(-)})} - 1\right] \quad (6.70)$$

In both cases the term in square brackets represents the reduced negative resistance. We see that the presence of $r_{(+)}$ reduces the value of the negative resistance. Writing (6.69) in the form:

$$\tilde{z}_e = r_e - \left[q'^2 \frac{1}{n_{(-)}(1 + r_{(-)})} - 1\right] \quad (6.71)$$

with

$$q'^2 = q^2\left(1 - \frac{n_{(-)}(1 + r_{(-)})}{(n_{(-)} + 2)(1 + r_{(+)})}\right) \quad (6.72)$$

we now see that the effect of $r_{(+)}$ is to cause a deterioration in the apparent quality of the varactor.

The input impedances, the gain and the instability factor may then be calculated by using the formulae given in Chapter 5 and by substituting q' for q. If we assume that the selectivity of the parasitic circuit at $\omega_{(+)}$ is low compared with that of the other circuits, the formula for the bandwidth is also applicable.

References

[1] Luksch, Matthews and Verwys, 'Design and operation of four-frequency parametric up-converters', *I.R.E. Trans. on Microwave Theory and Techniques*, January 1961.

[2] Anderson and Aukland, 'A general catalogue of gain, bandwidth and noise temperature, expressions for four-frequency parametric devices', *I.E.E.E. Trans. on Electronic Devices*, January 1963.

7

DEGENERATE AMPLIFIERS
by J. C. LIENARD

1. General study and classification

1.1. Classification of difference-frequency converters and amplifiers

In order to simplify the classification of difference-frequency converters and amplifiers we will replace the equivalent diagram of Fig. 5.12 by the more general circuit of Fig. 7.1.

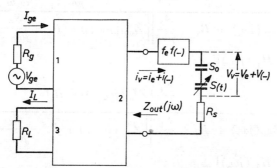

Figure 7.1. General equivalent circuit of difference-frequency circuits

This circuit consists of:

(1) a three-port which embodies the filters, the tuning reactances, and the circulators;
(2) an input circuit connected to terminals 1 of the three-port;
(3) an output circuit connected to terminals 3 of the three-port;
(4) a varactor circuit connected to terminals 2 of the three-port.

This last circuit comprises the varactor, or more precisely the elastance $S_0 + S(t)$, the series resistance R_s of the varactor and the filter which allows only currents of frequency f_e and $f_{(-)}$ to pass.

PARAMETRIC AMPLIFIERS

Let us consider once more the matrix equation (3.28). If we make $I_{(+)} = 0$ in this equation, we obtain the following equations:

$$V_e = \frac{S_0}{j\omega_e} I_e - \frac{S_1}{j\omega_{(-)}} I^*_{(-)} \tag{7.1}$$

$$V^*_{(-)} = -\frac{S_0}{j\omega_{(-)}} I^*_{(-)} + \frac{S_1}{j\omega_e} I_e \tag{7.2}$$

Let $H_{12}(j\omega)$ represent the voltage transfer function defined as the ratio of the voltage at terminals 2 (with port 2 open-circuited) and the voltage V_{ge} (with port 3 closed on the output circuit). Further, let $Z_{\text{out}}(j\omega)$ represent the impedance of the three-port, with its input and output terminations, seen from terminals 2. We then have:

$$V_e = H_{12}(j\omega_e)V_{ge} - [Z_{\text{out}}(j\omega_e) + R_s]I_e \tag{7.3}$$

$$V^*_{(-)} = -[Z^*_{\text{out}}(j\omega_{(-)}) + R_s]I^*_{(-)} \tag{7.4}$$

Solving the system (7.1), (7.2), (7.3), (7.4), we immediately get

$$V_e = \frac{\frac{S_0}{j\omega_e}\left[Z^*_{\text{out}}(j\omega_{(-)}) + R_s - \frac{S_0}{j\omega_{(-)}}\right] - \frac{|S_1|^2}{\omega_e \omega_{(-)}}}{\left[Z_{\text{out}}(j\omega_e) + R_s + \frac{S_0}{j\omega_e}\right]\left[Z^*_{\text{out}}(j\omega_{(-)}) + R_s - \frac{S_0}{j\omega_{(-)}}\right] - \frac{|S_1|^2}{\omega_e \omega_{(-)}}}$$
$$\times H_{12}(j\omega_e)V_{ge} \tag{7.5}$$

$$V^*_{(-)} = \frac{\frac{S_1}{j\omega_e}[Z^*_{\text{out}}(j\omega_{(-)}) + R_s]}{\left[Z_{\text{out}}(j\omega_e) + R_s + \frac{S_0}{j\omega_e}\right]\left[Z^*_{\text{out}}(j\omega_{(-)}) + R_s - \frac{S_0}{j\omega_{(-)}}\right] - \frac{|S_1|^2}{\omega_e \omega_{(-)}}}$$
$$\times H_{12}(j\omega_e)V_{ge} \tag{7.6}$$

$$I_e = \frac{Z^*_{\text{out}}(j\omega_{(-)}) + R_s - \frac{S_0}{j\omega_{(-)}}}{\left[Z_{\text{out}}(j\omega_e) + R_s + \frac{S_0}{j\omega_e}\right]\left[Z^*_{\text{out}}(j\omega_{(-)}) + R_s - \frac{S_0}{j\omega_{(-)}}\right] - \frac{|S_1|^2}{\omega_e \omega_{(-)}}}$$
$$\times H_{12}(j\omega_e)V_{ge} \tag{7.7}$$

$$I^*_{(-)} = \frac{-\frac{S_1}{j\omega_e}}{\left[Z_{\text{out}}(j\omega_e) + R_s + \frac{S_0}{j\omega_e}\right]\left[Z^*_{\text{out}}(j\omega_{(-)}) + R_s - \frac{S_0}{j\omega_{(-)}}\right] - \frac{|S_1|^2}{\omega_e \omega_{(-)}}}$$
$$\times H_{12}(j\omega_e)V_{ge} \tag{7.8}$$

DEGENERATE AMPLIFIERS 159

Note: relations (7.5) to (7.8) show that:

(1) With an appropriate choice of impedances, V_e, $V^*_{(-)}$, I_e, $I^*_{(-)}$ can be made infinite for a finite excitation V_{ge}. This is due to S_1 which causes a real negative term to appear in the denominators (negative-resistance effect). The quantities V_e, $V^*_{(-)}$; I_e and $I^*_{(-)}$ on the other hand become infinite simultaneously as they have the same denominator.
(2) When in the resonant condition, all the impedances are real and the initial phase of I_e with respect to $H_{12}(j\omega_e)V_{ge}$ is zero.
(3) When in the resonant condition, the initial phase of $I^*_{(-)}$ with respect to $H_{12}(j\omega_e)V_{ge}$ is equal to $\pi/2$.

Discussion

Let

$$\omega_e = \frac{\omega_p}{2} - \Delta \qquad (7.9)$$

It follows that:

$$\omega_{(-)} = \omega_p - \omega_e = \frac{\omega_p}{2} + \Delta \qquad (7.10)$$

Case 1: Δ/ω_e is comparable to unity. The two currents can easily be separated by filters in the three-port. If, then, only I_{Le} is detected, we have a reflection amplifier. If only $I_{L(-)}$ is detected, we have an inverting converter. These two devices were studied in Chapter 5.

Case 2: $\Delta/\omega_e \ll 1$. The currents at frequencies ω_e and $\omega_{(-)}$ can no longer be separated by filtering. The selection of the usable output signal must be made after frequency changing. For the calculation of the performance therefore, we need to know the amplification chain after the parametric amplifier. Thanks to filtering at the level of the intermediate frequency (IF), if $\Delta/\omega_{\text{IF}}$ is not too low, we can choose as the output signal either the IF transpose of ω_e, or the IF transpose of $\omega_{(-)}$, or the IF signal resulting from the synchronous detection of ω_e and $\omega_{(-)}$. Devices constructed in this way are degenerate amplifiers. If $\Delta/\omega_{\text{IF}} \ll 1$, we can no longer distinguish between the transposes of ω_e and $\omega_{(-)}$, even at the level of the IF frequency. The output signal will then be affected by a beat due to the addition of the two IF transposes. This case is of course of no practical interest.

Case 3: $\Delta = 0$. The two currents I_e and $I_{(-)}$ combine into a single one, and it can be shown that its value depends on the initial phase of V_{ge} with respect to S_1. This limiting case will not be considered in this book.

1.2. Practical form of the three-port

We saw in Chapter 5 that the use of non-reciprocal elements (circulators) produces a considerable improvement in the performance of a negative-resistance system. In what follows we will therefore consider the circuit shown in Fig. 7.2 with $\omega_e \approx \omega_{(-)} \approx \omega_p/2$. It will be assumed that the circulator may be considered as perfect within a frequency band including ω_e and $\omega_{(-)}$.

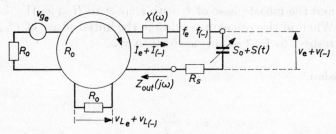

Figure 7.2. Circuit used for the representation of a degenerate amplifier

The calculation of the gains and the noise temperature for various types of degenerate amplifiers involves the calculation of the transfer function H_{12} for the various signal and noise sources, as well as the relations connecting the output voltage with the voltages and currents of the varactor. This calculation will be discussed in the following two sections.

1.3. Calculation of the transfer functions for the input signal and noise

If we apply relations (5.17) to the circuit in Fig. 7.2, we at once obtain:

$$H_{12}(j\omega_e) = 1 \qquad (7.11)$$
$$V_{Le} = \tfrac{1}{2}\{V_e - [R_0 - R_s - jX(\omega_e)]I_e\} \qquad (7.12)$$
$$V_{L(-)} = \tfrac{1}{2}\{V_{(-)} - [R_0 - R_s - jX(\omega_{(-)})]I_e\} \qquad (7.13)$$

1.4. Calculation of the transfer functions for the noise of R_s

The equivalent circuit for the noise of R_s is derived directly from Fig. 7.2 and is shown in Fig. 7.3, in which V_{se} designates the noise voltage of R_s at frequency f_e. We then easily obtain:

$$H_{12}(j\omega_e) = 1 \qquad (7.14)$$
$$V_{Le} = R_0 I_e \qquad (7.15)$$
$$V_{L(-)} = R_0 I_{(-)} \qquad (7.16)$$

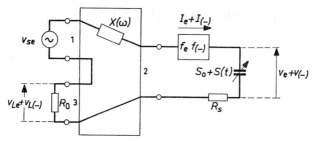

Figure 7.3. Equivalent circuit applicable to the calculation of the noise of R_s

1.5. Calculation of the impedances seen by the varactor

Figure 7.2 shows that we have:

$$Z_{\text{out}}(j\omega_e) = R_0 + jX(\omega_e) \tag{7.17}$$

$$Z_{\text{out}}(j\omega_{(-)}) = R_0 + jX(\omega_{(-)}) \tag{7.18}$$

1.6. Expressions for the output signals and noise

In order to describe all the possible cases we must consider two signal sources and four noise sources:

V_{ge} input signal at frequency f_e.

$V_{g(-)}$ input signal at frequency $f_{(-)}$.

V_{ne} input noise voltage at frequency f_e.

$V_{n(-)}$ input noise voltage at frequency $f_{(-)}$.

V_{se} noise voltage of R_s at frequency f_e.

$V_{s(-)}$ noise voltage of R_s at frequency $f_{(-)}$.

Each of these voltages produces two currents and voltages in the varactor at two frequencies f_e and $f_{(-)}$. We will mark these quantities with two indices describing the voltage from which they arise. For example, $I^{ne}_{(-)}$ is the current in the varactor at frequency $f_{(-)}$ due to the input noise voltage V_{ne}. Similarly, the output voltage will be marked with suffices corresponding to the voltage which produces them: $V^{s(-)}_{Le}$ is the output voltage at frequency f_e due to the noise voltage of R_s at frequency $f_{(-)}$.

Relations (7.4) to (7.8) and (7.11) to (7.18) give:

$$V^{ge}_{Le} = \frac{1}{2}\{V^{ge}_e - [R_0 - R_s - jX(\omega_e)]I^{ge}_e\} = -\frac{1}{2}\frac{A}{D}V_{ge} \tag{7.19}$$

$$V^{ge}_{L(-)} = \frac{1}{2}\{V^{ge}_{(-)} - [R_0 - R_s - jX(\omega_{(-)})]I^{ge}_{(-)}\} = j\frac{1}{2}\frac{B}{D^*}V^*_{ge} \tag{7.20}$$

$$V_{Le}^{ne} = \frac{1}{2}\{V_e^{ne} - [R_0 - R_s - jX(\omega_e)]I_e^{ne}\} = -\frac{1}{2}\frac{A}{D}V_{ne} \quad (7.21)$$

$$V_{L(-)}^{ne} = \frac{1}{2}\{V_{(-)}^{ne} - [R_0 - R_s - jX(\omega_{(-)})]I_{(-)}^{ne}\} = j\frac{1}{2}\frac{B}{D^*}V_{ne}^* \quad (7.22)$$

$$V_{Le}^{se} = R_0 I_e^{se} = R_0 \frac{C}{D}V_{se} \quad (7.23)$$

$$V_{L(-)}^{se} = R_0 I_{(-)}^{se} = -j\frac{1}{2}\frac{B}{D^*}V_{se}^* \quad (7.24)$$

with

$$A = \left\{R_0 + R_s - j\left[X(\omega_{(-)}) - \frac{S_0}{\omega_{(-)}}\right]\right\}$$
$$\times \left\{R_0 - R_s - j\left[X(\omega_e) - \frac{S_0}{\omega_e}\right]\right\} + \frac{S_1^2}{\omega_e \omega_{(-)}} \quad (7.25)$$

$$B = \frac{S_1}{\omega_e} 2R_0 \quad (7.26)$$

$$C = R_0 + R_s - j\left[X(\omega_{(-)}) - \frac{S_0}{\omega_{(-)}}\right] \quad (7.27)$$

$$D = \left\{R_0 + R_s + j\left[X(\omega_e) - \frac{S_0}{\omega_e}\right]\right\}$$
$$\times \left\{R_0 + R_s - j\left[X(\omega_{(-)}) - \frac{S_0}{\omega_{(-)}}\right]\right\} - \frac{S_1^2}{\omega_e \omega_{(-)}} \quad (7.28)$$

By exchanging the indices e and $(-)$ in the relations (7.19) to (7.24), we obtain:

$$V_{L(-)}^{g(-)} = -\frac{1}{2}\frac{E}{D^*}V_{g(-)} \quad (7.29)$$

$$V_{Le}^{g(-)} = j\frac{1}{2}\frac{F}{D}V_{g(-)}^* \quad (7.30)$$

$$V_{L(-)}^{n(-)} = -\frac{1}{2}\frac{E}{D^*}V_{n(-)} \quad (7.31)$$

$$V_{Le}^{n(-)} = j\frac{1}{2}\frac{F}{D}V_{n(-)}^* \quad (7.32)$$

$$V_{L(-)}^{s(-)} = R_0 \frac{H}{D^*}V_{s(-)} \quad (7.33)$$

$$V_{Le}^{s(-)} = j\frac{1}{2}\frac{F}{D}V_{s(-)}^* \quad (7.34)$$

with

$$E = \left\{ R_0 + R_s - j\left[X(\omega_e) - \frac{S_0}{\omega_e}\right]\right\}$$
$$\times \left\{ R_0 - R_s - j\left[X(\omega_{(-)}) - \frac{S_0}{\omega_{(-)}}\right]\right\} + \frac{|S_1|^2}{\omega_e\omega_{(-)}} \quad (7.35)$$

$$F = \frac{S_1}{\omega_{(-)}} 2R_0 \quad (7.36)$$

$$H = R_0 + R_s - j\left[X(\omega_e) - \frac{S_0}{\omega_e}\right] \quad (7.37)$$

Let us define

$$X_e = X(\omega_e) - \frac{S_0}{\omega_e} \quad (7.38)$$

$$X_{(-)} = X(\omega_{(-)}) - \frac{S_0}{\omega_{(-)}} \quad (7.39)$$

For a high-quality varactor, we may write:

$$\frac{|S_1|}{R_s} = \omega'_c \gg \omega_e \cong \omega_{(-)} \quad (7.40)$$

Hence

$$\frac{|S_1|}{\sqrt{\omega_e\omega_{(-)}}} \gg R_s \quad (7.41)$$

In order to ensure the stability of the system we must have:

$$(R_0 + R_s)^2 > \frac{|S_1|^2}{\omega_e\omega_{(-)}} \quad (7.42)$$

$$R_0 + R_s \gg R_s \quad (7.43)$$

We may write:

$$R_0 + R_s \cong R_0 - R_s \cong R_0 \quad (7.44)$$

Note, on the other hand, that we have:

$$\omega_e\omega_{(-)} = \left(\frac{\omega_p}{2} + \Delta\right)\left(\frac{\omega_p}{2} - \Delta\right) = \frac{\omega_p^2}{4} - \Delta^2 \cong \frac{\omega_p^2}{4} \quad \text{for} \quad \Delta \ll \frac{\omega_p}{2} \quad (7.45)$$

Let

$$r_0 = \frac{R_0}{R_s} \quad (7.46)$$

$$x_e = \frac{X_e}{R_s} \quad (7.47)$$

$$x_{(-)} = \frac{X_{(-)}}{R_s} \quad (7.48)$$

Taking account of relations (7.44) to (7.48), we obtain:

$$A = R_s^2\left[(r_0 - jx_{(-)})(r_0 - jx_e) + 4\left(\frac{\omega_c'}{\omega_p}\right)^2\right] \qquad (7.49)$$

$$B = R_s^2\left[2r_0\frac{\omega_c'}{\omega_e}\right] \simeq R_s^2\left[4r_0\frac{\omega_c'}{\omega_p}\right] \qquad (7.50)$$

$$C = R_s[r_0 - jx_{(-)}] \qquad (7.51)$$

$$D = R_s^2\left[(r_0 + jx_e)(r_0 - jx_{(-)}) - 4\left(\frac{\omega_c'}{\omega_p}\right)^2\right] \qquad (7.52)$$

$$E = R_s^2\left[(r_0 - jx_e)(r_0 - jx_{(-)}) + 4\left(\frac{\omega_c'}{\omega_p}\right)^2\right] \qquad (7.53)$$

$$F = R_s^2\left[2r_0\frac{\omega_c'}{\omega_{(-)}}\right] \simeq R_s^2\left[4r_0\frac{\omega_c'}{\omega_p}\right] \qquad (7.54)$$

$$H = R_s[r_0 - jx_e] \qquad (7.55)$$

1.7. Classification of degenerate amplifiers

The basic characteristic which distinguishes degenerate from other negative-resistance amplifiers lies in the fact that they produce identical amplification of the two bands of frequencies which are symmetrical with relation to half the pumping frequency. They may therefore be used to amplify single- or double-band signals, or signals consisting of two non-correlated bands (radiometry). In each case, the degenerate amplifier supplies two approximately equal signals of frequencies symmetrical with respect to half the pumping frequency.

In the following paragraphs we will assume that the difference between these two frequencies is large enough for their IF transposes to be separated. The classification is based on the method of detection of the output signals.

2. Single-band reception

2.1. The three methods of operation

The input signal consists of V_{ge} only. The degenerate amplifier supplies two output signals V_{Le}^{ge} and $V_{L(-)}^{ge}$. The mixer which follows the amplifier produces the two corresponding IF signals. The IF filter enables the required band to be selected. There are three possible cases:

Case 1: The IF transpose of V_{Le}^{ge} is chosen. The degenerate amplifier operates as a reflection amplifier.

Case 2: The IF transpose of $V_{L(-)}^{ge}$ is chosen. The degenerate amplifier operates as an inverting converter.

Case 3: There will be synchronous detection of the two signals V_{Le}^{ge} and $V_{L(-)}^{ge}$. We will see later that with this system we can obtain higher gains for the same stability.

These three cases will be studied in the following paragraphs. In order to simplify the presentation, each type of degenerate amplifier will be represented by the letter X, with indices for the input frequency and suffices for the output frequency. For example, the three above cases will be represented respectively by: X_e^e, $X_{(-)}^e$ and $X_{e(-)}^e$.

We will assume that the input signal has as its expression:

$$v_{ge} = V_{ge}\, e^{j\omega_e t} + V_{ge}^*\, e^{-j\omega_e t} \tag{7.56}$$

The noise performance may be studied by means of equations (1.152), (1.155) or (1.165) depending on the method of operation chosen. We will use here a system of notation analogous to that used in Chapters 4, 5 and 6: in the expressions for the gains G_{2k}^{1i} and G_{2k}^{1N}, we will omit the indices corresponding to the ports and replace the numerical indices corresponding to the frequencies by the indices e and $(-)$.

To enable the operation of a degenerate amplifier to be described fully, we must calculate the transducer gain (applicable to the signal), G; the noise gain G^N; and the effective noise temperature, T_{eff}. It should be remembered that the comparison of degenerate amplifiers amongst themselves and with non-degenerate amplifiers should be made by means of the operational noise temperature T_{op} defined by equation (1.174).

2.2. Amplifier X_e^e

(1) *General expression for the transducer gain*

The available power of the input signal is $|V_{ge2}|/2R_0$. The output power is equal to $2|V_{Le}^{ge}|^2/R_0$. The transducer gain is then equal to:

$$G_e^e = 4\left|\frac{V_{Le}^{ge}}{V_{ge}}\right|^2 \tag{7.57}$$

Equation (7.19) enables us to write:

$$G_e^e = \left|\frac{A}{D}\right|^2 \tag{7.58}$$

Taking (7.49) and (7.52) into account, we have finally:

$$G_e^e = \left| \frac{(r_0 - jx_e)(r_0 - jx_{(-)}) + 4\left(\dfrac{\omega'}{\omega_p}\right)^2}{(r_0 + jx_e)(r_0 - jx_{(-)}) - 4\left(\dfrac{\omega'_c}{\omega_p}\right)^2} \right|^2 \qquad (7.59)$$

(2) Gain at the nominal frequency and bandwidth

Expression (7.59) shows that it is desirable to make $x_e = x_{(-)} = 0$ for the nominal frequency to be amplified, especially in order to obtain as high a gain as required. Let us assume that the circuit traversed by the currents at frequencies ω_e and $\omega_{(-)}$ is a simple series tuned circuit. It is then impossible to obtain $x_e = x_{(-)} = 0$ simultaneously

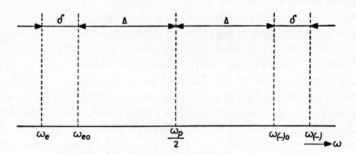

Figure 7.4. Relative positions of ω_e, $\omega_{(-)}$ and $\omega_p/2$

for $\omega_e \neq \omega_p/2$. It might be thought that the circuit could be tuned to a frequency ω_0 near to $\omega_p/2$ and such that the response curve of the amplifier is symmetrical with respect to the nominal frequency ω_{e_0}. Calculations show that this cannot be done except by strongly detuning the circuit for the usable frequencies (ω_{e_0} and $\omega_{(-)_0}$), so that in these conditions the transducer gain cannot exceed 3 dB. It is therefore impossible to obtain a symmetrical response curve by means of a simply tuned circuit. It can be shown that in order to obtain the maximum possible gain for $\omega = \omega_{e_0}$ the circuit must be tuned to the frequency $\omega_0 = \omega_p/2$, and we will proceed on this assumption. Let Q represent the quality factor of the simple circuit. Figure 7.4 shows the relative positions of ω_e, $\omega_{(-)}$ and $\omega_p/2$. The nominal input frequency ω_{e_0} corresponds to $\delta = 0$ and the nominal

DEGENERATE AMPLIFIERS

idler frequency $\omega_{(-)_0}$ corresponds to it. The introduction of δ enables the bandwidth to be calculated. We may write:

$$\frac{x_e}{r_0} = -4Q\frac{\Delta + \delta}{\omega_p} \qquad (7.60)$$

$$\frac{x_{(-)}}{r_0} = 4Q\frac{\Delta + \delta}{\omega_p} \qquad (7.61)$$

Again let:

$$\alpha = \left(\frac{2}{r_0} \cdot \frac{\omega_c'}{\omega_p}\right)^2 \qquad (7.62)$$

(7.59) becomes:

$$G_e^e = \left|\frac{1 + \alpha + 16Q^2\left(\frac{\Delta + \delta}{\omega_p}\right)^2}{1 - \alpha - 16Q^2\left(\frac{\Delta + \delta}{\omega_p}\right)^2 - j8Q\frac{\Delta + \delta}{\omega_p}}\right|^2 \qquad (7.63)$$

The gain at the nominal frequency is obtained from (7.63) if we make $\delta = 0$

$$(G_e^e)_{\delta=0} = \frac{\left[1 + \alpha + 16Q^2\left(\frac{\Delta}{\omega_p}\right)^2\right]^2}{\left[1 - \alpha - 16Q^2\left(\frac{\Delta}{\omega_p}\right)^2\right]^2 + 64Q^2\left(\frac{\Delta}{\omega_p}\right)^2} \qquad (7.64)$$

Note that we should always have $\alpha < 1$, since for $\alpha = 1$ the oscillation will occur at the frequency $\omega_p/2$ (corresponding to $\delta = -\Delta$). The maximum gain obtainable with $\delta = 0$ can therefore be calculated from (7.64) by putting $\alpha = 1 - 16Q^2(\Delta/\omega_p)^2$ in it.

$$|(G_e^e)_{\delta=0}|_{max} = \frac{1}{16Q^2(\Delta/\omega_p)^2} \qquad (7.65)$$

It is therefore desirable to choose low values for Q and Δ/ω_p. Equation (7.63) shows us that the maximum gain is obtained when $\delta = -\Delta$. The shape of the gain-frequency curve is shown in Fig. 7.5, but it should be noted that this curve applies only in the non-shaded area. The hypothesis on the basis of which relation (7.63) was calculated assumes that the stages after the amplifier select only the transposed frequencies corresponding to this region.

It is now required to calculate the values of $\delta_{-3\,dB}$ and $\delta_{+3\,dB}$ which limit the frequency band in which the gain does not vary by

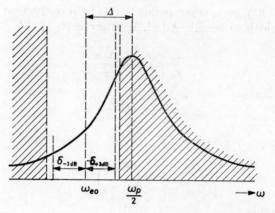

Figure 7.5. Response curve of the simply-tuned amplifier

more than 3 dB from the gain at the nominal frequency. For this purpose, we will write (7.63) in the following form:

$$G_e = \frac{1}{1 - 4\alpha \dfrac{1}{[1 + \alpha + 16Q^2((\Delta + \delta)/\omega_p)^2]^2}} \qquad (7.66)$$

The values $\delta_{-3\,\text{dB}}$ and $\delta_{+3\,\text{dB}}$ are solutions of the following equations:

$$1 - 4\alpha \frac{1}{[1 + \alpha + 16Q^2((\Delta + \delta)/\omega_p)^2]^2}$$
$$= 2\left\{1 - 4\alpha \frac{1}{[1 + \alpha + 16Q^2(\Delta/\omega_p)^2]^2}\right\} \qquad (7.67)$$

$$1 - 4\frac{1}{[1 + \alpha + 16Q^2((\Delta + \delta)/\omega_p)^2]^2}$$
$$= \frac{1}{2}\left\{1 - 4\alpha \frac{1}{[1 + \alpha + 16Q^2(\Delta/\omega_p)^2]^2}\right\} \qquad (7.68)$$

If in making our calculations we limit ourselves to first-order terms in $16Q^2((\Delta + \delta)/\omega_p)^2$, we obtain:

$$\frac{\delta_{-3\,\text{dB}}}{\omega_p/2} = -\frac{\Delta}{\omega_p/2} + \sqrt{2\left(\frac{\Delta}{\omega_p/2}\right)^2 + \frac{(1-\alpha)^2(1+\alpha)}{32Q^2\alpha}} \qquad (7.69)$$

$$\frac{\delta_{+3\,\text{dB}}}{\omega_p/2} = -\frac{\Delta}{\omega_p/2} + \sqrt{\frac{1}{2}\left(\frac{\Delta}{\omega_p/2}\right)^2 - \frac{(1-\alpha)^2(1+\alpha)}{64Q^2\alpha}} \qquad (7.70)$$

DEGENERATE AMPLIFIERS

The reduced bandwidth is then:

$$B_e^e = \frac{\delta_{-3\,dB}}{\omega_p/2} - \frac{\delta_{+3\,dB}}{\omega_p/2} = \sqrt{2\left(\frac{\Delta}{\omega_p/2}\right)^2 + \frac{(1-\alpha)^2(1+\alpha)}{32Q^2\alpha}}$$

$$- \sqrt{\frac{1}{2}\left(\frac{\Delta}{\omega_p/2}\right)^2 - \frac{(1-\alpha)^2(1+\alpha)}{64Q^2\alpha}} \quad (7.71)$$

For maximum gain ($\alpha \cong 1$) we have:

$$[B_e^e]_{\alpha \cong 1} \cong \frac{1}{\sqrt{2}} \cdot \frac{\Delta}{\omega_p} + 3\sqrt{2}\, Q^2 \left(\frac{\Delta}{\omega_p}\right)^3 \cong \frac{1}{\sqrt{2}} \frac{\Delta}{\omega_p} \quad (7.72)$$

It will be seen that the limit bandwidth depends very little on Q as the term $3\sqrt{2}\, Q^2(\Delta/\omega_p)^3$ is very small compared with the first term. In the same conditions ($\alpha \cong 1$), the voltage gain bandwidth product is equal to:

$$[B_e^e \sqrt{G_e^e}]_{\alpha \cong 1} \cong \frac{1}{4\sqrt{2}\, Q} \quad (7.73)$$

and we see that it is desirable to select as low a value of Q as possible. We may conclude from the above discussion that the use of a simply

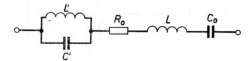

Figure 7.6. Equivalent circuit of the total impedance with double tuning

tuned circuit is quite unsuitable for a degenerate amplifier: the gain is limited (cf. formula (7.65)) and even if this first disadvantage is reduced to a minimum by the choice of a low Q-value, the pass band is always very unsymmetrical. We are thus led to consider a doubly tuned circuit which at the same time makes $x_e = x_{(-)} = 0$ for the nominal frequency to be amplified. The total circuit resistance, which has two zeros at ω_e and $\omega_{(-)}$ respectively, should have one pole between ω_e and $\omega_{(-)}$. It is clear, due to the symmetry, that this pole should be at $\omega_p/2$, which has the additional advantage that it eliminates the possibility of oscillation at this same frequency. In the range of frequencies with which we are concerned, the equivalent circuit of the total circuit impedance is that shown in Fig. 7.6. The anti-resonant $L'C'$ circuit and the resonant LC_0 circuit are both tuned to $\omega_p/2$. Figure 7.4 also shows the relative positions of ω_e, $\omega_{(-)}$, $\omega_p/2$, ω_{e_0} and $\omega_{(-)_0}$.

We may write:

$$\omega_e = \frac{\omega_p}{2} - \Delta - \delta \tag{7.74}$$

$$\omega_{(-)} = \frac{\omega_p}{2} + \Delta + \delta \tag{7.75}$$

$$\frac{x_e}{r_0} = -4Q\frac{\Delta + \delta}{\omega_p} + \frac{1}{2C'r_0} \cdot \frac{1}{\Delta + \delta} \tag{7.76}$$

$$\frac{x_{(-)}}{r_0} = -\frac{x_e}{r_0} \tag{7.77}$$

C' is chosen so as to equate x_e and $x_{(-)}$ to zero for $\omega_e = \omega_{e_0}$, that is for $\delta = 0$. We then have:

$$\frac{1}{2C'r_0} = \frac{4Q\Delta^2}{\omega_p} \tag{7.78}$$

$$\frac{x_e}{r_0} = -\frac{x_{(-)}}{r_0} = -\frac{4Q}{\omega_p}\left[(\Delta + \delta) - \frac{\Delta^2}{\Delta + \delta}\right] \tag{7.79}$$

(7.59) becomes, if we take (7.62) and (7.79) into account:

$$G_e^e = \left[\frac{1 + \alpha + \left(\frac{x_e}{r_0}\right)^2}{\left[1 - \alpha - \left(\frac{x_e}{r_0}\right)^2\right] + j\,2\frac{x_e}{r_0}}\right]^2 \tag{7.80}$$

At the nominal frequency we have:

$$\delta = 0$$
$$x_e = 0$$

$$G_e^e(\omega_{e_0}) = \left(\frac{1 + \alpha}{1 - \alpha}\right)^2 \tag{7.81}$$

The maximum gain varies from 1 to ∞ as α varies from 0 to 1. If we introduce (7.79) into (7.80), we see that G_e^e has at $\omega_{(-)_0}$ the same maximum as at ω_{e_0} and more usually assumes the same value for two frequencies symmetrical with respect to $\omega_p/2$. However, just as for the simple circuit, it is only the region in the neighbourhood of ω_{e_0} which must be taken into account. It should be noted that the equations obtained are valid for $\Delta < 0$ and that consequently it is immaterial whether ω_e is lower than $\omega_p/2$ (as shown in Fig. 7.4) or higher than $\omega_p/2$. The domain of permissible values for δ is such that $\Delta + \delta$ should be of the same sign as Δ.

DEGENERATE AMPLIFIERS

At the extremities of the 3 dB bandwidth we have:
$$G_e^e(\omega_e) = \tfrac{1}{2} G_e^e(\omega_{e0}) \tag{7.82}$$

(7.80) may be put in the form:
$$G_e^e(\omega_{e0}) = \frac{1}{1 - \{4\alpha/[1 + \alpha + (x_e/r_0)^2]^2\}} \tag{7.83}$$

In accordance with (7.82) we then have:
$$1 - \frac{4\alpha}{[1 + \alpha + (x_e/r_0)^2]} = 2\left(\frac{1-\alpha}{1+\alpha}\right)^2 \tag{7.84}$$

$$\left(\frac{x_e}{r_0}\right)^2 = (1 + \alpha)\left[-1 \pm \sqrt{\frac{4\alpha}{4\alpha - (1-\alpha)^2}}\right] \tag{7.85}$$

The expression (7.85) is meaningless unless it gives $(x_e/r_0)^2$ a real positive value. In order that $(x_e/r_0)^2$ should be real we must have $\alpha > (\sqrt{2} - 1)^2$. This condition is equivalent to $G_e^e(\omega_{e0}) > 2$. As on the other hand (7.81) gives us $G_e^e \geqslant 1$, it is clear that this latter condition is necessary in order that there should be a frequency for which G_e^e has decreased by 3 dB. In practice we have $G(\omega_{e0}) > 2$ and the expression is consequently real. Further, we have:

$$\frac{4\alpha}{4\alpha - (1-\alpha)^2} > 2 \tag{7.86}$$

and in order that (x_e/r_0) should be real we can only retain the positive sign in expression (7.85). Finally, we have:

$$\left(\frac{x_e}{r_0}\right)^2 = (1 + \alpha)\left[-1 + \sqrt{\frac{1}{1 - (1-\alpha)^2/4\alpha}}\right] \tag{7.87}$$

In the normal case in which α is near to 1, we have:

$$\left(\frac{x_e}{r_0}\right)^2 \cong (1 + \alpha)\frac{(1-\alpha)^2}{8\alpha} \tag{7.88}$$

This approximation gives rise to errors of less than 10% for $G > 10$ dB. The corresponding maximum error in the bandwidth will therefore be of the order of 5%.

Let
$$\beta = \frac{\omega_p}{16Q} \tag{7.89}$$

Introducing (7.79) into (7.88) we find:

$$(\Delta + \delta)^4 - 2\left[\Delta^2 + \frac{(1+\alpha)(1-\alpha)^2}{\alpha}\beta^2\right](\Delta + \delta)^2 + \Delta^4 = 0 \tag{7.90}$$

This equation gives the values of δ for which the gain has decreased by 3 dB. It is easily seen that the discriminant is always positive and that there are always two distinct and positive real values of $(\Delta + \delta)^2$ which satisfy (7.90). As already mentioned above, we need consider only those values of δ which give $(\Delta + \delta)$ the same sign as Δ. There are thus two values δ_1 and δ_2 which satisfy (7.90). They are given by the relations:

$$1 + \frac{\delta_1}{\Delta}$$
$$= \sqrt{1 + \frac{(1+\alpha)(1-\alpha)^2 \beta^2}{\alpha \Delta^2} + \sqrt{2\frac{(1+\alpha)(1-\alpha)^2 \beta^2}{\alpha \Delta^2} + \frac{(1+\alpha)^2(1-\alpha)^4 \beta^4}{\alpha^2 \Delta^4}}} \quad (7.91)$$

$$1 + \frac{\delta_2}{\Delta}$$
$$= \sqrt{1 + \frac{(1+\alpha)(1-\alpha)^2 \beta^2}{\alpha \Delta^2} - \sqrt{2\frac{(1+\alpha)(1-\alpha)^2 \beta^2}{\alpha \Delta^2} + \frac{(1+\alpha)^2(1-\alpha)^4 \beta^4}{\alpha^2 \Delta^4}}} \quad (7.92)$$

The reduced bandwidth is equal to:

$$B_e^e = \left|\frac{\delta_1 - \delta_2}{\omega_p/2}\right| = \frac{\Delta}{\omega_p/2} \cdot \left|\left(1 + \frac{\delta_1}{\Delta}\right) - \left(1 + \frac{\delta_2}{\Delta}\right)\right| \quad (7.93)$$

It will be seen that this bandwidth is asymmetrical as $\delta_1 \neq -\delta_2$. We will therefore calculate the relative dissymmetry D defined by:

$$D_e^e = \left|\frac{\delta_1 + \delta_2}{\delta_1 - \delta_2}\right| = \frac{1}{B_e^e} \cdot \frac{|\delta_1 + \delta_2|}{\omega_p/2} = \frac{1}{B_e^e} \cdot \frac{\Delta}{\omega_p/2} \left|\frac{\delta_1}{\Delta} + \frac{\delta_2}{\Delta}\right| \quad (7.94)$$

Let us first calculate the bandwidth. We may write:

$$\left[\left(1 + \frac{\delta_1}{\Delta}\right) - \left(1 + \frac{\delta_2}{\Delta}\right)\right]^2 = 2\frac{(1+\alpha)(1-\alpha)^2 \beta^2}{\alpha \Delta^2} \quad (7.95)$$

$$B_e^e = \sqrt{\frac{1+\alpha}{2\alpha}} \frac{1-\alpha}{4Q} \simeq \frac{1-\alpha}{4Q} \quad (7.96)$$

The voltage gain at resonance is

$$\sqrt{G_e^e(\omega_{e_0})} = \frac{1+\alpha}{1-\alpha} \simeq \frac{2}{1-\alpha} \quad (7.97)$$

The voltage gain bandwidth product is then equal to:

$$B_e^e \sqrt{G_e^e(\omega_{e_0})} = \frac{1}{2Q} \quad (7.98)$$

DEGENERATE AMPLIFIERS 173

The next step is to calculate the relative dissymmetry. We may write:

$$\left(1 + \frac{\delta_1}{\Delta}\right)\left(1 + \frac{\delta_2}{\Delta}\right) = 1 + \frac{\delta_1 + \delta_2}{\Delta} + \frac{\delta_1\delta_2}{\Delta^2} \qquad (7.99)$$

$$\frac{\delta_1\delta_2}{\Delta^2} = \frac{1}{4}\left(\frac{\delta_1 + \delta_2}{\Delta}\right)^2 - \frac{1}{4}\left(\frac{\delta_1 - \delta_2}{\Delta}\right)^2 \qquad (7.100)$$

$$\left(\frac{\delta_1 + \delta_2}{\Delta}\right)^2 + 4\left(\frac{\delta_1 + \delta_2}{\Delta}\right) - 2\frac{(1 + \alpha)(1 - \alpha)^2\beta^2}{\alpha\Delta^2} = 0 \qquad (7.101)$$

$$\frac{\delta_1 + \delta_2}{\Delta} = -2 \pm \sqrt{4 + 2\frac{(1 + \alpha)(1 - \alpha)^2\beta^2}{\alpha\Delta^2}} \qquad (7.102)$$

The root corresponding to the negative sign is excluded since we must have:

$$1 + \frac{\delta_1}{\Delta} > 0 \qquad (7.103)$$

$$1 + \frac{\delta_2}{\Delta} > 0 \qquad (7.104)$$

$$\frac{\delta_1}{\Delta} + \frac{\delta_2}{\Delta} > -2 \qquad (7.105)$$

The relative dissymmetry is then:

$$D_e^e = \left|\frac{\delta_1 + \delta_2}{\delta_1 - \delta_2}\right| = -\frac{4\Delta}{\omega_p B_e^e} + \sqrt{1 + \left(\frac{4\Delta}{\omega_p B_e^e}\right)^2} \qquad (7.106)$$

To sum up, the bandwidth depends only on Q (for a fixed gain). The symmetry of this band will improve as the value of Δ increases. The bandwidth decreases, but its symmetry improves as the gain increases.

(3) *Noise gain*

The contribution made to the output noise by the noise of the signal generator consists of two terms V_{Le}^{ne} and $V_{Le}^{n(-)}$. The input noises V_{ne} and $V_{n(-)}$ are not phase correlated. The same applies to V_{Le}^{ne} and $V_{Le}^{n(-)}$. The power of the output noise is thus the sum of the powers corresponding to V_{Le}^{ne} and $V_{Le}^{n(-)}$, namely:

$$\Delta f[p]_{\text{ideal}} = \frac{2|V_{Le}^{ne}|^2}{R_0} + \frac{2|V_{Le}^{n(-)}|^2}{R_0} \qquad (7.107)$$

From (7.21) and (7.32) we have:

$$\Delta f[p]_{\text{ideal}} = \frac{1}{2R_0}\left[\left|\frac{A}{D}\right|^2 |V_{ne}|^2 + \left|\frac{F}{D}\right|^2 |V_{n(-)}|^2\right] \quad (7.108)$$

The available power of the input noise in each of the two frequency bands Δf centered on f_e and $f_{(-)}$ is equal to:

$$\frac{|V_{ne}|^2}{2R_0} = \frac{|V_{n(-)}|^2}{2R_0} = kT_1\Delta f \quad (7.109)$$

where T_1 is the noise temperature of the signal source. (7.108) and (7.109) give us:

$$\Delta f[p]_{\text{ideal}} = kT_1\Delta F\left[\left|\frac{A}{D}\right|^2 + \left|\frac{F}{D}\right|^2\right] \quad (7.110)$$

The available power of the input noise is the sum of the available powers of each band, namely $2kT_1\Delta f$. The noise gain is therefore:

$$G_e^{Ne} = \frac{1}{2}\frac{|A|^2 + |F|^2}{|D|^2} \quad (7.111)$$

If we assume that the circuits are tuned (by means of the doubly tuned circuit discussed earlier) and if we admit the approximations introduced when we were studying the gain, we obtain:

$$G_e^{Ne} = \frac{(1+\alpha)^2 + 4\alpha}{(1-\alpha)^2} \quad (7.112)$$

(4) *Effective noise temperature*

The contribution made to the noise by the noise of R_s consists of two terms V_{Le}^{se} and $V_{Le}^{s(-)}$ which are not phase-correlated. The same applies to V_{Le}^{se} and $V_{Le}^{s(-)}$. The power of the output noise is therefore the sum of the powers corresponding to V_{Le}^{se} and $V_{Le}^{s(-)}$, namely:

$$\Delta f\{[p]_{\text{actual}} - [p]_{\text{ideal}}\} = \frac{2|V_{Le}^{se}|^2}{R_0} + \frac{2|V_{Le}^{s(-)}|^2}{R_0} \quad (7.113)$$

From (7.23) and (7.34), we have:

$$\Delta f\{[p]_{\text{actual}} - [p]_{\text{ideal}}\} = \frac{1}{2R_0}\left[4R_0^2\left|\frac{C}{D}\right|^2|V_{se}|^2 + \left|\frac{F}{D}\right|^2|V_{s(-)}|^2\right] \quad (7.114)$$

The available power of the noise of R_s in the frequency bands Δf centered on f_e and $f_{(-)}$ is equal to:

$$\frac{|V_{se}|^2}{2R_s} = \frac{|V_{s(-)}|^2}{2R_s} = kT_v\Delta f \quad (7.115)$$

where T_v is the temperature of the varactor. Equations (7.114) and (7.115) give us:

$$[p]_{\text{actual}} - [p]_{\text{ideal}} = kT_v - \frac{R_s}{R_0}\left[4R_0^2\left|\frac{C}{D}\right|^2 + \left|\frac{F}{D}\right|^2\right] \quad (7.116)$$

Equation (1.143) tells us that the effective noise temperature is equal to :†

$$T_e^e = \frac{\omega_e}{\omega_c'} T_v \frac{4\sqrt{\alpha}(1+\alpha)}{(1+\alpha)^2 + 4\alpha} \quad (7.117)$$

Here again we have assumed tuning and we have used the approximations made when studying the gain. When the gain tends to infinity, α tends to unity and we have:

$$[T_e^e]_{G \to \infty} = \frac{\omega_e}{\omega_c'} T_v \quad (7.118)$$

The effective temperature corresponding to any given finite gain is always less than this limiting value, since, for $\alpha < 1$ we always have:

$$\frac{4\sqrt{\alpha}(1+\alpha)}{(1+\alpha)^2 + 4\alpha} < 1 \quad (7.119)$$

(5) *Gain instability factor*

By definition we have:

$$S_e^e = \frac{dG_e^e/G_e^e}{d(\omega_c'^2)/\omega_c'^2} = \frac{dG_e^e/G_e^e}{d\alpha/\alpha} = \frac{\alpha}{G_e^e} \times \frac{dG_e^e}{d\alpha} \quad (7.120)$$

where G_e^e is the maximum gain for a fixed value of α (reactive tuning). It is easily shown that this definition coincides with that adopted in Chapters 4 and 5. We are assuming that tuning has been effected by means of the doubly tuned circuit discussed earlier. We then at once have:

$$S_e^e = \frac{4\alpha}{1-\alpha^2} \quad (7.121)$$

2.3. Amplifier $X_{(-)}^e$

The same method of calculation is used as in the previous section, and we will therefore give the approximate results at once. We are assuming that the circuit which is traversed by currents of frequencies ω_e and $\omega_{(-)}$ is the doubly tuned circuit studied in the previous section.

† As no confusion is possible, we will in future omit the index 'eff'.

PARAMETRIC AMPLIFIERS

(1) *General expression for the transducer gain*

$$G^e_{(-)} = \left(\frac{\omega_p}{\omega_e}\right)^2 \left|\frac{2r_0(\omega'_c/\omega_p)}{(r_0 - jx_e)(r_0 + jx_{(-)}) - 4(\omega'_c/\omega_p)^2}\right|^2 \quad (7.122)$$

(2) *Gain at the nominal frequency and bandwidth*

$$G^e_{(-)}(\omega_{e_0}) = \frac{4\alpha}{(1 - \alpha)^2} \quad (7.123)$$

$$B^e_{(-)} = \frac{4\alpha}{\sqrt{2(1 + \alpha)}} \cdot \frac{1 - \alpha}{4Q} \simeq \frac{1 - \alpha}{4Q} \quad (7.124)$$

$$B^e_{(-)}\sqrt{G^e_{(-)}(\omega_{e_0})} = \frac{1}{2Q} \quad (7.125)$$

$$D^e_{(-)} = \frac{-4\Delta}{\omega_p B^e_{(-)}} + \sqrt{1 + \left(\frac{4\Delta}{\omega_p B^e_{(-)}}\right)^2} \quad (7.126)$$

(3) *Noise gain*

$$G^{Ne}_{(-)} = \frac{(1 + \alpha)^2 + 4\alpha}{(1 - \alpha)^2} \quad (7.127)$$

(4) *Effective noise temperature*

$$T^e_{(-)} = \frac{\omega_e}{\omega'_c} T_v \frac{4\sqrt{\alpha}\,(1 + \alpha)}{(1 + \alpha)^2 + 4\alpha} \quad (7.128)$$

When the gain tends to infinity, α tends to unity and we have:

$$[T^e_{(-)}]_{G \to \infty} = \frac{\omega_e}{\omega'_c} T_v \quad (7.129)$$

The effective temperature corresponding to any given finite gain is always lower than this limiting value, since, for $\alpha < 1$ we always have:

$$\frac{4\sqrt{\alpha}\,(1 + \alpha)}{(1 + \alpha)^2 + 4\alpha} < 1 \quad (7.130)$$

(5) *Instability factor of the gain*

$$S^e_{(-)} = \frac{1 + \alpha}{1 - \alpha} \quad (7.131)$$

2.4. Amplifier $X^e_{e(-)}$

We will assume that the circuit traversed by currents of frequencies ω_e and $\omega_{(-)}$ is the doubly tuned circuit already studied in the previous paragraphs.

DEGENERATE AMPLIFIERS

(1) *General expression for the transducer gain*
In order to calculate the available output power it is necessary to calculate the IF power that would be supplied by synchronous loss-free detection of the two signals V_{Le}^{ge} and $V_{L(-)}^{ge}$. Representing the instantaneous values of the voltages by lower case letters, we have:

$$v_{ge} = V_{ge} e^{j\omega_e t} + V_{ge}^* e^{-j\omega_e t} \tag{7.132}$$

$$v_{Le}^{ge} = V_{Le}^{ge} e^{j\omega_e t} + V_{Le}^{ge*} e^{-j\omega_e t}$$
$$= -\frac{1}{2}\left[\frac{A}{D} V_{ge} e^{j\omega_e t} + \frac{A^*}{D^*} V_{ge}^* e^{-j\omega_e t}\right] \tag{7.133}$$

$$v_{L(-)}^{ge} = V_{L(-)}^{ge} e^{j\omega_{(-)}t} + V_{L(-)}^{ge*} e^{-j\omega_{(-)}t}$$
$$= j\frac{1}{2}\left[\frac{B}{D^*} V_{ge}^* e^{j\omega_{(-)}t} - \frac{B^*}{D} V_{ge} e^{-j\omega_{(-)}t}\right] \tag{7.134}$$

A local oscillator is used with a frequency ω_{OL} equal to half the pumping frequency and in phase with it:

$$\omega_{OL} = \frac{\omega_p}{2} = \frac{\omega_e + \omega_{(-)}}{2} = \omega_e + \Delta = \omega_{(-)} - \Delta \tag{7.135}$$

The voltage of the local oscillator can then be expressed by:

$$v_{OL} = V_{OL} e^{j\omega_{OL} t} + V_{OL}^* e^{-j\omega_{OL} t} \tag{7.136}$$

Synchronous loss-free demodulation is obtained by multiplying V_{Le}^{ge} and $V_{L(-)}^{ge}$ by v_{OL} (with $V_{OL} = 1/\sqrt{2}\, e^{j\phi}$), by filtering to eliminate all frequencies except Δ, and by adding the two voltages in the load resistance R_0. We obtain:

$$v_{IF} = -\frac{1}{2\sqrt{2}}$$
$$\times \left[\frac{A e^{-j\phi} + jB^* e^{j\phi}}{D} V_{ge} e^{-j\Delta t} + \frac{A^* e^{j\phi} - jB e^{-j\phi}}{D^*} V_{ge}^* e^{j\Delta t}\right]$$
$$= V_{IF} e^{j\Delta t} + V_{IF}^* e^{-j\Delta t} \tag{7.137}$$

$$V_{IF} = -\frac{1}{2\sqrt{2}} \frac{A^* e^{j\phi} - jB e^{-j\phi}}{D^*} V_{ge}^* \tag{7.138}$$

$$V_{IF}^* = -\frac{1}{2\sqrt{2}} \frac{A e^{-j\phi} + jB^* e^{j\phi}}{D} V_{ge} \tag{7.139}$$

The output power is equal to:

$$\frac{2|V_{IF}|^2}{R_0} = \frac{2 V_{IF} V_{IF}^*}{R_0} = \frac{|V_{ge}|^2}{4 R_0} \frac{|A|^2 + |B|^2 + 2 \operatorname{Re}\{jA^* B^* e^{j2\phi}\}}{|D|^2} \tag{7.140}$$

It is required to bring this power to a maximum when the input frequency is the nominal frequency ($x_e = x_{(-)} = 0$). Relations (7.49), (7.50) and (7.140) show that for this to happen we must have $\phi = -\pi/4$. Assuming that this condition is satisfied, the output power will be equal to:

$$\frac{2|V_{\text{IF}}|^2}{R_0} = \frac{|V_{ge}|^2}{4R_0} \cdot \left|\frac{A+B}{D}\right|^2 \tag{7.141}$$

The available input power is equal to $|V_{ge}|^2/2R_0$. The transducer gain is then:

$$G^e_{e(-)} = \frac{1}{2}\left|\frac{A+B}{D}\right|^2 \tag{7.142}$$

Finally, taking account of (7.49), (7.50) and (7.52), we have:

$$G^e_{e(-)} = \frac{1}{2}\left|\frac{(r_0 - jx_{(-)})(r_0 - jx_e) + 4(\omega'_c/\omega_p)^2 + 2r_0(\omega'_c/\omega_e)}{(r_0 + jx_e)(r_0 - jx_{(-)}) - 4(\omega'_c/\omega_p)^2}\right|^2 \tag{7.143}$$

(2) *Gain at the nominal frequency and bandwidth*

The method of calculation is the same as that used in the previous paragraphs, and we will therefore at once set out the approximate results for the case when $\phi = -\pi/4$.

$$G^e_{e(-)}(\omega_{e0}) = \frac{1}{2}\left(\frac{1+\sqrt{\alpha}}{1-\sqrt{\alpha}}\right)^2 \tag{7.144}$$

$$B^e_{e(-)} = \frac{1}{\sqrt[4]{\alpha}}\frac{1-\alpha}{4Q} \simeq \frac{1-\alpha}{4Q} \tag{7.145}$$

$$B^e_{e(-)}\sqrt{G^e_{e(-)}(\omega_{e0})} = \frac{(1+\sqrt{\alpha})^2}{4\sqrt{2}Q} \simeq \frac{1}{Q\sqrt{2}} \tag{7.146}$$

$$D^e_{e(-)} = \frac{-4\Delta}{\omega_p B^e_{e(-)}} + \sqrt{1+\left(\frac{4\Delta}{\omega_p B^e_{e(-)}}\right)^2} \tag{7.147}$$

(3) *Noise gain*

The calculation is analogous to that for the transducer gain. The approximate results for the case when $\phi = -\pi/4$ are given below.

$$G^{Ne}_{e(-)} = \left(\frac{1+\sqrt{\alpha}}{1-\sqrt{\alpha}}\right)^2 \tag{7.148}$$

(4) *Effective noise temperature*

In the same way we find:

$$T^e_{e(-)} = \frac{\omega_e}{\omega'_c} T_v \frac{4\sqrt{\alpha}}{(1+\sqrt{\alpha})^2} \tag{7.149}$$

DEGENERATE AMPLIFIERS

When the gain tends to infinity, α tends to unity and we have:

$$[T^e_{e(-)}]_{G \to \infty} = \frac{\omega_e}{\omega'_c} T_v \qquad (7.150)$$

The effective temperature corresponding to any given value of finite gain is always lower than this limiting value, since, for $\alpha < 1$ we always have:

$$\frac{4\sqrt{\alpha}}{(1 + \sqrt{\alpha})^2} < 1 \qquad (7.151)$$

(5) *Gain instability factor*

By differentiation of (7.144) we obtain:

$$S^e_{e(-)} = \frac{2\sqrt{\alpha}}{1 - \alpha} \qquad (7.152)$$

2.5. Comparison between the three amplifiers and the reflection amplifiers

(1) *Note: Performance of the reflection amplifier*

In order to avoid confusion in notation, we will in this paragraph denote all the characteristics of the reflection amplifier by the index r. Only the case in which $r_{(-)} = 0$ will be considered, since this is the optimum case from the point of view of noise. In order to simplify the comparison we will put:

$$\alpha = \frac{q^2}{n_{(-)} r_e} \qquad (7.153)$$

Expressions (5.136), (5.138) and (5.149) can then be written approximately in the form:

$$G_r = \left(\frac{1 + \alpha}{1 - \alpha}\right)^2 \qquad (7.154)$$

$$T_{r \min} = \frac{4\alpha}{(1 + \alpha)^2} \cdot \left(\frac{\omega'_c}{\omega_{(-)}} + \frac{\omega_{(-)}}{\omega'_c}\right) \frac{\omega_e}{\omega'_c} T_v \qquad (7.155)$$

$$S_r = \frac{4\alpha}{1 - \alpha^2} \qquad (7.156)$$

When the idler frequency is optimized ($\omega_{(-)} = \omega'_c$) we have:

$$T_{r \text{ opt}} = \frac{8\alpha}{(1 + \alpha)^2} \cdot \frac{\omega_e}{\omega'_c} T_v \qquad (7.157)$$

Further, as the reflection amplifier is of the non-degenerate type, its noise gain is the same as its transducer gain.

PARAMETRIC AMPLIFIERS

(2) *Criterion for the comparison*

As was done in Chapter 5, we will compare the various systems at given stability. Strictly speaking, we should then, in the expressions for the gains and noise temperatures, replace α by its value as a function of the instability factor. However, under normal operating conditions (adequate gain) the values of the different instability factors corresponding to the same value of α differ very little. The comparison will therefore be made between the various systems with a constant value of α.

(3) *Transducer gain and noise gain*

Table 7.1 and Fig. 7.7 give the transducer gain and the noise gain as

Table 7.1

α	$G_e^e = G_r$, dB	$G_{(-)}^e$, dB	$G_{e(-)}^e$, dB	$G_e^{Ne} = G_{(-)}^{Ne}$, dB	$G_{e(-)}^{Ne}$, dB
0·50	9·55	9·03	12·29	12·30	15·30
0·55	10·74	10·35	13·58	13·56	16·59
0·60	12·04	11·76	14·93	14·91	17·94
0·65	13·46	13·26	16·36	16·37	19·37
0·70	15·07	14·93	18·03	18·01	21·04
0·75	16·90	16·81	19·86	19·87	22·87
0·80	19·08	19·03	22·11	22·07	25·12
0·85	21·82	21·79	24·93	24·81	27·94
0·90	25·57	25·56	28·81	28·58	31·82
0·95	31·82	31·81	34·93	34·83	37·94

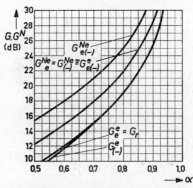

Figure 7.7. Transducer gain and noise gain of single-band degenerate amplifiers

functions of α for the three degenerate single-band amplifiers and for the reflection amplifier.

(4) Effective noise temperature

Table 7.2 and Fig. 7.8 give the effective noise temperature as a function of α for the three degenerate single-band amplifiers and for the reflection amplifier with the idler frequency optimized. It will be seen that the columns headed T_e^e and $T_{(-)}^e$ contain only the number 1000. It can be proved by means of (7.128) that for $0.5 \leqslant \alpha \leqslant 1$, $T_{(-)}^e/[(\omega_e/\omega_c')T_v]$ differs from unity by an amount less than 0.5×10^{-3}.

Table 7.2

α	$\dfrac{T_e^e}{(\omega_c'/\omega_e)T_v} = \dfrac{T_{(-)}^e}{(\omega_e/\omega_c')T_v}$	$\dfrac{T_{e(-)}^e}{(\omega_e/\omega_c')T_v}$	$\dfrac{T_{r\,\text{opt}}}{(\omega_e/\omega_c')T_v}$
0·50	1·000	0·984	1·875
0·55	1·000	0·987	1·899
0·60	1·000	0·990	1·920
0·65	1·000	0·992	1·939
0·70	1·000	0·994	1·955
0·75	1·000	0·996	1·969
0·80	1·000	0·998	1·980
0·85	1·000	0·999	1·989
0·90	1·000	0·999	1·995
0·95	1·000	1·000	1·999
1·00	1·000	1·000	2·000

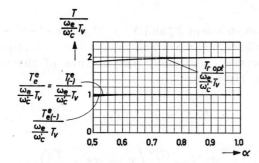

Figure 7.8. Noise temperature of degenerate single-band amplifiers

Conclusions

We have just seen that a degenerate amplifier of the type required to amplify a signal at frequency f_e also amplifies with a comparable gain the image signal $f_{(-)}$. The use of an amplifier of this type is contra-indicated unless one is sure that any disturbing signal in the image band is absent. We will base our study of the comparison of the amplifiers X_e^e, $X_{(-)}^e$, $X_{e(-)}^e$ and X_r on this assumption. (The non-degenerate reflection amplifier will be designated by X_r.) It should be noted first that the amplifier $X_{(-)}^e$ has the same effective temperature and the same noise gain as the amplifier X_e^e. Conversely, $G_{(-)}^e$ is always lower than G_e^e. All that is needed, therefore, is to compare the amplifiers X_e^e, $X_{e(-)}^e$ and X_r.

Amplifier $X_{e(-)}^e$ requires a balanced mixer with a wide band. In order to reduce this band it is therefore useful to use for Δ a value only slightly higher than the maximum value of δ, Fig. 7.5. In any case the circuit which ensures equilibrium (for example a hybrid ring), must be centred on the frequency of the local oscillator. It will operate less satisfactorily than in other cases where it is possible to centre the circuits on the signal frequency. This disadvantage becomes more marked as the intermediate frequency increases. Further, amplifier $X_{e(-)}^e$ requires phase correlation between the pumping voltage and the local oscillator.

This being so, we will compare two cases. In the first, the use of the amplifier $X_{e(-)}^e$ is excluded for various technical reasons (for example we may wish to design a universal amplifier to work with any existing mixer). There remains, therefore, the comparison between amplifiers X_e^e and X_r. In the second case it is possible to use amplifier $X_{e(-)}^e$. The curves in Figs. 7.7 and 7.8 show that for useful values of gain (higher than 10 dB) the noise temperatures and the ratios G^N/G of the amplifiers X_e^e and $X_{e(-)}^e$ are very close, while amplifier $X_{e(-)}^e$ has by far the best gain. It will be enough therefore to compare $X_{e(-)}^e$ and X_r.

Case 1: Comparison of X_e^e and X_r. If T_2 is the effective temperature of the amplification chain which follows the first amplifier, the expressions for the operational temperatures of the two types of amplifiers are given by:

$$T_{\text{op } r} = T_1 + T_{r \min} + \frac{T_2}{G_r} \tag{7.158}$$

$$T_{\text{op } e} = \frac{G_e^{Ne}}{G_e^e} \left(T_1 + T_e^e + \frac{T_2}{G_e^{Ne}} \right) \tag{7.159}$$

DEGENERATE AMPLIFIERS

The numerical calculation performed above shows that we have with a good approximation:

$$G_e^{Ne} \simeq 2G_e^e = 2G_r \tag{7.160}$$

$$T_{r\,\min} = \frac{1}{2}\left(\frac{\omega_c'}{\omega_{(-)}} + \frac{\omega_{(-)}}{\omega_c'}\right) T_{r\,\text{opt}} \simeq \left(\frac{\omega_c'}{\omega_{(-)}} + \frac{\omega_{(-)}}{\omega_c'}\right) T_e^e \tag{7.161}$$

We then have:

$$T_{\text{op}\,r} \simeq T_1 + \left(\frac{\omega_c'}{\omega_{(-)}} + \frac{\omega_{(-)}}{\omega_c'}\right) T_e^e + \frac{T_2}{G_e^e} \tag{7.162}$$

$$T_{\text{op}\,e}^e \simeq 2T_1 + 2T_e^e + \frac{T_2}{G_e^e} \tag{7.163}$$

When the idler frequency of X_r is optimized ($\omega_{(-)} = \omega_c'$) we always have:

$$T_{\text{op}\,e}^e > T_{\text{op}\,r} \tag{7.164}$$

If $\omega_{(-)}$ is not optimized, the inequality (7.164) is satisfied when:

$$T_1 + \left(2 - \frac{\omega_c'}{\omega_{(-)}} - \frac{\omega_{(-)}}{\omega_c'}\right) T_e^e > 0 \tag{7.165}$$

Consequently, we must choose the degenerate amplifier X_e^e for 'cold' sources (T_1 low) when we cannot obtain the optimum idler frequency for X_r. Conversely, if this optimum frequency is obtained we must always choose the reflection amplifier. This would still be preferable even for $\omega_{(-)} \neq \omega_c'$ if the signal source is 'hot' (T_1 high).

Case 2: Comparison of $X_{e(-)}^e$ and X_r. The operational temperature of $X_{e(-)}^e$ is:

$$T_{\text{op}\,e(-)}^e = \frac{G_{e(-)}^{Ne}}{G_{e(-)}^e}\left(T_1 + T_{e(-)}^e + \frac{T_2}{G_{e(-)}^{Ne}}\right) \tag{7.166}$$

The numerical calculation performed above shows that we have, with a very good approximation:

$$G_{e(-)}^{Ne} = 2G_{e(-)}^e \simeq 4G_r \tag{7.167}$$

$$T_{r\,\min} \simeq \left(\frac{\omega_c'}{\omega_{(-)}} + \frac{\omega_{(-)}}{\omega_c'}\right) T_{e(-)}^e \tag{7.168}$$

We then have:

$$T_{\text{op}\,r} \simeq T_1 + \left(\frac{\omega_c'}{\omega_{(-)}} + \frac{\omega_{(-)}}{\omega_c'}\right) T_{e(-)}^e + \frac{2T_2}{G_{e(-)}^e} \tag{7.169}$$

$$T_{\text{op}\,e(-)}^e \simeq 2T_1 + 2T_{e(-)}^e + \frac{T_2}{G_{e(-)}^e} \tag{7.170}$$

The reflection amplifier will be preferable if:

$$T_1 + \left(2 - \frac{\omega_c'}{\omega_{(-)}} - \frac{\omega_{(-)}}{\omega_c'}\right) T^e_{e(-)} - \frac{T_2}{G^e_{e(-)}} \geqslant 0 \qquad (7.171)$$

The degenerate amplifier $X^e_{e(-)}$ is therefore suitable for a low value of T_1 and a high value of T_2.

3. Double-band reception, correlated signals

3.1. The two methods of operation

The input signal consists of v_{ge} and $v_{g(-)}$. The degenerate amplifier produces four output signals: v^{ge}_{Le}, $v^{ge}_{L(-)}$, $v^{g(-)}_{Le}$ and $v^{g(-)}_{L(-)}$. The mixer which follows produces four corresponding IF signals. The IF circuit enables us to select the band which it is desired to use. There are two possible cases:

Case 1: A single band is chosen, for example the transpose of $v^{ge}_{Le} + v^{g(-)}_{Le}$.

Case 2: A synchronous detection of the two signals, $v^{ge}_{Le} + v^{g(-)}_{Le}$ and $v^{ge}_{L(-)} + v^{g(-)}_{L(-)}$, is effected.

The following sections will be devoted to the study of these two cases. In the presentation we have adopted, the first case is represented by the symbol $X^{e(-)}_e$, and the second by $X^{e(-)}_{e(-)}$. It will be assumed that the expressions for the two input signals will be:

$$v_{ge} = V_{ge}\, e^{j\omega_e t} + V^*_{ge}\, e^{-j\omega_e t} \qquad (7.172)$$

$$v_{g(-)} = V_{g(-)}\, e^{j\omega_{(-)} t} + V^*_{g(-)}\, e^{-j\omega_{(-)} t} \qquad (7.173)$$

with

$$|V_{ge}| = |V_{g(-)}| \qquad (7.174)$$

The calculation of the transducer gains described below will require a knowledge of the expression for the available power of the input signal. It is obtained by a loss-free synchronous detection of the signal. The calculation is analogous to that performed in section 2.4(1). If we optimize the phase of the local oscillator, we find the value: $|V_{ge}|^2/R_0$. On the other hand, if we designate by $\phi/2$ the phase at the origin of the carrier from which the signals v_{ge} and $v_{g(-)}$ are produced by LF modulation, we know that we have:

$$V_{ge} = V^*_{g(-)}\, e^{-j\phi} \qquad (7.175)$$

3.2. Amplifier $X^e_{e(-)}$

(1) *General expression for the transducer gain*

The available power of the input signal is $|V_{ge}|^2/R_0$. The output power is equal to:

$$\frac{2|V^{ge}_{Le} + V^{g(-)}_{Le}|^2}{R_0}$$

The transducer gain is then:

$$G^{e(-)}_e = 2\left|\frac{V^{ge}_{Le} + V^{g(-)}_{Le}}{V_{ge}}\right|^2 \qquad (7.176)$$

Equations (7.19) and (7.30) enable us to write:

$$G^{e(-)}_e = \frac{1}{2}\left|\frac{A - jF\,e^{+j\phi}}{D}\right|^2 \qquad (7.177)$$

Finally, taking (7.49), (7.54) and (7.52) into account we have:

$$G^{e(-)}_e = \frac{1}{2}\frac{\left|\left[r_0^2 + 4r_0\frac{\omega'_c}{\omega_p}\sin\phi + 4\left(\frac{\omega'_c}{\omega_p}\right)^2 - x_e x_{(-)}\right] - j\left[r_0(x_e + x_{(-)}) + 4r_0\frac{\omega'_c}{\omega_p}\cos\phi\right]\right|^2}{\left|(r_0 + jx_e)(r_0 - jx_{(-)}) - 4\left[\frac{\omega'_c}{\omega_p}\right]^2\right|^2} \qquad (7.178)$$

ϕ will be chosen so as to maximize $G^{e(-)}_e$. It can easily be shown that we should have:

$$\cotan\phi = +\frac{r_0(x_e + x_{(-)})}{r_0^2 + 4(\omega'_c/\omega_p)^2 - x_e x_{(-)}} \qquad (7.179)$$

We will assume that the circuit traversed by currents at frequencies ω_e and $\omega_{(-)}$ is the doubly tuned circuit studied in the previous chapter. We then have:

$$x_e + x_{(-)} = 0 \qquad (7.180)$$

$$\cotan\phi = 0 \qquad (7.181)$$

The maximum gain corresponds to $\phi = \pi/2$ and is equal to:

$$G^{e(-)}_e = \frac{1}{2}\frac{\left[\left(r_0 + 2\frac{\omega'_c}{\omega_p}\right)^2 - x_e x_{(-)}\right]^2}{\left[(r_0 + jx_e)(r_0 - jx_{(-)}) - 4\left(\frac{\omega'_c}{\omega_p}\right)^2\right]^2} \qquad (7.182)$$

(2) *Gain at the nominal frequency and bandwidth*

We will, as before, put:

$$\alpha = \left(\frac{2}{r_0} \cdot \frac{\omega_c'}{\omega_p}\right)^2 \qquad (7.183)$$

We have at once:

$$G_e^{e(-)}(\omega_{e_0}, \omega_{(-)_0}) = \frac{1}{2}\left(\frac{1 + \sqrt{\alpha}}{1 - \sqrt{\alpha}}\right)^2 \qquad (7.184)$$

If Q is the quality factor of the circuit, the expressions for the bandwidth and for the relative dissymmetry are:

$$B_e^{e(-)} = \frac{2}{\sqrt{4\sqrt{\alpha} - (1 - \sqrt{\alpha})^2}} \cdot \frac{1 - \alpha}{4Q} \simeq \frac{1 - \alpha}{4Q} \qquad (7.185)$$

$$D_e^{e(-)} = \frac{-4\Delta}{\omega_p B_e^{e(-)}} + \sqrt{1 + \left(\frac{4\Delta}{\omega_p B_e^{e(-)}}\right)^2} \qquad (7.186)$$

The voltage gain bandwidth product is equal to:

$$B_e^{e(-)}\sqrt{G_e^{e(-)}(\omega_{e_0}, \omega_{(-)_0})} \simeq \frac{(1 + \sqrt{\alpha})^2}{4\sqrt{2}\,Q} \simeq \frac{1}{\sqrt{2}\,Q} \qquad (7.187)$$

(3) *Noise gain*

It is clear that the noise gain is the same as that of the amplifier X_e^e (or $X_{(-)}^e$). We then have:

$$G_e^{Ne(-)} = \frac{(1 + \alpha)^2 + 4\alpha}{(1 - \alpha)^2} \qquad (7.188)$$

(4) *Effective noise temperature*

This is the same as that of X_e^e (or $X_{(-)}^e$). We then have:

$$T_e^{e(-)} = \frac{\omega_e}{\omega_c'} T_v \frac{4\sqrt{\alpha}(1 + \alpha)}{(1 + \alpha)^2 + 4\alpha} \qquad (7.189)$$

(5) *Instability factor of the gain*

Differentiation of (7.184) gives at once:

$$S_e^{e(-)} = \frac{2\sqrt{\alpha}}{1 - \alpha} \qquad (7.190)$$

3.3. Amplifier $X^{e(-)}_{e(-)}$

(1) General expression for the transducer gain

The available power of the input signal is $|V_{ge}|^2/R_0$. The output power arises from the loss-free detection of the two coherent signals $v^{ge}_{Le} + v^{g(-)}_{Le}$ and $v^{ge}_{L(-)} + v^{g(-)}_{L(-)}$. The voltage of the local oscillator is of the form:

$$v_{OL} = \frac{1}{\sqrt{2}} [e^{j\psi} e^{j(\omega_p/2)t} + e^{-j\psi} e^{-j(\omega_p/2)t}] \quad (7.191)$$

A maximum value for the output power will be obtained by an appropriate choice of the two initial phases of ϕ and ψ. As these calculations are similar to those performed in the previous paragraphs, we will not discuss them in detail. It can be shown that the output power is a maximum for $\phi = \pi/2$ and $\psi = -\pi/4$, and is then equal to:

$$\frac{|V_{ge}|^2}{R_0} \cdot \frac{|A + F|^2}{|D|^2} \quad (7.192)$$

The transducer gain may then be written:

$$G^{e(-)}_{e(-)} = \left|\frac{A + F}{D}\right|^2 \quad (7.193)$$

Finally, taking account of (7.49), (7.52) and (7.54), we have:

$$G^{e(-)}_{e(-)} = \left|\frac{(r_0 - jx_e)(r_0 - jx_{(-)}) + 4\left(\frac{\omega'_c}{\omega_p}\right)^2 + 4r_0 \frac{\omega'_c}{\omega_p}}{(r_0 + jx_e)(r_0 - jx_{(-)}) - 4\left(\frac{\omega'_c}{\omega_p}\right)^2}\right|^2 \quad (7.194)$$

We will assume in all that follows that the circuit traversed by the currents at frequencies ω_e and $\omega_{(-)}$ is the doubly tuned circuit described earlier.

(2) Gain at the nominal frequency and bandwidth

Let us, as before, make:

$$\alpha = \left(\frac{2}{r_0} \cdot \frac{\omega'_c}{\omega_p}\right)^2 \quad (7.195)$$

We then have at once:

$$G^{e(-)}_{e(-)}(\omega_{e_0}\omega_{(-)_0}) = \left(\frac{1 + \sqrt{\alpha}}{1 - \sqrt{\alpha}}\right)^2 \quad (7.196)$$

The expressions for the bandwidth and for the relative dissymmetry are:

$$B_{e(-)}^{e(-)} = \frac{1}{\sqrt[4]{\alpha}} \cdot \frac{1-\alpha}{4Q} \cong \frac{1-\alpha}{4Q} \qquad (7.197)$$

$$D_{e(-)}^{e(-)} = \frac{-4\Delta}{\omega_p B_{e(-)}^{e(-)}} + \sqrt{1 + \left(\frac{4\Delta}{\omega_p B_{e(-)}^{e(-)}}\right)^2} \qquad (7.198)$$

The voltage gain bandwidth product is equal to:

$$B_{e(-)}^{e(-)} \sqrt{G_{e(-)}^{e(-)}(\omega_{e_0}\omega_{(-)_0})} = \frac{(1+\sqrt{\alpha})^2}{4Q} \cong \frac{1}{Q} \qquad (7.199)$$

(3) *Noise gain*

It is clear that the noise gain is the same as that of the amplifier $X_{e(-)}^e$. We then have:

$$G_{e(-)}^{Ne(-)} = \left(\frac{1+\sqrt{\alpha}}{1-\sqrt{\alpha}}\right)^2 \qquad (7.200)$$

(4) *Effective noise temperature*

This is the same as that of $X_{e(-)}^e$. We then have:

$$T_{e(-)}^{e(-)} = \frac{\omega_e}{\omega_c'} T_v \frac{4\sqrt{\alpha}}{(1+\sqrt{\alpha})^2} \qquad (7.201)$$

(5) *Instability factor of the gain*

Differentiation of (7.196) gives:

$$S_{e(-)}^{e(-)} = \frac{2\sqrt{\alpha}}{1-\alpha} \qquad (7.202)$$

3.4. Conclusions

It should first be noted that we have:

$$\left.\begin{array}{l} G_e^{e(-)} = G_{e(-)}^e \\ G_{e(-)}^{e(-)} = 2G_{e(-)}^e \\ G_e^{Ne(-)} = G_e^{Ne} \\ G_{e(-)}^{Ne(-)} = G_{e(-)}^{Ne} \\ T_e^{e(-)} = T_e^e \\ T_{e(-)}^{e(-)} = T_{e(-)}^e \end{array}\right\} \qquad (7.203)$$

DEGENERATE AMPLIFIERS

For the numerical data it will suffice if we refer to Tables 7.1 and 7.2 and to Figs. 7.7 and 7.8. It is clear that amplifier $X_{e\{-\}}^{e\{-\}}$ is preferable to $X_e^{e(-)}$.

4. Double band reception, random signals (radiometry)

4.1. The two methods of operation

The input signal consists of v_{ge} and $v_{g(-)}$. There is no phase correlation between v_{ge} and $v_{g(-)}$. This can only happen if the two signals contain no information and consist therefore of noise. The degenerate amplifier supplies four output signals: the two signals v_{Le}^{ge} and $v_{L(-)}^{ge}$ on the one hand, and the two signals $v_{Le}^{g(-)}$ and $v_{L(-)}^{g(-)}$ on the other hand. The initial phase of $v_{L(-)}^{ge}$ with relation to v_{Le}^{ge} depends on the initial phase of the pumping frequency. The same applies to the initial phase of $v_{Le}^{g(-)}$ and $v_{L(-)}^{g(-)}$. Conversely, there is no phase correlation between v_{Le}^{ge} and $v_{L(-)}^{ge}$ on the one hand, and $v_{Le}^{g(-)}$ and $v_{L(-)}^{g(-)}$ on the other hand. There are two possible cases.

Case 1: The IF circuit is designed so as to select only one of the bands produced by the mixer. For example, it selects the IF transposes of v_{Le}^{ge} and $v_{Le}^{g(-)}$. As these two signals are not correlated, the time mean of the resulting power of the signal is equal to the sum of the individual powers.

Case 2: The frequency of the local oscillator is chosen equal to half the pumping frequency and is correlated in phase with it. There is thus phase correlation between the IF transposes of v_{Le}^{ge} and $v_{L(-)}^{ge}$. If a correct choice of the initial phase of the local oscillator with respect to the pump is made, we can then add the amplitudes of these signals. The same applies to the IF transposes of $v_{Le}^{g(-)}$ and $v_{L(-)}^{g(-)}$. As these two resulting signals are not correlated, the time mean of the power of the resulting total signal is the sum of the powers of these two signals.

In order to represent the two types of amplifiers described above, the abbreviated presentation adopted earlier must be made more precise. In order to remind ourselves of the absence of phase correlation between the two bands of the input signal, we will add an index a (random). The above two cases will then be represented by the following symbols:

Case 1: $X_e^{e(-)a}$

Case 2: $X_{e\{-\}}^{e\{-\}a}$

In the following paragraphs we must know the time mean of

the available power of the input signal in order to calculate the transducer gains. This is of course equal to the sum of the available powers in each band. It will be assumed that the expressions for the two input signals are:

$$v_{ge} = V_{ge}\, e^{j\omega_e t} + V_{ge}^*\, e^{-j\omega_e t} \qquad (7.204)$$

$$v_{g(-)} = V_{g(-)}\, e^{j\omega_{(-)} t} + V_{g(-)}^*\, e^{-j\omega_{(-)} t} \qquad (7.205)$$

with

$$|V_{ge}| = |V_{g(-)}| \qquad (7.206)$$

The available power of the input signal is then equal to $|V_{ge}|^2/R_0$.

4.2 Amplifier $X_e^{e'(-)a}$

(1) *General expression for the transducer gain*

The available input power is $|V_{ge}|^2/R_0$. The output power is equal to

$$\frac{2|V_{Le}^{ge}|^2}{R_0} + \frac{2|V_{Le}^{g(-)}|^2}{R_0}$$

The transducer gain is then:

$$G_e^{e(-)a} = 2\, \frac{|V_{Le}^{ge}|^2 + |V_{Le}^{g(-)}|^2}{|V_{ge}|^2} \qquad (7.207)$$

(7.19) and (7.30) enable us to write:

$$G_e^{e(-)a} = \frac{1}{2}\, \frac{|A|^2 + |F|^2}{|D|^2} \qquad (7.208)$$

Finally, taking (7.49), (7.52) and (7.54) into account, we have:

$$G_e^{e(-)a} = \frac{\left|(r_0 - jx_e)(r_0 - jx_{(-)}) + 4\left(\dfrac{\omega_c'}{\omega_p}\right)^2\right|^2 + \left|4 r_0 \dfrac{\omega_c'}{\omega_p}\right|^2}{2\left|(r_0 + jx_e)(r_0 - jx_{(-)}) - 4\left(\dfrac{\omega_c'}{\omega_p}\right)^2\right|^2} \qquad (7.209)$$

(2) *Gain at the nominal frequency and bandwidth*

We will, as before, make:

$$\alpha = \left(\frac{2}{r_0} \cdot \frac{\omega_c'}{\omega_p}\right)^2 \qquad (7.210)$$

We then at once have:

$$G_e^{e(-)a}(\omega_{e_0}, \omega_{(-)_0}) = \frac{(1+\alpha)^2 + 4\alpha}{2(1-\alpha)^2} \qquad (7.211)$$

DEGENERATE AMPLIFIERS

Similarly, we can calculate the bandwidth and the relative dissymmetry for the doubly tuned circuit used in the previous chapters.

$$B_e^{e(-)a} = \frac{1-\alpha}{4Q}\sqrt{\frac{8}{1+\alpha}\cdot\frac{8\alpha+(1-\alpha)^2}{8\alpha-(1-\alpha)^2}} \simeq \frac{1-\alpha}{4Q} \quad (7.212)$$

$$D_e^{e(-)a} = \frac{-4\Delta}{\omega_p B_e^{e(-)a}} + \sqrt{1+\left(\frac{4\Delta}{\omega_p B_e^{e(-)a}}\right)^2} \quad (7.213)$$

The voltage gain bandwidth product is equal to:

$$B_e^{e(-)a}\sqrt{G_e^{e(-)a}(\omega_{e_0},\omega_{(-)_0})} = \frac{1}{2Q}\sqrt{\frac{(1+\alpha)^2+4\alpha}{8}} \simeq \frac{1}{2Q} \quad (7.214)$$

(3) *Noise gain*

It is clear that the noise gain is the same as for the amplifier X_e^e. We then have:

$$G_e^{Ne(-)a} = \frac{(1+\alpha)^2+4\alpha}{(1-\alpha)^2} \quad (7.215)$$

(4) *Effective noise temperature*

This is the same as that of X_e^e. We then have:

$$T_e^{e(-)a} = \frac{\omega_e}{\omega_c'}T_v\frac{4\sqrt{\alpha}(1+\alpha)}{(1+\alpha)^2+4\alpha} \quad (7.216)$$

(5) *Instability factor of the gain*

Differentiation of (7.211) gives:

$$S_e^{e(-)a} = \frac{8\alpha(1+\alpha)}{(1-\alpha)[(1+\alpha)^2+4\alpha]} \quad (7.217)$$

4.3. Amplifier $X_{e(-)}^{e(-)a}$

(1) *General expression for the transducer gain*

If a correct choice is made of the initial phase of the voltage of the local oscillator ($\phi = \pi/2$), we can add the amplitudes of the two IF transposed signals of v_{Le}^{ge} and $v_{L(-)}^{ge}$. The condition $\phi = \pi/2$ also ensures the addition of the amplitudes of the IF transposed signals of $v_{Le}^{g(-)}$ and $v_{L(-)}^{g(-)}$. It should be remembered that the sums of the two signals are added vectorially. Thanks to relations (7.19), (7.20), (7.29) and (7.30) it can be shown that the transducer gain is equal to:

$$G_{e(-)}^{e(-)a} = \frac{1}{2}\left|\frac{A+B}{D}\right|^2 \quad (7.218)$$

Taking account of (7.49), (7.50) and (7.52), we have:

$$G_{e(-)}^{e(-)a} = \frac{1}{2} \left| \frac{r_0^2 + 4r_0 \frac{\omega_c'}{\omega_p} + 4\left(\frac{\omega_c'}{\omega_p}\right)^2 - x_e x_{(-)} - jr_0(x_e + x_{(-)})}{(r_0 + jx_e)(r_0 - jx_{(-)}) - 4\left(\frac{\omega_c'}{\omega_p}\right)^2} \right|^2 \quad (7.219)$$

(2) *Gain at the nominal frequency and bandwidth*

Putting as before:

$$\alpha = \left(\frac{2}{r_0} \cdot \frac{\omega_c'}{\omega_p}\right)^2 \quad (7.220)$$

We then at once have:

$$G_{e(-)}^{e(-)a}(\omega_{e_0}, \omega_{(-)_0}) = \frac{1}{2}\left(\frac{1 + \sqrt{\alpha}}{1 - \sqrt{\alpha}}\right)^2 \quad (7.221)$$

The same method of calculation can be used for the bandwidth and for the relative dissymmetry corresponding to the doubly tuned circuit used in the previous chapters.

$$B_{e(-)}^{e(-)a} = \frac{1 - \alpha}{4Q} \sqrt{\frac{4}{4\sqrt{\alpha} - (1 - \sqrt{\alpha})^2}} \simeq \frac{1 - \alpha}{4Q} \quad (7.222)$$

$$D_{e(-)}^{e(-)a} = \frac{-4\Delta}{\omega_p B_{e(-)}^{e(-)a}} + \sqrt{1 + \left(\frac{4\Delta}{\omega_p B_{e(-)}^{e(-)a}}\right)^2} \quad (7.223)$$

The voltage gain bandwidth product is equal to:

$$B_{e(-)}^{e(-)a} \sqrt{G_{e(-)}^{e(-)a}(\omega_{e_0}, \omega_{(-)_0})} = \frac{(1 + \sqrt{\alpha})^2}{2\sqrt{2}\,Q} \simeq \frac{1}{Q\sqrt{2}} \quad (7.224)$$

(3) *Noise gain*

It is clear that the noise gain is the same as for the amplifier $X_{e(-)}^e$. We then have:

$$G_{e(-)}^{Ne(-)a} = \left(\frac{1 + \sqrt{\alpha}}{1 - \sqrt{\alpha}}\right)^2 \quad (7.225)$$

(4) *Effective noise temperature*

This is the same as that of $X_{e(-)}^e$. We then have:

$$T_{e(-)}^{e(-)a} = \frac{\omega_e}{\omega_c'} T_v \frac{4\sqrt{\alpha}}{(1 + \sqrt{\alpha})^2} \quad (7.226)$$

(5) *Instability factor of the gain*

Differentiation of (7.221) gives:

$$S_{e(-)}^{e(-)a} = \frac{2\sqrt{\alpha}}{1-\alpha} \qquad (7.227)$$

4.4. Conclusions

Note first of all that we have:

$$\left.\begin{array}{l} G_e^{e(-)a} = \tfrac{1}{2} G_e^{Ne} \\ G_{e(-)}^{e(-)a} = G_{e(-)}^{e} \\ G_e^{Ne(-)a} = G_e^{Ne} \\ G_{e(-)}^{Ne(-)a} = G_{e(-)}^{Ne} \\ T_e^{e(-)a} = T_e^{e} \\ T_{e(-)}^{e(-)a} = T_{e-)}^{e} \end{array}\right\} \qquad (7.228)$$

For the numerical data it will suffice to refer back to Tables 7.1 and 7.2 and to Figs. 7.7 and 7.8. It is clear that the amplifier $X_{e(-)}^{e(-)a}$ is preferable to $X_e^{e(-)a}$.

References

[1] Angot, A., 'Complements de mathématiques à l'usage des ingénieurs de l'electrotechnique et des télécommunications', *Edition de la Revue d'Optique*, p. 735 (4th edn.), 1957.
[2] Marton, L., *Methods of Experimental Physics*, Academic Press.
[3] 'A.C. capacitance, dielectric constant, and loss characteristics of electrical insulating materials', *A.S.T.M. D150–59T*.
[4] 'Electrical resistance of insulating materials', *A.S.T.M. D257–61*.
[5] 'Cleaning plastic specimens for insulation resistance testing', *A.S.T.M. D1371–59*.
[6] Brookshier, W. K., 'Electrometer circuit design for extended band widths', *Nuclear Instruments and Methods*, vol. 25, pp. 317–327, 1964.
[7] Fowler, E. P., 'A New Reactor Log. Power and Periodmeter, Atomic Energy Establishment, Winfrith, Dorchester, Dorset.

8

PUMPING OF PARAMETRIC DIODES FOR MICROWAVE FREQUENCIES

by J. C. LIENARD

1. General

1.1. Description of the pumped varactor

Let us consider a semiconductor junction. We know that the capacitance of the junction depends on the voltage applied to it. In general, the differential capacitance is expressed by the relation:

$$C = \frac{C_{0v}}{\sqrt[n]{1 + V/\phi}} \qquad (8.1)$$

where C_{0v} is the differential capacitance with zero voltage at the terminals of the junction; V is a voltage applied to the terminals of the junction (positive in the cut-off direction); and ϕ is a characteristic parameter which depends on the nature of the semiconductors used. The degree, n, of the root is determined by the method of construction of the junction; n lies between 2 and 3. In order to simplify the study of parametric systems, it is often practical to use the differential elastance (inverse of the differential capacitance) the expression for which is:

$$S = S_{0v}\sqrt[n]{1 + V/\phi} \qquad (8.2)$$

In addition to the non-linear capacitance, the varactor contains the following components:

(1) A parasitic resistance, R_s, in series with C, due to the finite conductivity of the semiconductors used.
(2) A cartridge capacitance C_c in parallel with R_s and C.
(3) An inductance L_c in series with the combination of the other elements, due to the connecting wires.

PUMPING OF DIODES FOR MICROWAVE FREQUENCIES 195

The equivalent circuit of the varactor may be represented by Fig. 8.1.

The only influence of the parasitic elements, L_c and C_c, on the performance of a parametric amplifier is of a secondary nature. They can therefore be considered as part of the exterior circuitry. Conversely, R_s affects all the fundamental limitations of the performance. This will be taken into account in the following discussion.

In a parametric amplifier, the varactor has applied to it a voltage, v_p, of large amplitude (pumping voltage) of fundamental frequency, ω_p. The voltages of frequencies other than ω_p and their harmonics

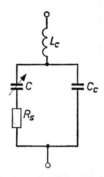

Figure 8.1. Equivalent circuit of a varactor

are assumed to have a very low amplitude compared with that of the pumping voltage. We can then, in equivalent circuits applicable to the case of small signals, replace the non-linear capacitance (elastance) by a linear capacitance (elastance) dependent on time:

$$C(t) = \sum_{k=-\infty}^{+\infty} C_k \, e^{jk\omega_p t} \qquad (8.3)$$

$$S(t) = \sum_{k=-\infty}^{+\infty} S_k \, e^{jk\omega_p t} \qquad (8.4)$$

The object of this study is the calculation of the coefficients C_k and S_k, of the power P required for pumping, and the polarizing voltage, V_0, that has to be applied to the varactor.

1.2. Maximization of $|S_1|$

Generally, the performance of parametric devices improves as the value of $|S_1|$ increases. It is obvious that this requires a large amplitude for the pumping voltage, but also for a given amplitude there is an optimum waveform for the pumping voltage. We shall assume that the pumping voltage, v_p, varies periodically from $V_{\min}$ to $V_{\max}$; the time origin is chosen so that $v_p(0)$ is a maximum; and the function $v_p(t)$ in the frequency interval $(0, 2\pi/\omega_p)$ has a vertical axis of symmetry, $t = \pi/\omega_p$.

The voltage, $v_p(t)$, has then the shape shown in Fig. 8.2.

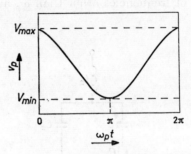

Figure 8.2. General shape of the pumping voltage

Since there is a bi-univocal and monotonic relation (8.2) between S and v_p, the curve of S as a function of $\omega_p t$ has the same symmetry as the curve for v_p. From this symmetry with respect to a vertical axis $\omega_p t = \pi$, we derive:

$$2|S_1| = S_1 + S_1^* = S_1 + S_{-1} \tag{8.5}$$

Let us now determine the function $S(\omega_p t)$ that, for given $S_{\max}$ and $S_{\min}$, leads to a maximum value of S_1. We may write:

$$2|S_1| = S_1 + S_{-1} = \frac{1}{\pi} \int_0^{2\pi} S(\omega_p t) \cos \omega_p t \, d(\omega_p t) \tag{8.6}$$

$$|S_1| = \frac{1}{2\pi}$$
$$\times \left[\int_0^{\pi/2} S(x) \cos x \, dx + \int_{\pi/2}^{3\pi/2} S(x) \cos x \, dx + \int_{3\pi/2}^{2\pi} S(x) \cos x \, dx \right] \tag{8.7}$$

$$|S_1| = \frac{1}{2\pi} \left| \int_{-\pi/2}^{+\pi/2} S(x) \cos x \, dx + \int_{\pi/2}^{3\pi/2} S(x) \cos x \, dx \right| \tag{8.8}$$

PUMPING OF DIODES FOR MICROWAVE FREQUENCIES

Since S is necessarily positive, we have:

$$0 \leqslant \int_{-\pi/2}^{\pi/2} S(x) \cos x \, dx \leqslant \int_{-\pi/2}^{\pi/2} S_{\max} \cos x \, dx = 2S_{\max} \quad (8.9)$$

$$0 \geqslant \int_{\pi/2}^{3\pi/2} S(x) \cos x \, dx \geqslant \int_{\pi/2}^{3\pi/2} S_{\min} \cos x \, dx = -2S_{\min} \quad (8.10)$$

On the other hand, with our choice of the time origin and since S is a decreasing function of x in the interval $(0, \pi)$, and increasing in the interval $(\pi, 2\pi)$, we have:

$$\int_{-\pi/2}^{\pi/2} S(x) \cos x \, dx \geqslant \left| \int_{\pi/2}^{3\pi/2} S(x) \cos x \, dx \right| \quad (8.11)$$

and we may write:

$$|S_1| = \frac{1}{2\pi} \left[\int_{-\pi/2}^{\pi/2} S(x) \cos x \, dx - \left| \int_{\pi/2}^{3\pi/2} S(x) \cos x \, dx \right| \right] \quad (8.12)$$

In order to give S_1 a maximum value, we must have

$$S(x) = S_{\max} \quad \text{for } 0 \leqslant \omega_p t \leqslant \pi/2 \text{ and}$$
$$\text{for } 3\pi/2 \leqslant \omega_p t \leqslant 2\pi$$
$$S(x) = S_{\min} \quad \text{for } \pi/2 \leqslant \omega_p t \leqslant 3\pi/2$$

The elastance, and consequently the voltage v_p, should have a square waveform as shown in Fig. 8.3.

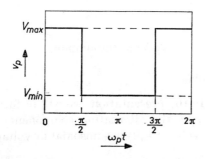

Figure 8.3. Optimum waveform of the pumping voltage

1.3. Discussion: waveform of the pumping voltage

It was seen in the previous section that the optimum pumping voltage is a square wave. However, parametric amplifiers operate chiefly in the microwave field in which it is almost impossible to obtain a waveform of this type. On the other hand, it is easy to connect to the terminals of the varactor (consisting of C and R_s in series) a circuit with zero or infinite impedance for the harmonic of ω_p.

If we decide to make the impedance zero, the voltage at the terminals of the varactor is zero for the harmonic frequencies of ω_p, but the current can flow through C and R_s at these same frequencies. The voltage across C will be the sinusoidal voltage present at the terminals of the varactor, decreased by the non-sinusoidal voltage drop in R_s. The impedance of R_s for the harmonics of appreciable amplitude is usually very low compared to that of C; thus, the assumption that the voltage at the terminals of C is sinusoidal, will be approximately true; this is called voltage pumping. If we decide to make the impedance infinite, the current in the varactor is sinusoidal; this is called current pumping.

We will deal only with voltage and current pumping, as they alone ensure satisfactory operation in the microwave field. The range of the pumping voltage at the terminals of the varactor is limited by the Zener voltage, V_B, in the cut-off direction and by the voltage ϕ in the conducting direction. The pumping is said to be full if v_p varies from $-\phi$ to V_B; in this case, the d.c. polarization, V_0, should be made equal to the mean value of $v_p(t)$. The pumping is called partial if the field of variation of v_p is narrower; in this case the d.c. polarization V_0 should be chosen in a range such that at every instant we have:

$$-\phi \leqslant v_p(t) \leqslant V_B \tag{8.13}$$

2. Voltage pumping

2.1. General formulae

In order to simplify the presentation, we will in future only use the symbol $v_p(t)$ to define the alternating component of the pumping voltage. Since the voltage $v_p(t)$ is sinusoidal in voltage pumping, we may write:

$$v_p = V_p(e^{j\omega_p t} + e^{-j\omega_p t}) = 2V_p \cos \omega_p t \tag{8.14}$$

PUMPING OF DIODES FOR MICROWAVE FREQUENCIES 199

where the time origin is chosen in such a way that the initial phase is zero.

The voltage applied to the varactor is:

$$v_p + V_0 = V_0 + 2V_p \cos \omega_p t \qquad (8.15)$$

The expressions for the elastance and the differential capacitance may be written:

$$C(t) = \frac{C_{0v}}{\sqrt[n]{1 + (V_0 + 2V_p \cos \omega_p t)/\phi}} \qquad (8.16)$$

$$S(t) = S_{0v} \sqrt[n]{1 + \frac{V_0 + 2V_p \cos \omega_p t}{\phi}} \qquad (8.17)$$

This gives the coefficients C_k and S_k by the relations:

$$C_k = \frac{1}{T} \int_0^T C(t)\, e^{-jk\omega_p t}\, dt \qquad (8.18)$$

$$S_k = \frac{1}{T} \int_0^T S(t)\, e^{-jk\omega_p t}\, dt \qquad (8.19)$$

In the time interval from 0 to T, v_p is symmetrical with respect to the ordinate $t = T/2$. The same applies to $C(t)$ and $S(t)$, and C_k and S_k are therefore real and we may write:

$$2|C_k| = |C_k + C_{-k}| = \left|\frac{2}{T} \int_0^T C(t) \cos k\omega_p t\, dt\right| \qquad (8.20)$$

$$2|S_k| = |S_k + S_{-k}| = \left|\frac{2}{T} \int_0^T S(t) \cos k\omega_p t\, dt\right| \qquad (8.21)$$

(8.16), (8.17), (8.20) and (8.21) give us:

$$|C_k| = \left|\frac{C_{0v}}{T} \int_0^T \frac{\cos k\omega_p t}{\sqrt[n]{1 + (V_0/\phi) + (2V_p/\phi) \cos \omega_p t}}\, dt\right|$$

$$= \left|\frac{C_{0v}}{2\pi} \int_0^{2\pi} \frac{\cos kx}{\sqrt[n]{1 + (V_0/\phi) + (2V_p/\phi) \cos x}}\, dx\right| \qquad (8.22)$$

$$|S_k| = \left| \frac{S_{0v}}{T} \int_0^T \cos k\omega_p t \sqrt[n]{1 + \frac{V_0}{\phi} + \frac{2V_p}{\phi} \cos \omega_p t}\, dt \right|$$

$$= \left| \frac{S_{0v}}{2\pi} \int_0^{2\pi} \cos kx \sqrt[n]{1 + \frac{V_0}{\phi} + \frac{2V_p}{\phi} \cos x}\, dx \right| \quad (8.23)$$

The current in the varactor is given by the relation:

$$i(t) = \frac{1}{S(t)} \cdot \frac{dv_p(t)}{dt} = \frac{-2V_p\omega_p \sin \omega_p t}{S_{0v}\sqrt[n]{1 + (V_0/\phi) + (2V_p/\phi) \cos \omega_p t}} \quad (8.24)$$

The pumping power may be calculated with the aid of this expression:

$$P = \frac{R_s}{T} \int_0^T [i(t)]^2\, dt$$

$$= \frac{2}{\pi} R_s \frac{V_p^2 \omega_p^2}{S_{0v}^2} \int_0^{2\pi} \frac{\sin^2 x}{\sqrt[n/2]{1 + (V_0/\phi) + (2V_p/\phi) \cos x}}\, dx \quad (8.25)$$

Figure 8.4 shows the sum of the pumping voltage v_p and the direct polarization V_0 as a function of time.

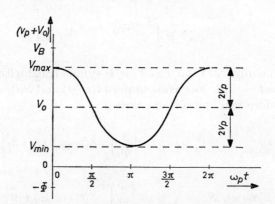

Figure 8.4. Pump voltage for voltage pumping

The maximum peak-to-peak amplitude of $v_p + V_0$ is, by virtue of (8.13), $V_B + \phi$. In the general case, the peak-to-peak amplitude of $v_p + V_0$ is $V_{\max} - V_{\min}$.

Let:

$$\alpha_v = \frac{V_{\max} - V_{\min}}{V_B + \phi} = \frac{4V_p}{V_B + \phi} \quad (8.26)$$

We of course have:
$$0 \leqslant \alpha_v \leqslant 1 \tag{8.27}$$

Once the amplitude, $2V_p$, of the pumping voltage v_p is fixed, coefficient α_v is determined. Any value may be chosen for the d.c. polarization V_0, provided we always have

$$-\phi \leqslant V_{\min} < V_{\max} \leqslant V_B \tag{8.28}$$

The polarization, V_0, will be a minimum if we make:

$$V_{\min} = -\phi \tag{8.29}$$

It is then equal to:

$$V_0 = \frac{V_{\max} + V_{\min}}{2} = \frac{V_{\max} - \phi}{2} = \frac{V_{\max} + \phi}{2} - \phi \tag{8.30}$$

hence:

$$1 + \frac{V_0}{\phi} = \alpha_v \frac{1}{2}\left(1 + \frac{V_B}{\phi}\right) \tag{8.31}$$

In the general case we may make:

$$1 + \frac{V_0}{\phi} = \frac{\alpha_v}{\beta_v} \cdot \frac{1}{2}\left(1 + \frac{V_B}{\phi}\right) \tag{8.32}$$

where β_v is less than unity.

Taking (8.28) into account, we may write:

$$\frac{\alpha_v}{2 - \alpha_v} \leqslant \beta_v \leqslant 1 \tag{8.33}$$

Formulae (8.22), (8.23) and (8.25) become:

$$|C_k| = \frac{C_{0v}}{2\pi \sqrt[n]{(\alpha_v/2\beta_v)(1 + V_B/\phi)}} \left| \int_0^{2\pi} \frac{\cos kx}{\sqrt[n]{1 + \beta_v \cos x}} dx \right| \tag{8.34}$$

$$|S_k| = \frac{S_{0v}}{2\pi} \sqrt[n]{\frac{\alpha_v}{2\beta_v}\left(1 + \frac{V_B}{\phi}\right)} \left| \int_0^{2\pi} (\sqrt[n]{1 + \beta_v \cos x}) \cos kx \, dx \right| \tag{8.35}$$

$$P = \frac{2^{2/n} R_s \omega_p^2 \alpha_v^2 (V_B + \phi)^2}{8\pi S_{0v}^2 \sqrt[n/2]{(\alpha_v/\beta_v)(1 + V_B/\phi)}} \int_{-\pi}^{+\pi} \frac{\sin^2 x}{\sqrt[n/2]{1 + \beta_v \cos x}} dx \tag{8.36}$$

We will now introduce the following definitions:

Minimum capacitance:

$$C_{\min} = \frac{C_{0v}}{\sqrt[n]{1 + V_B/\phi}} \tag{8.37}$$

Maximum elastance:

$$S_{\max} = S_{0v} \sqrt[n]{1 + \frac{V_B}{\phi}} \qquad (8.38)$$

Maximum static cut-off frequency:

$$\omega_{c\,\max} = \frac{S_{\max}}{R_s} = \frac{1}{C_{\min} R_s} \qquad (8.39)$$

Normalization power:

$$P_{\text{norm}} = \frac{(V_B + \phi)^2}{R_s} \qquad (8.40)$$

Index of the law of variation of the capacitance:

$$\nu = \frac{1}{n} \qquad (8.41)$$

Relations (8.34), (8.35) and (8.36) may be written:

$$\frac{|C_k|}{C_{\min}} = 2\nu \left(\frac{\beta_v}{\alpha_v}\right)^\nu \left| \frac{1}{2\pi} \int_0^{2\pi} \frac{\cos kx}{(1 + \beta_v \cos x)^\nu} \, dx \right| \qquad (8.42)$$

$$\frac{|S_k|}{S_{\max}} = \left(\frac{1}{2}\right)^\nu \left(\frac{\alpha_v}{\beta_v}\right)^\nu \left| \frac{1}{2\pi} \int_0^{2\pi} (1 + \beta_v \cos x)^\nu \cos kx \, dx \right| \qquad (8.43)$$

$$\frac{P}{P_{\text{norm}}(\omega_p/\omega_{c\,\max})^2} = \frac{1}{4^{1-\nu}} \alpha_v^{2(1-\nu)} \beta_v^{2\nu} \frac{1}{2\pi} \int_{-\pi}^{+\pi} \frac{\sin^2 x}{(1 + \beta_v \cos x)^{2\nu}} \, dx \qquad (8.44)$$

The calculation of the integrals which appear in (8.42), (8.43) and (8.44) will be dealt with in the following paragraphs.

2.2. Calculation of the $|C_\kappa|$'s.

We may write:

$$(1 + \beta_v \cos x)^{-\nu} = \sum_{p=0}^{\infty} \frac{\Gamma(1 - \nu)}{p! \, \Gamma(1 - \nu - p)} \beta_v^p \cos^p x \qquad (8.45)$$

where Γ is the gamma function or factorial function or Euler function of the first order.

Let:

$$I_{kp} = \frac{1}{2\pi} \int_0^{2\pi} \cos^p x \cos kx \, dx \qquad (8.46)$$

PUMPING OF DIODES FOR MICROWAVE FREQUENCIES

(8.42) gives us:

$$\frac{|C_k|}{C_{min}} = 2^\nu \left(\frac{\beta_v}{\alpha_v}\right)^\nu \Gamma(1-\nu) \left| \sum_{p=0}^{\infty} \frac{\beta_v^p}{p!\,\Gamma(1-\nu-p)} I_{kp} \right| \quad (8.47)$$

On the other hand, we know the relation:

$$\cos^p x = 2^{-p} \sum_{q=0}^{p} \frac{p!}{q!(p-q)!} \cos(p-2q)x \quad (8.48)$$

We may therefore write:

$$I_{kp} = 2^{-p} p! \sum_{q=0}^{p} \frac{1}{q!(p-q)!} I_{kpq} \quad (8.49)$$

in which we made:

$$I_{kpq} = \frac{1}{2\pi} \int_0^{2\pi} \cos(p-2q)x \cos kx \, dx \quad (8.50)$$

The integral, I_{kpq}, is clearly zero except for $p - 2q = \pm k$. We then have $I_{kp} = 0$ for $p < k$ and for $p = k + 2s + 1$ (s is a positive integer). On the other hand, $I_{kp} \neq 0$ for $p = k + 2s$.

Formula (8.47) may therefore be written:

$$\frac{|C_k|}{C_{min}} = 2^\nu \left(\frac{\beta_v}{\alpha_v}\right)^\nu \Gamma(1-\nu) \left| \sum_{s=0}^{\infty} \frac{\beta_v^{k+2s}}{(k+2s)!\,\Gamma(1-\nu-k-2s)} I_{kp} \right| \quad (8.51)$$

The integral, I_{kpq}, is different from zero for $q = (p+k)/2 = k + 2s$ and for $q = (p-k)/2 = s$. For these two values of q, it is easily shown that I_{kpq} is equal to $\frac{1}{2}$. We then have:

$$I_{kp} = 2^{-(k+2s)} \frac{(k+2s)!}{s!(k+s)!} \quad (8.52)$$

$$\frac{|C_k|}{C_{min}} = 2^\nu \left(\frac{\beta_v}{\alpha_v}\right)^\nu \Gamma(1-\nu) \left| \sum_{s=0}^{\infty} \frac{1}{(k+s)!\,s!\,\Gamma(1-\nu-k-2s)} \left(\frac{\beta_v}{2}\right)^{k+2s} \right| \quad (8.53)$$

We know the relations:

$$\Gamma(\nu + k + 2s) = \frac{(-1)^k \pi}{\sin \pi\nu \, \Gamma(1-\nu-k-2s)} \quad (8.54)$$

$$\Gamma(\nu + k + 2s) = \frac{2^{\nu+k+2s-1}}{\sqrt{\pi}} \Gamma\left(s + \frac{\nu+k}{2}\right) \Gamma\left(s + \frac{\nu+k+1}{2}\right) \quad (8.55)$$

These enable us to write (8.53) in the form:

$$\frac{|C_k|}{C_{\min}} = \frac{4^v \sin \pi v}{2\pi \sqrt{\pi}} \left(\frac{\beta_v}{\alpha_v}\right)^v \Gamma(1 - v)$$

$$\left| \sum_{s=0}^{\infty} \frac{\Gamma[(v + k)/2 + s]\Gamma[(v + k + 1)/2 + s]}{s!(k + s)!} \beta_v^{k+2s} \right| \quad (8.56)$$

The hypergeometric function is defined by:

$$F(\alpha, \beta|\gamma|x) = \frac{\Gamma(\gamma)}{\Gamma(\alpha)\Gamma(\beta)} \sum_{n=0}^{\infty} \frac{\Gamma(\alpha + n)\Gamma(\beta + n)}{n!\Gamma(\gamma + n)} x^n \quad (8.57)$$

We then have:

$$\frac{|C_k|}{C_{\min}} = \frac{4^v \sin \pi v}{2\pi \sqrt{\pi}} \left(\frac{\beta_v}{\alpha_v}\right)^v \beta_v^k \frac{\Gamma(1 - v)\Gamma[(v + k)/2]\Gamma[(v + k + 1)/2]}{k!}$$

$$\times F\left(\frac{v + k}{2}, \frac{v + k + 1}{2} \bigg| 1 + k | \beta_v^2 \right) \quad (8.58)$$

Applying the well-known formulae:

$$\Gamma\left(\frac{v + k}{2}\right)\Gamma\left(\frac{v + k + 1}{2}\right) = \frac{\sqrt{\pi}\,\Gamma(v + k)}{2^{v+k-1}} \quad (8.59)$$

$$\Gamma(v + k)\Gamma(1 - v - k) = \frac{(-1)^k \pi}{\sin \pi v} \quad (8.60)$$

$$\Gamma(1 - 2v) = \frac{2^{-2v}}{\sqrt{\pi}} \Gamma(\tfrac{1}{2} - v)\Gamma(1 - v) \quad (8.61)$$

$$\Gamma(1 - v + k) = \frac{2^{k-v}}{\sqrt{\pi}} \Gamma\left(\frac{1 - v + k}{2}\right)\Gamma\left(1 - \frac{v - k}{2}\right) \quad (8.62)$$

we obtain:

$$\frac{|C_k|}{C_{\min}} = \frac{4^v}{\alpha_v^v} \left| \frac{\Gamma(1 - 2v)}{\Gamma(1 - v + k)\Gamma(1 - v - k)} \right| \beta_v^{v+k}$$

$$\times \frac{\Gamma[(1 + k - v)/2]\Gamma[1 + (k - v)/2]}{k!\,\Gamma(\tfrac{1}{2} - v)} \quad (8.63)$$

$$\times F\left(\frac{k + v}{2}, \frac{1 + k + v}{2} \bigg| 1 + k | \beta_v^2 \right)$$

The properties of the factorial and hypergeometric functions as well as the numerous formulae used in this and the following paragraphs are taken from the references [1], [2], [3] and [4].

There exists, between $|C_{k+2}|$, $|C_{k+1}|$ and $|C_k|$ a recurrence formula, quoted by S. Sensiper and R. D. Weglein [5], which we will discuss hereunder.

This formula may be written:

$$|C_{k+2}| = \frac{2}{\beta_v} \cdot \frac{k+1}{k-\nu+2} |C_{k+1}| - \frac{k+\nu}{k-\nu+2} |C_k| \qquad (8.64)$$

Relation (8.53) enables us to write (8.64) in the form:

$$\frac{|C_{k+2}|}{C_{\min}} = 2^\nu \left(\frac{\beta_v}{\alpha_v}\right)^\nu \Gamma(1-\nu) \frac{1}{k-\nu+2}$$

$$\times \left[(k+1) \left| \sum_{s=0}^{\infty} \frac{1}{(k+1+s)!s!\Gamma(-\nu-k-2s)} \left(\frac{\beta_v}{2}\right)^{k+2s} \right| \right.$$

$$\left. - (k+\nu) \left| \sum_{s=0}^{\infty} \frac{1}{(k+s)!s!\Gamma(1-\nu-k-2s)} \left(\frac{\beta_v}{2}\right)^{k+2s} \right| \right] \qquad (8.65)$$

It can easily be shown, thanks to the properties of the gamma function, that all the terms of a single series in s are of the same sign and that two terms belonging to two different series are of opposite sign. We then have:

$$\frac{|C_{k+2}|}{C_{\min}} = 2^\nu \left(\frac{\beta_v}{\alpha_v}\right)^\nu \Gamma(1-\nu) \frac{1}{k-\nu+2}$$

$$\times \left| \sum_{s=0}^{\infty} \left[\frac{(k+1)(-\nu-k-2s)}{(k+1+s)} + (k+\nu) \right] \right.$$

$$\left. \times \frac{(\beta_v/2)^{k+2s}}{(k+s)!s!\Gamma(1-\nu-k-2s)} \right| \qquad (8.66)$$

$$\frac{|C_{k+2}|}{C_{\min}} = 2^\nu \left(\frac{\beta_v}{\alpha_v}\right)^\nu \Gamma(1-\nu)$$

$$\times \left| \sum_{s=0}^{\infty} \frac{s}{k+1+s} \cdot \frac{1}{(k+s)!s!\Gamma(1-\nu-k-2s)} \left(\frac{\beta_v}{2}\right)^{k+2s} \right| \qquad (8.67)$$

Putting $r = s - 1$, we obtain:

$$\frac{|C_{k+2}|}{C_{\min}} = 2^\nu \left(\frac{\beta_v}{\alpha_v}\right)^\nu \Gamma(1-\nu)$$

$$\times \left| \sum_{r=0}^{\infty} \frac{1}{(k+2+r)!r!\Gamma(1-\nu-k-2-2r)} \left(\frac{\beta_v}{2}\right)^{k+2+2r} \right| \qquad (8.68)$$

This last relation is the same as (8.53) written for the suffix $k+2$ instead of k. Formula (8.64) is thus verified.

2.3. Calculation of the $|S_\kappa|$'s

If we compare equations (8.42) and (8.43), the formula which gives

$|S_k|$ is the same as that which gives $|C_k|$ except that ν has been changed into $-\nu$. We then at once obtain:

$$\frac{|S_k|}{S_{\max}} = \frac{\alpha_v^\nu}{4^\nu} \left| \frac{\Gamma(1+2\nu)}{\Gamma(1+\nu+k)\Gamma(1+\nu-k)} \beta_v^{-\nu+k} \right.$$

$$\times \frac{\Gamma[(1+k+\nu)/2]\Gamma[1+(k+\nu)/2]}{k!\,\Gamma(\tfrac{1}{2}+\nu)}$$

$$\times \left| F\!\left(\frac{k-\nu}{2}, \frac{1+k-\nu}{2} \right| 1 + k|\beta_v^2\right) \right| \quad (8.69)$$

The following recurrence formulae, like those in the previous paragraph, are proved:

$$|S_2| = \frac{2}{\beta_v} \cdot \frac{1}{2+\nu} |S_1| - \frac{\nu}{\nu+2} |S_0| \quad (8.70)$$

$$|S_{k+2}| = \frac{2}{\beta_v} \cdot \frac{k+1}{k+\nu+2} |S_{k+1}| - \frac{k-\nu}{k+\nu+2} |S_k| \quad \text{with } k > 0 \quad (8.71)$$

2.4. Calculation of the pumping power

We may write:

$$(1 + \beta_v \cos x)^{-2\nu} = \sum_{p=0}^{\infty} \frac{\Gamma(1-2\nu)}{p!\,\Gamma(1-2\nu-p)} \beta_v^p \cos^p x \quad (8.72)$$

(8.44) becomes:

$$\frac{P}{P_{\text{norm}}(\omega_p/\omega_{c\,\max})^2} = \frac{1}{4^{1-\nu}} \alpha_v^{2(1-\nu)} \beta_v^{2\nu}$$

$$\times \sum_{p=0}^{\infty} \beta_v^p \frac{\Gamma(1-2\nu)}{p!\,\Gamma(1-2\nu-p)} \cdot \frac{1}{2\pi} \int_0^{2\pi} \cos^p x \sin^2 x \, \mathrm{d}x \quad (8.73)$$

We know the integral:

$$\int_0^{\pi/2} \cos^p x \sin^2 x \, \mathrm{d}x = \tfrac{1}{2} B\!\left(\frac{3}{4}, \frac{p+1}{2}\right) \quad (8.74)$$

in which $B(m, n)$ is the beta function of Euler function of the second class. This gives us:

$$\int_{3\pi/2}^{2\pi} \cos^p x \sin^2 x \, \mathrm{d}x = \tfrac{1}{2} B\!\left(\frac{3}{2}, \frac{p+1}{2}\right) \quad (8.75)$$

$$\int_{\pi/2}^{\pi} \cos^p x \sin^2 x \, \mathrm{d}x$$

$$= \int_{\pi}^{3\pi/2} \cos^p x \sin^2 x \, \mathrm{d}x = (-1)^p \tfrac{1}{2} B\!\left(\frac{3}{2}, \frac{p+1}{2}\right) \quad (8.76)$$

$$\int_0^{2\pi} \cos^p x \sin^2 x \, dx = 0 \quad \text{for } p = 2s + 1 \quad (8.77)$$

$$\int_0^{2\pi} \cos^p x \sin^2 x \, dx = 2B\left(\frac{3}{2}, \frac{p+1}{2}\right) \quad \text{for } p = 2s \quad (8.78)$$

where s represents a positive integer.

We know the relations:

$$B\left(\frac{3}{2}, \frac{p+1}{2}\right) = \frac{\Gamma\left(\frac{3}{2}\right)\Gamma\left(\frac{p+1}{2}\right)}{\Gamma\left(\frac{3}{2} + \frac{p+1}{2}\right)} = \frac{0{\cdot}5\sqrt{\pi}\,\Gamma\left(0{\cdot}5 + \frac{p}{2}\right)}{\Gamma\left(2 + \frac{p}{2}\right)} \quad (8.79)$$

$$\frac{\Gamma\left(0{\cdot}5 + \frac{p}{2}\right)}{\sqrt{\pi}\, p!} = \frac{1}{\left(\frac{p}{2}\right)! \, 4^{p/2}} \quad (8.80)$$

(8.73) may then be written:

$$\frac{P}{P_{\text{norm}}(\omega_p/\omega_{c\,\text{max}})^2} = \alpha_v^{2(1-\nu)} \beta_v^{2\nu} \frac{1}{2^{3-2\nu}}$$

$$\times \sum_{s=0}^{\infty} \frac{\Gamma(1 - 2\nu)}{s!(s+1)!\,\Gamma(1 - 2\nu - 2s)} \left(\frac{\beta_v}{2}\right)^{2s} \quad (8.81)$$

As in the previous paragraphs, this expression can be transformed thanks to the properties of the factorial and hyper-geometric functions. We obtain:

$$\frac{P}{P_{\text{norm}}(\omega_p/\omega_{c\,\text{max}})^2} = \frac{\alpha_v^{2(1-\nu)}}{(1-\nu)2^{4(1-\nu)}} \cdot \frac{\Gamma(3 - 4\nu)}{[\Gamma(2 - 2\nu)]^2} \beta_v^{2\nu}$$

$$\times \frac{\Gamma(2 - \nu)\Gamma\left(\dfrac{3 - 2\nu}{2}\right)}{\Gamma\left(\dfrac{3 - 4\nu}{2}\right)} F(\nu, \nu + \tfrac{1}{2}|2|\beta_v^2) \quad (8.82)$$

2.5. Special cases

In future we will set out the quantities for which we have just found the expressions in the form $f(\alpha_v, \beta_v)$. Reverting to the general formulae:

$$\frac{|C_k(\alpha_v, \beta_v)|}{C_{\min}} = \frac{4^\nu}{\alpha_v^\nu} \left| \frac{\Gamma(1 - 2\nu)}{\Gamma(1 - \nu + k)\Gamma(1 - \nu - k)} \right| \beta_v^{\nu+k}$$

$$\times \frac{\Gamma[(1 + k - \nu)/2]\Gamma[1 + (k - \nu)/2]}{k!\,\Gamma(\tfrac{1}{2} - \nu)}$$

$$\times F\left(\frac{k + \nu}{2}, \frac{1 + k + \nu}{2} \middle| 1 + k \middle| \beta_v^2\right) \quad (8.83)$$

$$\frac{|S_k(\alpha_v, \beta_v)|}{S_{\max}} = \frac{\alpha_v^\nu}{4^\nu} \left| \frac{\Gamma(1 + 2\nu)}{\Gamma(1 + \nu + k)\Gamma(1 + \nu - k)} \right| \beta_v^{-\nu+k}$$

$$\times \frac{\Gamma[(1 + k + \nu)/2]\Gamma[1 + (k + \nu)/2]}{k!\,\Gamma(\tfrac{1}{2} + \nu)}$$

$$\times \left| F\left(\frac{k - \nu}{2}, \frac{1 + k - \nu}{2} \middle| 1 + k \middle| \beta_v^2\right) \right| \quad (8.84)$$

$$\frac{P(\alpha_v, \beta_v)}{P_{\text{norm}}(\omega_p/\omega_{c\,\max})^2} = \frac{\alpha_v^{2(1-\nu)}}{(1 - \nu)2^{4(1-\nu)}}$$

$$\times \frac{\Gamma(3 - 4\nu)}{[\Gamma(2 - 2\nu)]^2} \beta_v^{2\nu} \frac{\Gamma(2 - \nu)\Gamma\!\left(\dfrac{3 - 2\nu}{2}\right)}{\Gamma\!\left(\dfrac{3 - 4\nu}{2}\right)} F(\nu, \nu + \tfrac{1}{2}|2|\beta_v^2) \quad (8.85)$$

$$1 + \frac{V_0(\alpha_v, \beta_v)}{\phi} = \frac{\alpha_v}{\beta_v} \cdot \frac{1}{2}\left(1 + \frac{V_B}{\phi}\right) \quad (8.86)$$

When the pumping is partial with minimum polarization, we have $\beta_v = 1$ and the formulae (8.83) to (8.86) become:

$$\frac{|C_k(\alpha_v, 1)|}{C_{\min}} = \frac{4^\nu}{\alpha_v^\nu} \left| \frac{\Gamma(1 - 2\nu)}{\Gamma(1 - \nu + k)\Gamma(1 - \nu - k)} \right| \quad (8.87)$$

$$\frac{|S_k(\alpha_v, 1)|}{S_{\max}} = \frac{\alpha_v^\nu}{4^\nu} \left| \frac{\Gamma(1 + 2\nu)}{\Gamma(1 + \nu + k)\Gamma(1 + \nu - k)} \right| \quad (8.88)$$

$$\frac{P(\alpha_v, 1)}{P_{\text{norm}}(\omega_p/\omega_{c\,\max})^2} = \frac{\alpha_v^{2(1-\nu)}}{(1 - \nu)2^{4(1-\nu)}} \cdot \frac{\Gamma(3 - 4\nu)}{[\Gamma(2 - 2\nu)]^2} \quad (8.89)$$

$$1 + \frac{V_0(\alpha_v, 1)}{\phi} = \alpha_v \frac{1}{2}\left(1 + \frac{V_B}{\phi}\right) \quad (8.90)$$

When the pumping is full, we have $\alpha_v = \beta_v = 1$ and the formulae (8.83) to (8.86) become:

$$\frac{|C_k(1, 1)|}{C_{\min}} = 4^\nu \left| \frac{\Gamma(1 - 2\nu)}{\Gamma(1 - \nu + k)\Gamma(1 - \nu - k)} \right| \quad (8.91)$$

$$\frac{|S_k(1,1)|}{S_{\max}} = \frac{1}{4^\nu} \left| \frac{\Gamma(1+2\nu)}{\Gamma(1+\nu+k)\Gamma(1+\nu-k)} \right| \tag{8.92}$$

$$\frac{P(1,1)}{P_{\text{norm}}(\omega_p/\omega_{c\,\max})^2} = \frac{1}{(1-\nu)2^{4(1-\nu)}} \cdot \frac{\Gamma(3-4\nu)}{[\Gamma(2-2\nu)]^2} \tag{8.93}$$

$$1 + \frac{V_0(1,1)}{\phi} = \frac{1}{2}\left(1 + \frac{V_B}{\phi}\right) \tag{8.94}$$

Formulae (8.88) to (8.90) are quoted in the work of Penfield and Rafuse [6]. For a direct proof of formulae (8.87) to (8.94) see reference [7].

For the numerical calculation, it is of advantage to proceed in three stages. To begin with, we calculate the values for full pumping by formulae (8.91) to (8.94). The choice of α_v then determines the ratios between the values for partial pumping with minimum polarization and those with full pumping. Finally, the choice of β_v fixes the ratios between the values with partial pumping with any given polarization and those with partial pumping and minimum polarization. This last ratio does not depend on α_v and may therefore be tabulated once for all. The three-stage calculation just mentioned can be performed with the aid of the tables and monographs described later. The following formulae explain the ratios $f(\alpha_v, 1)/f(1,1)$ and $f(\alpha_v, \beta_v)/f(\alpha_v, 1)$:

$$\frac{|C_k(\alpha_v, 1)|}{|C_k(1,1)|} = \frac{1}{\alpha_v^\nu} \tag{8.95}$$

$$\frac{|S_k(\alpha_v, 1)|}{|S_k(1,1)|} = \alpha_v^\nu \tag{8.96}$$

$$\frac{P(\alpha_v, 1)}{P(1,1)} = \alpha_v^{2(1-\nu)} \tag{8.97}$$

$$\frac{1 + V_0(\alpha_v, 1)/\phi}{1 + V_0(1,1)/\phi} = \alpha_v \tag{8.98}$$

$$\frac{|C_k(\alpha_v, \beta_v)|}{|C_k(\alpha_v, 1)|} = \beta_v^{\nu+k} \frac{\Gamma[(1+k-\nu)/2]\Gamma[1+(k-\nu)/2]}{k!\,\Gamma(\tfrac{1}{2}-\nu)}$$
$$\times F\left(\frac{k+\nu}{2}, \frac{1+k+\nu}{2} \Big| 1 + k|\beta_v^2\right) \tag{8.99}$$

$$\frac{|S_k(\alpha_v, \beta_v)|}{|S_k(\alpha_v, 1)|} = \beta_v^{-\nu+k} \frac{\Gamma[(1+k+\nu)/2]\Gamma[1+(k+\nu)/2]}{k!\,\Gamma(\tfrac{1}{2}-\nu)}$$
$$\times \left| F\left(\frac{k-\nu}{2}, \frac{1+k-\nu}{2} \Big| 1 + k|\beta_v^2\right) \right| \tag{8.100}$$

$$\frac{P(\alpha_v, \beta_v)}{P(\alpha_v, 1)} = \beta_v^{2\nu} \frac{\Gamma(2 - \nu)\Gamma[(3 - 2\nu)/2]}{\Gamma[(3 - 4\nu)/2]} F(\nu, \nu + \tfrac{1}{2}|2|\beta_v^2) \qquad (8.101)$$

$$\frac{1 + V_0(\alpha_v, \beta_v)/\phi}{1 + V_0(\alpha_v, 1)/\phi} = \frac{1}{\beta_v} \qquad (8.102)$$

3. Current pumping

3.1. General formulae

The pumping current is of the form:

$$i_p = jI_p(e^{j\omega_p t} - e^{-j\omega_p t}) = -2I_p \sin \omega_p t = 2I_p \cos\left(\omega_p t + \frac{\pi}{2}\right) \qquad (8.103)$$

in which the time origin was chosen so as to make the initial phase of this current equal to $\pi/2$.

By definition of the differential capacitance, we have:

$$i_p(t) = C \frac{dv_p(t)}{dt} \qquad (8.104)$$

(8.1), (8.103) and (8.104) give us:

$$-\frac{2I_p}{\phi} \sin \omega_p t = C_{0v}\left[1 + \frac{V_0}{\phi} + \frac{v_p}{\phi}\right]^{-1/n} d\left[1 + \frac{V_0}{\phi} + \frac{v_p}{\phi}\right] \qquad (8.105)$$

Taking the indefinite integral of each term of this equation:

$$\frac{2I_p}{C_{0v}\omega_p \phi}\left[\frac{1}{\beta_I} + \cos \omega_p t\right] = \frac{n}{n-1}\left[1 + \frac{V_0}{\phi} + \frac{v_p}{\phi}\right]^{(n-1)/n} \qquad (8.106)$$

The constant β_I is determined by the pumping conditions (full or partial) and by the polarization (minimum or of any given value). The parameters will be $V_{\max}$ and $V_{\min}$, extreme values of $(v_p + V_0)$, which should of course lie between $-\phi$ and V_B. We see that $(v_p + V_0) = V_{\max}$ for $\omega_p t = 2k\pi$ and that $(v_p + V_0) = V_{\min}$ for $\omega_p t = (2k + 1)\pi$.

This being so, (8.106) gives us:

$$\frac{n}{n-1}\left[1 + \frac{V_{\max}}{\phi}\right]^{(n-1)/n} = \frac{2I_p}{C_{0v}\omega_p \phi}\left(\frac{1}{\beta_I} + 1\right) \qquad (8.107)$$

$$\frac{n}{n-1}\left[1 - \frac{V_{\min}}{\phi}\right]^{(n-1)/n} = \frac{2I_p}{C_{0v}\omega_p \phi}\left(\frac{1}{\beta_I} - 1\right) \qquad (8.108)$$

Hence:

$$\frac{1 + \beta_I}{1 - \beta_I} = \left[\frac{1 + V_{\max}/\phi}{1 + V_{\min}/\phi}\right]^{(n-1)/n} \qquad (8.109)$$

$$\beta_I = \frac{[1 + V_{max}/\phi]^{(n-1)/n} - [1 + V_{min}/\phi]^{(n-1)/n}}{[1 + V_{max}/\phi]^{(n-1)/n} + [1 + V_{min}/\phi]^{(n-1)/n}} \quad (8.110)$$

$$I_p = \frac{n}{n-1} \cdot \frac{C_{0v}\omega_p\phi}{4} \left[\left(1 + \frac{V_{max}}{\phi}\right)^{(n-1)/n} - \left(1 + \frac{V_{min}}{\phi}\right)^{(n-1)/n} \right] \quad (8.111)$$

From (8.106) we can also obtain the expressions for $(v_p + V_0)$ and V_0:

$$1 + \frac{V_0 + v_p}{\phi} = \left[\frac{n-1}{n} \cdot \frac{2I_p}{C_{0v}\omega_p\phi} \right]^{n/(n-1)} \left[\frac{1}{\beta_I} + \cos \omega_p t \right]^{n/(n-1)} \quad (8.112)$$

$$1 + \frac{V_0}{\phi} = \overline{1 + \frac{V_0 + v_p}{\phi}}$$

$$= \left[\frac{n-1}{n} \cdot \frac{2I_p}{C_{0v}\omega_p\phi} \right]^{n/(n-1)} \frac{1}{2\pi} \int_0^{2\pi} \left(\frac{1}{\beta_I} + \cos x \right)^{n/(n-1)} dx \quad (8.113)$$

The differential capacitance is given by:

$$C(t) = C_{0v} \left[1 + \frac{V_0 + v_p}{\phi} \right]^{-1/n}$$

$$= C_{0v} \left[\frac{n-1}{n} \cdot \frac{2I_p}{C_{0v}\omega_p\phi} \right]^{-1/(n-1)} \left[\frac{1}{\beta_I} + \cos \omega_p t \right]^{-1/(n-1)} \quad (8.114)$$

The coefficients $|C_k|$ are then given by:

$$|C_k| = C_{0v} \left[\frac{n-1}{n} \cdot \frac{2I_p}{C_{0v}\omega_p\phi} \right]^{-1/(n-1)} \beta_I^{+1/(n-1)}$$

$$\times \left| \frac{1}{2\pi} \int_{-\pi}^{+\pi} (1 + \beta_I \cos x)^{-1/(n-1)} \cos kx \, dx \right| \quad (8.115)$$

Similarly the expressions for $S(t)$ and $|S_k|$ are:

$$S(t) = S_{0v} \left[\frac{n-1}{n} \cdot \frac{2S_{0v}I_p}{\phi\omega_p} \right]^{1/(n-1)} \left[\frac{1}{\beta_I} + \cos \omega_p t \right]^{1/(n-1)} \quad (8.116)$$

$$|S_k| = S_{0v} \left[\frac{n-1}{n} \cdot \frac{2S_{0v}I_p}{\phi\omega_p} \right]^{1/(n-1)} \left(\frac{1}{\beta_I} \right)^{1/(n-1)}$$

$$\times \left| \frac{1}{2\pi} \int_{-\pi}^{+\pi} (1 + \beta_I \cos x)^{1/(n-1)} \cos kx \, dx \right| \quad (8.117)$$

The pumping power is given by:

$$P = 2R_s I_p^2 \quad (8.118)$$

Equation (8.111) shows that the maximum amplitude of the pumping current is given by:

$$I_{p\,\text{max}} = \frac{n}{n-1} \cdot \frac{C_{0v}\omega_p \phi}{4}\left[1 + \frac{V_B}{\phi}\right]^{(n-1)/n} \qquad (8.119)$$

Coefficient α_I will be defined by:

$$\alpha_I = \frac{I_p}{I_{p\,\text{max}}} = \frac{\left(1 + \dfrac{V_{\text{max}}}{\phi}\right)^{(n-1)/n} - \left(1 + \dfrac{V_{\text{min}}}{\phi}\right)^{(n-1)/n}}{\left(1 + \dfrac{V_B}{\phi}\right)^{(n-1)/n}} \qquad (8.120)$$

When α_I is fixed, the pumping power is determined and in accordance with (8.111), and (8.118) is equal to:

$$P = \frac{(V_B + \phi)^2}{R_s}\left(\frac{R_s C_{0v}}{\sqrt[n]{1 + V_B/\phi}}\right)^2 \omega_p^2 \alpha_I^2 \tfrac{1}{8}\left(\frac{n}{n-1}\right)^2 \qquad (8.121)$$

Taking account of the definitions (8.37) to (8.41), (8.121) may be written:

$$\frac{P}{P_{\text{norm}}(\omega_p/\omega_{c\,\text{max}})^2} = \alpha_I^2 \tfrac{1}{8}\frac{1}{(1-\nu)^2} \qquad (8.122)$$

We see that, for a given α_I, the pumping power does not depend on the polarization chosen, which is fixed if the parameter β_I is also fixed. From (8.109) we see at once that:

$$\frac{\alpha_I}{2 - \alpha_I} \leqslant \beta_I \leqslant 1 \qquad (8.123)$$

When α_I is equal to unity, (8.124) requires that $\beta_I = 1$; this is the case in which there is full pumping. When α_I has any given value but when we make $\beta_I = 1$, the polarization is a minimum as (8.109) gives $V_{\text{min}} = -\phi$. Parameters α_I and β_I for current pumping have the same role as that of α_v and β_v for voltage pumping.

From definitions (8.37) to (8.41) we can put (8.113), (8.115) and (8.117) in the form:

$$\frac{1 + V_0/\phi}{1 + V_B/\phi}$$

$$= \left(\frac{1}{2}\right)^{1/(1-\nu)}\left(\frac{\alpha_I}{\beta_I}\right)^{1/(1-\nu)} \frac{1}{2\pi}\int_0^{2\pi}(1 + \beta_I \cos x)^{1/(1-\nu)}\,dx \qquad (8.124)$$

PUMPING OF DIODES FOR MICROWAVE FREQUENCIES 213

$$\frac{|C_k|}{C_{\min}} = 2^{\nu/(1-\nu)} \left(\frac{\beta_I}{\alpha_I}\right)^{\nu/(1-\nu)}$$

$$\times \left|\frac{1}{2\pi}\int_0^{2\pi} (1 + \beta_I \cos x)^{-\nu/(1-\nu)} \cos kx\, dx\right| \quad (8.125)$$

$$\frac{|S_k|}{S_{\max}} = \left(\frac{1}{2}\right)^{\nu/(1-\nu)} \left(\frac{\alpha_I}{\beta_I}\right)^{\nu/(1-\nu)}$$

$$\times \left|\frac{1}{2\pi}\int_0^{2\pi} (1 + \beta_I \cos x)^{\nu/(1-\nu)} \cos kx\, dx\right| \quad (8.126)$$

3.2. Calculation of the $|C_k|$'s

A comparison of (8.125) and (8.42) shows that $|C_k|$ for current pumping is obtained by replacing α_v, β_v and ν respectively by α_I, β_I and $\nu/(1-\nu)$ in the expression for $|C_k|$ with voltage pumping. We then at once have:

$$\frac{|C_k|}{C_{\min}} = \frac{4^{\nu/(1-\nu)}}{\alpha_I^{\nu/(1-\nu)}} \left| \frac{\Gamma\left(1 - \frac{2\nu}{1-\nu}\right)}{\Gamma\left(1 - \frac{\nu}{1-\nu} + k\right)\Gamma\left(1 - \frac{\nu}{1-\nu} - k\right)}\right| \beta_I^{\nu/(1-\nu)+k}$$

$$\times \frac{\Gamma\left[\frac{1 + k - \nu/(1-\nu)}{2}\right]\Gamma\left[1 + \frac{k - \nu/(1-\nu)}{2}\right]}{k!\,\Gamma\left(\frac{1}{2} - \frac{\nu}{1-\nu}\right)}$$

$$\times F\left[\frac{k + \nu/(1-\nu)}{2}, \frac{1 + k + \nu/(1-\nu)}{2}\bigg|1 + k|\beta_v^2\right] \quad (8.127)$$

The recurrence formula is obtained in the same way:

$$(C_{k+2}) = \frac{2}{\beta_I} \cdot \frac{k+1}{k - \frac{\nu}{1-\nu} + 2} |C_{k+1}| - \frac{k + \frac{\nu}{1-\nu}}{k - \frac{\nu}{1-\nu} + 2} |C_k| \quad (8.128)$$

3.3. Calculation of the $|S_k|$'s

A comparison of (8.126) and (8.125) shows that $|S_k|$ is obtained by replacing $\nu/(1-\nu)$ by $-\nu/(1-\nu)$ in the expression for $|C_k|$. We then have at once:

$$\frac{|S_k|}{S_{\max}} = \frac{\alpha_I^{\nu/(1-\nu)}}{4^{\nu/(1-\nu)}} \left| \frac{\Gamma\left(1 + \frac{2\nu}{1-\nu}\right)}{\Gamma\left(1 + \frac{\nu}{1-\nu} + k\right)\Gamma\left(1 + \frac{\nu}{1-\nu} - k\right)} \right|$$

$$\times \beta_I^{-\nu/(1-\nu)+k} \frac{\Gamma\left[\frac{1+k+\nu/(1-\nu)}{2}\right]\Gamma\left[1 + \frac{k+\nu/(1-\nu)}{2}\right]}{k!\,\Gamma\left(\frac{1}{2} + \frac{\nu}{1-\nu}\right)}$$

$$\times \left| F\left(\frac{k - \frac{\nu}{1-\nu}}{2}, \frac{1+k-\frac{\nu}{1-\nu}}{2} \middle| 1 + k|\beta_I^2\right) \right| \quad (8.129)$$

The recurrence formulae are obtained in the same way:

$$|S_2| = \frac{2}{\beta_I} \cdot \frac{1}{2 + \frac{\nu}{1-\nu}} |S_1| - \frac{\frac{\nu}{1-\nu}}{2 + \frac{\nu}{1-\nu}} |S_0| \quad (8.130)$$

$$|S_{k+2}| = \frac{2}{\beta_I} \cdot \frac{k+1}{k + \frac{\nu}{1-\nu} + 2} |S_{k+1}| - \frac{k - \frac{\nu}{1-\nu}}{k + \frac{\nu}{1-\nu} + 2} |S_k|$$

$$\text{with } k > 0 \quad (8.131)$$

3.4. Calculation of the bias

Comparison of equations (8.124) and (8.126) shows that we obtain $(1 + V_0/\phi)/(1 + V_\beta/\phi)$ if we replace $\nu/(1-\nu)$ by $1/(1-\nu)$ and if we make $k = 0$ in the expression $|S_k|/S_{\max}$. We then have at once:

$$\frac{1 + \frac{V_0}{\phi}}{1 + \frac{V_B}{\phi}} = \frac{\alpha_I^{1/(1-\nu)}}{4^{1/(1-\nu)}} \frac{\Gamma\left(1 + \frac{2}{1-\nu}\right)}{\left[\Gamma\left(1 + \frac{1}{1-\nu}\right)\right]^2}$$

$$\times \beta_I^{-1/(1-\nu)} \frac{\Gamma\left[\frac{1 + 1/(1-\nu)}{2}\right]\Gamma\left[1 + \frac{1}{2(1-\nu)}\right]}{\Gamma\left(\frac{1}{2} + \frac{1}{1-\nu}\right)}$$

$$\times F\left(\frac{-1}{2(1-\nu)}, \frac{1 - 1/(1-\nu)}{2} \middle| 1|\beta_I^2\right) \quad (8.132)$$

PUMPING OF DIODES FOR MICROWAVE FREQUENCIES 215

3.5. Special cases

We will in future write the quantities for which the expressions have just been found in the form $f(\alpha_I, \beta_I)$.

Reminding ourselves of the general formulae:

$$\frac{|C_k(\alpha_I, \beta_I)|}{C_{\min}} = \frac{4^{\nu/(1-\nu)}}{\alpha_I^{\nu/(1-\nu)}} \left| \frac{\Gamma\left(1 - \dfrac{2\nu}{1-\nu}\right)}{\Gamma\left(1 - \dfrac{\nu}{1-\nu} + k\right)\Gamma\left(1 - \dfrac{\nu}{1-\nu} - k\right)} \right|$$

$$\times \beta_I^{\nu/(1-\nu)+k} \frac{\Gamma\left[\dfrac{(1+k-\nu/(1-\nu)}{2}\right]\Gamma\left[1 + \dfrac{k - \nu/(1-\nu)}{2}\right]}{k!\,\Gamma\left(\dfrac{1}{2} - \dfrac{\nu}{1-\nu}\right)}$$

$$\times F\left[\dfrac{k + \nu/(1-\nu)}{2}, \dfrac{1 + k + \nu/(1-\nu)}{2} \,\middle|\, 1 + k \middle| \beta_I^2\right] \quad (8.133)$$

$$\frac{|S_k(\alpha_I, \beta_I)|}{S_{\max}} = \frac{\alpha_I^{\nu/(1-\nu)}}{4^{\nu/(1-\nu)}} \left| \frac{\Gamma\left(1 + \dfrac{2\nu}{1-\nu}\right)}{\Gamma\left(1 + \dfrac{\nu}{1-\nu} + k\right)\Gamma\left(1 + \dfrac{\nu}{1-\nu} - k\right)} \right|$$

$$\times \beta_I^{-\nu/(1-\nu)+k} \frac{\Gamma\left[\dfrac{1 + k + \nu/(1+\nu)}{2}\right]\Gamma\left[1 + \dfrac{k + \nu/(1-\nu)}{2}\right]}{k!\,\Gamma\left(\dfrac{1}{2} + \dfrac{\nu}{1-\nu}\right)}$$

$$\times \left| F\left[\dfrac{k - \nu/(1-\nu)}{2}, \dfrac{1 + k - \nu/(1-\nu)}{2} \,\middle|\, 1 + k \middle| \beta_I^2\right] \right| \quad (8.134)$$

$$\frac{P(\alpha_I, \beta_I)}{P_{\text{norm}}(\omega_p/\omega_{c\,\max})^2} = \alpha_I^2 \frac{1}{8(1-\nu)^2} \quad (8.135)$$

$$\frac{1 + \dfrac{V_0(\alpha_I, \beta_I)}{\phi}}{1 + \dfrac{V_B}{\phi}} = \frac{\alpha_I^{1/(1-\nu)}}{4^{1/(1-\nu)}} \cdot \frac{\Gamma\left(1 + \dfrac{2}{1-\nu}\right)}{\left[\Gamma\left(1 + \dfrac{1}{1-\nu}\right)\right]^2}$$

$$\times \beta_v^{-1/(1-\nu)} \frac{\Gamma\left[\dfrac{1 + 1/(1-\nu)}{2}\right]\Gamma\left[1 + \dfrac{1}{2(1-\nu)}\right]}{\Gamma\left(\dfrac{1}{2} + \dfrac{1}{1-\nu}\right)}$$

$$\times F\left[\dfrac{-1}{2(1-\nu)}, \dfrac{1 - 1/(1-\nu)}{2} \,\middle|\, 1 \middle| \beta_I^2\right] \quad (8.136)$$

With partial pumping ($\alpha_I < 1$) and a minimum polarization ($\beta_I = 1$) we have:

$$\frac{|C_k(\alpha_I, 1)|}{C_{\min}} = \frac{4^{\nu/(1-\nu)}}{\alpha_I^{\nu/(1-\nu)}} \left| \frac{\Gamma\left(1 - \dfrac{2\nu}{1-\nu}\right)}{\Gamma\left(1 - \dfrac{\nu}{1-\nu} + k\right)\Gamma\left(1 - \dfrac{\nu}{1-\nu} - k\right)} \right| \quad (8.137)$$

$$\frac{|S_k(\alpha_I, 1)|}{S_{\max}} = \frac{\alpha_I^{\nu/(1-\nu)}}{4^{\nu/(1-\nu)}} \left| \frac{\Gamma\left(1 + \dfrac{2\nu}{1-\nu}\right)}{\Gamma\left(1 + \dfrac{\nu}{1-\nu} + k\right)\Gamma\left(1 + \dfrac{\nu}{1-\nu} - k\right)} \right| \quad (8.138)$$

$$\frac{P(\alpha_I, 1)}{P_{\text{norm}}(\omega_p/\omega_{c\,\max})^2} = \alpha_I^2 \frac{1}{8(1-\nu)^2} \quad (8.139)$$

$$\frac{1 + \dfrac{V_0(\alpha_I, 1)}{\phi}}{1 + \dfrac{V_B}{\phi}} = \frac{\alpha_I^{1/(1-\nu)}}{4^{1/(1-\nu)}} \cdot \frac{\Gamma\left(1 + \dfrac{2}{1-\nu}\right)}{\left[\Gamma\left(1 + \dfrac{1}{1-\nu}\right)\right]^2} \quad (8.140)$$

For full pumping ($\alpha_I = \beta_I = 1$) we have:

$$\frac{|C_k(1, 1)|}{C_{\min}} = 4^{\nu/(1-\nu)} \left| \frac{\Gamma\left(1 - \dfrac{2\nu}{1-\nu}\right)}{\Gamma\left(1 - \dfrac{\nu}{1-\nu} + k\right)\Gamma\left(1 - \dfrac{\nu}{1-\nu} - k\right)} \right| \quad (8.141)$$

$$\frac{|S_k(1, 1)|}{S_{\max}} = \frac{1}{4^{\nu/(1-\nu)}} \left| \frac{\Gamma\left(1 + \dfrac{2\nu}{1-\nu}\right)}{\Gamma\left(1 + \dfrac{\nu}{1-\nu} + k\right)\Gamma\left(1 + \dfrac{\nu}{1-\nu} - k\right)} \right| \quad (8.142)$$

$$\frac{P(1, 1)}{P_{\text{norm}}(\omega_p/\omega_{c\,\max})^2} = \frac{1}{8(1-\nu)^2} \quad (8.143)$$

$$\frac{1 + \dfrac{V_0^{(1,1)}}{\phi}}{1 + \dfrac{V_B}{\phi}} = \frac{1}{4^{1/(1-\nu)}} \cdot \frac{\Gamma\left(1 + \dfrac{2}{1-\nu}\right)}{\left[\Gamma\left(1 + \dfrac{1}{1-\nu}\right)\right]^2} \quad (8.144)$$

Formulae (8.138) to (8.140) and (8.142) to (8.144) are quoted in the work of Penfield and Rafuse [6]. A direct proof of formulae (8.137) to (8.144) will be found in [7].

PUMPING OF DIODES FOR MICROWAVE FREQUENCIES 217

As in the case of voltage pumping, the numerical calculation is performed, most simply, in three stages. A rapid calculation can be made by means of the tables and charts given below. The following formulae explain the ratios $f(\alpha_I, 1)/f(1, 1)$ and $f(\alpha_I, \beta_I)/f(\alpha_I, 1)$.

$$\frac{|C_k(\alpha_I, 1)|}{|C_k(1, 1)|} = \frac{1}{\alpha_I^{\nu/(1-\nu)}} \tag{8.145}$$

$$\frac{|S_k(\alpha_I, 1)|}{|S_k(1, 1)|} = \alpha_I^{\nu/(1-\nu)} \tag{8.146}$$

$$\frac{P(\alpha_I, 1)}{P(1, 1)} = \alpha_I^2 \tag{8.147}$$

$$\frac{1 + \dfrac{V_0(\alpha_I, 1)}{\phi}}{1 + \dfrac{V_0(1, 1)}{\phi}} = \alpha_I^{1/(1-\nu)} \tag{8.148}$$

$$\frac{|C_k(\alpha_I, \beta_I)|}{|C_k(\alpha_I, 1)|} = \beta_I^{\nu/(1-\nu)+k}$$

$$\times \frac{\Gamma\left[\dfrac{1 + k - \nu/(1-\nu)}{2}\right]\Gamma\left[1 + \dfrac{k - \nu/(1-\nu)}{2}\right]}{k!\,\Gamma\left(\dfrac{1}{2} - \dfrac{\nu}{1-\nu}\right)}$$

$$\times F\left(\frac{k + \dfrac{\nu}{1-\nu}}{2}, \frac{1 + k + \dfrac{\nu}{1-\nu}}{2}\bigg| 1 + k \big| \beta_I^2\right) \tag{8.149}$$

$$\frac{|S_k(\alpha_I, \beta_I)|}{|S_k(\alpha_I, 1)|} = \beta_I^{-\nu/(1-\nu)+k}$$

$$\times \frac{\Gamma\left[\dfrac{1 + k + \nu/(1-\nu)}{2}\right]\Gamma\left[1 + \dfrac{k + \nu/(1-\nu)}{2}\right]}{k!\,\Gamma\left(\dfrac{1}{2} + \dfrac{\nu}{1-\nu}\right)}$$

$$\times \left| F\left(\frac{k - \dfrac{\nu}{1-\nu}}{2}, \frac{1 + k - \dfrac{\nu}{1-\nu}}{2}\bigg| 1 + k \big| \beta_I^2\right) \right| \tag{8.150}$$

$$\frac{P(\alpha_I, \beta_I)}{P(\alpha_I, 1)} = 1 \tag{8.151}$$

$$\frac{1 + \dfrac{V_0(\alpha_I, \beta_I)}{\phi}}{1 + \dfrac{V_0(\alpha_I, 1)}{\phi}} = \beta_I^{-1/(1-\nu)} \frac{\Gamma\left[\dfrac{1 + 1/(1-\nu)}{2}\right]\Gamma\left[1 + \dfrac{1}{2(1-\nu)}\right]}{\Gamma\left(\dfrac{1}{2} + \dfrac{1}{1-\nu}\right)}$$

$$\times F\left[\frac{-1}{2(1-\nu)}, \frac{1 - 1/(1-\nu)}{2} \,\bigg|\, 1 \,\bigg|\, \beta_I^2\right] \quad (8.152)$$

The above formulae can be presented in a particularly simple form when $n = 2$ ($\nu = \frac{1}{2}$). We then have:

$$\frac{|S_0(\alpha_I, \beta_I)|}{S_{\max}} = \frac{1}{2} \cdot \frac{\alpha_I}{\beta_I} \quad (8.153)$$

$$\frac{|S_1(\alpha_I, \beta_I)|}{S_{\max}} = \frac{1}{4} \alpha_I \quad (8.154)$$

$$\frac{|S_k(\alpha_I, \beta_I)|}{S_{\max}} = 0 \quad \text{for } k \geqslant 2 \quad (8.155)$$

$$\frac{P(\alpha_I, \beta_I)}{P_{\text{norm}}(\omega_p/\omega_{c\,\max})^2} = \frac{1}{2} \alpha_I^2 \quad (8.156)$$

$$\frac{1 + V(\alpha_I, \beta_I)/\phi}{1 + V_B/\phi} = \frac{1}{4}\left(\frac{\alpha_I}{\beta_I}\right)^2 \left(1 + \frac{1}{2} \beta_I^2\right) \quad (8.157)$$

4. Tables and charts for voltage pumping

4.1. Introduction

The numerical values of the $|S_k|$'s, of V_0 and of P are given in the following sections. These are in fact the quantities which occur in the design of parametric amplifiers for microwave frequencies. The values of $|C_k|$, on the other hand, are used in the design of low-frequency amplifiers. The tables and charts for these amplifiers will therefore be dealt with in Part III of the book. We will give the complete tables for $|S_0|$, $|S_1|$, V_0 and P, as well as the values for full pumping for the $|S_k|$ with index greater than 1. Let us consider, for example, the case of $|S_0|$. The results are shown as follows:

(1) Table 8.1 gives the value for the full pumping $|S_0(1, 1)|/S_{\max}$ as a function of $n = 1/\nu$.
(2) Table 8.2 gives the value for partial pumping with minimum

PUMPING OF DIODES FOR MICROWAVE FREQUENCIES

polarization referred to the value for full pumping, $|S_0(\alpha_v, 1)|/|S_0(1, 1)|$ as a function of α_v with n as parameter.

(3) Table 8.3 gives the value for any given type of pumping refixed to the value for partial pumping with minimum bias corresponding to the same value of α_v, $|S_0(\alpha_v, \beta_v)|/|S_0(\alpha_v, 1)|$ as a function of β_v with n as parameter.

4.2. Mean value of the elastance

Table 8.1. Full pumping

n	$\dfrac{|S_0(1, 1)|}{S_{max}}$	n	$\dfrac{|S_0(1, 1)|}{S_{max}}$
2·0	0·637	2·6	0·687
2·2	0·654	2·8	0·699
2·4	0·670	3·0	0·715

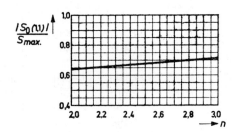

Figure 8.5. Mean value of the elastance for full pumping

Table 8.2. Partial pumping with minimum bias $|S_0(\alpha_v, 1)|/|S_0(1, 1)|$

α_v	$n = 2·0$	$n = 2·2$	$n = 2·4$	$n = 2·6$	$n = 2·8$	$n = 3·0$
0·0	0·000	0·000	0·000	0·000	0·000	0·000
0·1	0·316	0·351	0·383	0·412	0·439	0·464
0·2	0·447	0·481	0·512	0·539	0·563	0·585
0·3	0·548	0·578	0·606	0·630	0·651	0·670
0·4	0·632	0·659	0·682	0·703	0·721	0·737
0·5	0·707	0·730	0·749	0·766	0·781	0·794
0·6	0·774	0·793	0·809	0·822	0·833	0·843
0·7	0·837	0·850	0·862	0·872	0·880	0·887
0·8	0·894	0·903	0·911	0·918	0·923	0·928
0·9	0·949	0·953	0·957	0·960	0·963	0·965
1·0	1·000	1·000	1·000	1·000	1·000	1·000

It should be noted that formula (8.96) gives us:

$$\frac{|S_k(\alpha_v, 1)|}{|S_k(1, 1)|} = \frac{|S_0(\alpha_v, 1)|}{|S_0(1, 1)|}$$

The above table therefore applies to all the k indices.

Table 8.3. Arbitrary bias partial pumping $|S_0(\alpha_v, \beta_v)|/|S_0(\alpha_v, 1)|$

β_v	$n = 2 \cdot 0$	$n = 2 \cdot 2$	$n = 2 \cdot 4$	$n = 2 \cdot 6$	$n = 2 \cdot 8$	$n = 3 \cdot 0$
0·00	∞	∞	∞	∞	∞	∞
0·05	4·967	4·347	3·885	3·529	3·248	3·020
0·10	3·510	3·170	2·909	2·702	2·535	2·396
0·15	2·864	2·635	2·455	2·310	2·191	2·092
0·20	2·477	2·309	2·175	2·066	1·975	1·899
0·25	2·213	2·084	1·979	1·893	1·822	1·760
0·30	2·016	1·914	1·831	1·762	1·704	1·654
0·35	1·863	1·781	1·714	1·657	1·609	1·568
0·40	1·738	1·672	1·617	1·571	1·531	1·496
0·45	1·634	1·580	1·535	1·497	1·464	1·435
0·50	1·545	1·501	1·464	1·433	1·405	1·381
0·55	1·467	1·432	1·402	1·376	1·353	1·333
0·60	1·398	1·370	1·346	1·325	1·306	1·290
0·65	1·337	1·315	1·296	1·279	1·264	1·250
0·70	1·281	1·264	1·249	1·236	1·224	1·213
0·75	1·229	1·217	1·206	1·196	1·186	1·178
0·80	1·181	1·173	1·165	1·157	1·151	1·145
0·85	1·136	1·130	1·125	1·121	1·116	1·112
0·90	1·092	1·089	1·087	1·084	1·082	1·079
0·95	1·049	1·048	1·047	1·047	1·046	1·045
1·00	1·000	1·000	1·000	1·000	1·000	1·000

Note that the above table applies only to values of β_v higher than $\alpha_v/(2 - \alpha_v)$.

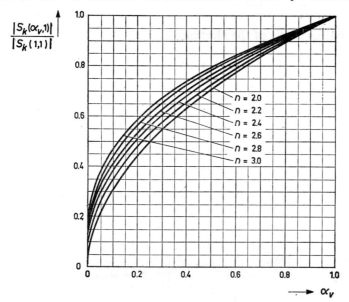

Figure 8.6. Development coefficients of the elastance with partial pumping and minimum bias

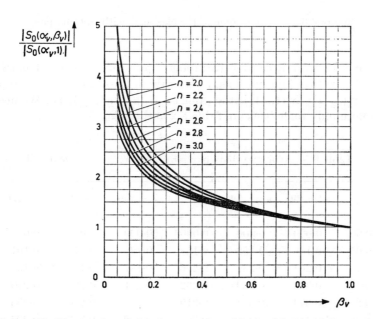

Figure 8.7. Mean value of the elastance for arbitrary bias partial pumping

4.3. First harmonic of the elastance

Table 8.4. Full pumping

n	$|S_1(1, 1)|/S_{\max}$	n	$|S_1(1, 1)|/S_{\max}$
2·0	0·212	2·6	0·191
2·2	0·204	2·8	0·184
2·4	0·197	3·0	0·179

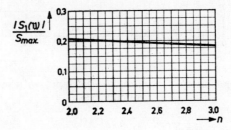

Figure 8.8. First harmonic of the elastance with full pumping

It is no longer necessary to tabulate $|S_1|$ for the case of partial pumping with minimum bias, as we saw in the previous paragraph that $|S_1(\alpha_v, 1)|/|S_1(1, 1)|$ is the same as $|S_0(\alpha_v, 1)|/|S_0(1, 1)|$. We need therefore only refer back to Table 8.2 and to Fig. 8.6.

Table 8.5. Arbitrary bias partial pumping $S_1(\alpha_v, \beta_v)/S_1(\alpha_v, 1)$

β_v	$n = 2 \cdot 0$	$n = 2 \cdot 2$	$n = 2 \cdot 4$	$n = 2 \cdot 6$	$n = 2 \cdot 8$	$n = 3 \cdot 0$
0·00	0·000	0·000	0·000	0·000	0·000	0·000
0·05	0·186	0·158	0·138	0·122	0·110	0·101
0·10	0·264	0·231	0·206	0·187	0·172	0·160
0·15	0·323	0·289	0·262	0·241	0·224	0·210
0·20	0·374	0·338	0·310	0·288	0·270	0·255
0·25	0·419	0·383	0·354	0·332	0·313	0·297
0·30	0·460	0·424	0·396	0·372	0·353	0·337

PUMPING OF DIODES FOR MICROWAVE FREQUENCIES

Table 8.5. (*continued*)

β_v	$n = 2 \cdot 0$	$n = 2 \cdot 2$	$n = 2 \cdot 4$	$n = 2 \cdot 6$	$n = 2 \cdot 8$	$n = 3 \cdot 0$
0·35	0·499	0·463	0·435	0·411	0·392	0·375
0·40	0·535	0·500	0·472	0·449	0·429	0·412
0·45	0·570	0·536	0·508	0·485	0·466	0·449
0·50	0·604	0·571	0·544	0·521	0·502	0·486
0·55	0·638	0·606	0·580	0·557	0·538	0·522
0·60	0·671	0·640	0·615	0·593	0·575	0·559
0·65	0·704	0·675	0·651	0·630	0·612	0·597
0·70	0·737	0·710	0·687	0·668	0·651	0·636
0·75	0·771	0·746	0·725	0·707	0·691	0·677
0·80	0·806	0·784	0·765	0·748	0·733	0·720
0·85	0·844	0·824	0·808	0·793	0·780	0·768
0·90	0·884	0·869	0·855	0·843	0·832	0·822
0·95	0·931	0·921	0·912	0·903	0·895	0·887
1·00	1·000	1·000	1·000	1·000	1·000	1·000

Note that the above table applies only to values of β_v higher than $\alpha_v/(2 - \alpha_v)$.

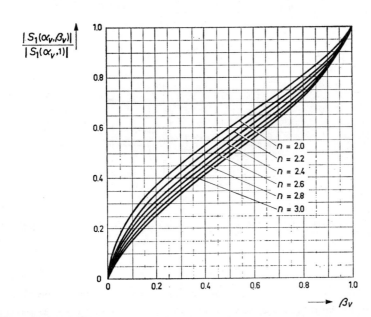

Figure 8.9. First harmonic of the elastance for arbitrary bias partial pumping

4.4. Higher harmonics of the elastance

With partial pumping and minimum bias we saw above that $|S_k(\alpha_v, 1)|/|S_k(1, 1)|$ is the same as $|S_0(\alpha_v, 1)|/|S_0(1, 1)|$. All that is necessary therefore is to refer to Table 8.2 and Fig. 8.6.

Table 8.6. Full pumping. Value of $|S_k(1, 1)|/S_{\max}$

k	$n = 2 \cdot 0$	$n = 2 \cdot 2$	$n = 2 \cdot 4$	$n = 2 \cdot 6$	$n = 2 \cdot 8$	$n = 3 \cdot 0$
2	0·0424	0·0452	0·0476	0·0491	0·0501	0·0510
3	0·0182	0·0202	0·0220	0·0235	0·0246	0·0255
4	0·0101	0·0116	0·0129	0·0141	0·0149	0·0157
5	0·0064	0·0075	0·0085	0·0095	0·0101	0·0108

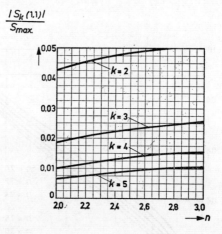

Figure 8.10. Higher harmonics of the elastance with full pumping

The values of $|S_k|$ with k greater than 1 are so seldom used that it would not be justified to draw up a table for the arbitrary bias partial pumping. However, if necessary these values may be obtained

PUMPING OF DIODES FOR MICROWAVE FREQUENCIES 225

quickly from the values of $|S_0|$ and $|S_1|$ thanks to the recurrence formulae (8.70) and (8.71).

4.5. Polarization

The expressions for the values of the bias with full and partial pumping with minimum bias are very simple, and this makes tabulation unnecessary. We have:

$$\frac{1 + V_0(1, 1)/\phi}{1 + V_B/\phi} = \frac{1}{2}; \quad \frac{1 + V_0(\alpha_v, 1)/\phi}{1 + V_0(1, 1)/\phi} = \alpha_v$$

The case of arbitrary bias partial pumping is tabulated later. We see that, whatever the degree of pumping, V_0 does not depend on n.

Table 8.7. Arbitrary bias partial pumping

β_v	$\dfrac{1 + V_0(\alpha_v \beta_v)/\phi}{1 + V_0(\alpha_v, 1)/\phi}$	β_v	$\dfrac{1 + V_0(\alpha_v \beta_v)/\phi}{1 + V_0(\alpha_v, 1)/\phi}$
0·0	∞	0·6	1·67
0·1	10·00	0·7	1·43
0·2	5·00	0·8	1·25
0·3	3·33	0·9	1·11
0·4	2·50	1·0	1·00
0·5	2·00		

Note that the above table applies only to values of β_v greater than $\alpha_v/(2 - \alpha_v)$.

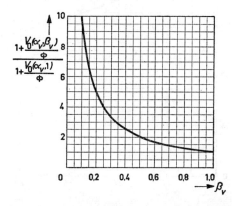

Figure 8.11. Arbitrary bias partial pumping

4.6. Pumping power

Table 8.8. Full pumping

n	$\dfrac{P(1,1)}{P_{\text{norm}}(\omega_p/\omega_{c\max})^2}$	n	$\dfrac{P(1,1)}{P_{\text{norm}}(\omega_p/\omega_{c\max})^2}$
2·0	0·500	2·6	0·316
2·2	0·408	2·8	0·289
2·4	0·353	3·0	0·267

Table 8.9. Partial pumping with minimum bias $P(\alpha_v, 1)/P(1, 1)$

α_v	$n = 2\cdot0$	$n = 2\cdot2$	$n = 2\cdot4$	$n = 2\cdot6$	$n = 2\cdot8$	$n = 3\cdot0$
0·0	0·000	0·000	0·000	0·000	0·000	0·000
0·1	0·100	0·081	0·068	0·059	0·052	0·046
0·2	0·200	0·172	0·152	0·138	0·126	0·116
0·3	0·300	0·268	0·245	0·227	0·212	0·200
0·4	0·400	0·368	0·343	0·324	0·308	0·294
0·5	0·500	0·469	0·445	0·426	0·410	0·396
0·6	0·600	0·572	0·551	0·534	0·519	0·506
0·7	0·700	0·677	0·659	0·645	0·632	0·621
0·8	0·800	0·784	0·770	0·760	0·750	0·742
0·9	0·900	0·891	0·884	0·878	0·873	0·869
1·0	1·000	1·000	1·000	1·000	1·000	1·000

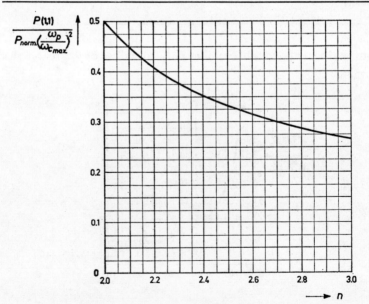

Figure 8.12. Pump power for full pumping

PUMPING OF DIODES FOR MICROWAVE FREQUENCIES

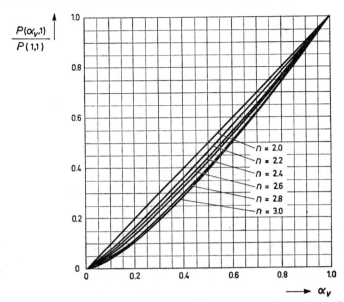

Figure 8.13. Pump power for partial pumping with minimum bias

Table 8.10. Arbitrary bias partial pumping $P(\alpha_v, \beta_v)/P(\alpha_v, 1)$

β_v	$n = 2\cdot0$	$n = 2\cdot2$	$n = 2\cdot4$	$n = 2\cdot6$	$n = 2\cdot8$	$n = 3\cdot0$
0·00	0·000	0·000	0·000	0·000	0·000	0·000
0·05	0·025	0·038	0·052	0·068	0·084	0·101
0·10	0·050	0·071	0·093	0·115	0·138	0·160
0·15	0·075	0·103	0·130	0·158	0·184	0·210
0·20	0·101	0·134	0·166	0·197	0·227	0·255
0·25	0·127	0·165	0·201	0·235	0·267	0·297
0·30	0·154	0·196	0·236	0·272	0·306	0·337
0·35	0·181	0·227	0·270	0·308	0·343	0·375
0·40	0·209	0·259	0·304	0·344	0·380	0·412
0·45	0·238	0·291	0·338	0·380	0·416	0·449
0·50	0·268	0·325	0·373	0·416	0·453	0·486
0·55	0·300	0·359	0·409	0·452	0·490	0·522
0·60	0·333	0·395	0·447	0·490	0·527	0·559
0·65	0·369	0·433	0·485	0·529	0·566	0·597
0·70	0·408	0·474	0·527	0·570	0·606	0·636
0·75	0·451	0·518	0·571	0·613	0·648	0·677
0·80	0·500	0·568	0·619	0·660	0·693	0·720
0·85	0·557	0·624	0·674	0·713	0·743	0·768
0·90	0·627	0·692	0·739	0·773	0·800	0·822
0·95	0·724	0·783	0·822	0·851	0·871	0·887
1·00	1·000	1·000	1·000	1·000	1·000	1·000

Note that the above table applies only for values of β_v greater than $\alpha_v/(2 - \alpha_v)$.

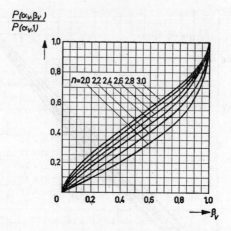

Figure 8.14. Pump power for arbitrary bias partial pumping

5. Tables and charts for current pumping

5.1. Introduction

The presentation of these results is the same as that used for voltage pumping. Here again we do no more than tabulate $|S_0|$, $|S_1|$, V_0 and P as functions of α_I and β_I. We give also the higher harmonics $|S_k|$ for full pumping.

5.2. Mean value of the elastance

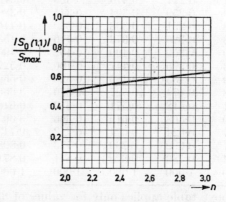

Figure 8.15. Mean value of the elastance with full pumping

PUMPING OF DIODES FOR MICROWAVE FREQUENCIES 229

Table 8.11. Full pumping

n	$\dfrac{\|S_0(1,1)\|}{S_{\max}}$	n	$\dfrac{\|S_0(1,1)\|}{S_{\max}}$
2·0	0·500	2·6	0·592
2·2	0·535	2·8	0·618
2·4	0·566	3·0	0·637

Table 8.12. Minimum bias partial pumping $|S_0(\alpha_I, 1)|/|S_0(1, 1)|$

α_I	$n = 2·0$	$n = 2·2$	$n = 2·4$	$n = 2·6$	$n = 2·8$	$n = 3·0$
0·0	0·000	0·000	0·000	0·000	0·000	0·000
0·1	0·100	0·147	0·193	0·238	0·278	0·316
0·2	0·200	0·262	0·317	0·366	0·409	0·448
0·3	0·300	0·367	0·423	0·471	0·513	0·548
0·4	0·400	0·466	0·520	0·564	0·602	0·632
0·5	0·500	0·561	0·610	0·648	0·681	0·707
0·6	0·600	0·654	0·694	0·726	0·753	0·774
0·7	0·700	0·743	0·775	0·800	0·820	0·837
0·8	0·800	0·830	0·853	0·870	0·883	0·894
0·9	0·900	0·916	0·927	0·936	0·943	0·949
1·0	1·000	1·000	1·000	1·000	1·000	1·000

Note that formula (8.146) gives us:

$$\frac{|S_k(\alpha_I, 1)|}{|S_k(1, 1)|} = \frac{|S_0(\alpha_I, 1)|}{|S_0(1, 1)|}$$

The above table therefore applies for all the k indices

Table 8.13. Arbitrary bias partial pumping $S_0(\alpha_I, \beta_I)/S_0(\alpha_I, 1)$

β_I	$n = 2·0$	$n = 2·2$	$n = 2·4$	$n = 2·6$	$n = 2·8$	$n = 3·0$
0·00	∞	∞	∞	∞	∞	∞
0·05	20·000	12·719	9·148	7·115	5·834	4·967
0·10	10·000	7·137	5·574	4·611	3·968	3·510
0·15	6·667	5·088	4·170	3·576	3·165	2·864
0·20	5·000	4·001	3·392	2·985	2·694	2·477
0·25	4·000	3·319	2·889	2·593	2·377	2·213
0·30	3·333	2·849	2·532	2·310	2·144	2·016
0·35	2·857	2·502	2·264	2·093	1·964	1·863

Table 8.13. (*continued*)

β_I	$n = 2 \cdot 0$	$n = 2 \cdot 2$	$n = 2 \cdot 4$	$n = 2 \cdot 6$	$n = 2 \cdot 8$	$n = 3 \cdot 0$
0·40	2·500	2·236	2·054	1·921	1·819	1·738
0·45	2·222	2·024	1·884	1·780	1·699	1·634
0·50	2·000	1·850	1·743	1·661	1·597	1·545
0·55	1·818	1·705	1·623	1·560	1·509	1·467
0·60	1·667	1·583	1·520	1·471	1·431	1·398
0·65	1·538	1·477	1·430	1·393	1·363	1·337
0·70	1·429	1·384	1·350	1·323	1·300	1·281
0·75	1·333	1·303	1·279	1·260	1·244	1·229
0·80	1·250	1·230	1·215	1·202	1·191	1·181
0·85	1·176	1·165	1·156	1·148	1·142	1·136
0·90	1·111	1·105	1·101	1·098	1·095	1·092
0·95	1·052	1·051	1·050	1·050	1·049	1·049
1.00	1·000	1·000	1·000	1·000	1·000	1·000

It should be remembered that the above table applies only to values of β_I higher than $\alpha_I/(2 - \alpha_I)$.

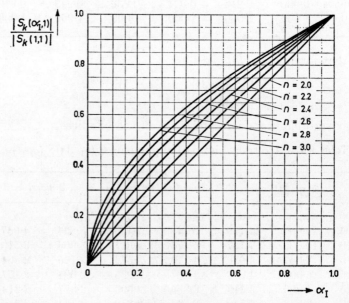

Figure 8.16. Development coefficients of the elastance for partial pumping with minimum bias

PUMPING OF DIODES FOR MICROWAVE FREQUENCIES

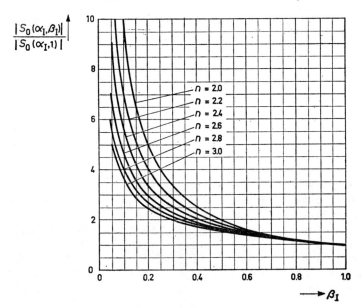

Figure 8.17. Mean value of the elastance for arbitrary bias partial pumping

5.3. First harmonic of the elastance

Table 8.14. Full pumping

n	$|S_1(1, 1)|/S_{\max}$	n	$|S_1(1, 1)|/S_{\max}$
2·0	0·250	2·6	0·228
2·2	0·243	2·8	0·221
2·4	0·236	3·0	0·212

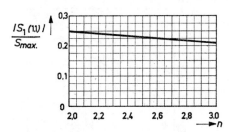

Figure 8.18. First harmonic of the elastance

It is no longer necessary to tabulate $|S_1|$ for the case of partial pumping with minimum polarization. We saw, in fact, in the previous paragraph that $|S_1(\alpha_I, 1)|/|S_1(1, 1)|$ is identical with $|S_0(\alpha_I, 1)|/$

$|S_0(1, 1)|$. All we need, therefore, is to refer to Table 8.12 and Fig. 8.16.

Table 8.15. Arbitrary bias partial pumping $|S_1(\alpha_I, \beta_I)|/|S_1(\alpha_I, 1)|$

β_I	$n = 2{\cdot}0$	$n = 2{\cdot}2$	$n = 2{\cdot}4$	$n = 2{\cdot}6$	$n = 2{\cdot}8$	$n = 3{\cdot}0$
0·00	1·000	0·000	0·000	0·000	0·000	0·000
0·05	1·000	0·583	0·392	0·289	0·227	0·186
0·10	1·000	0·655	0·478	0·375	0·309	0·264
0·15	1·000	0·701	0·537	0·437	0·370	0·323
0·20	1·000	0·735	0·584	0·487	0·422	0·374
0·25	1·000	0·764	0·623	0·531	0·466	0·419
0·30	1·000	0·788	0·657	0·569	0·507	0·460
0·35	1·000	0·809	0·688	0·605	0·544	0·499
0·40	1·000	0·828	0·716	0·637	0·580	0·535
0·45	1·000	0·845	0·742	0·668	0·613	0·570
0·50	1·000	0·861	0·767	0·698	0·645	0·604
0·55	1·000	0·876	0·790	0·726	0·677	0·638
0·60	1·000	0·891	0·813	0·754	0·708	0·671
0·65	1·000	0·905	0·834	0·781	0·738	0·704
0·70	1·000	0·918	0·856	0·808	0·769	0·737
0·75	1·000	0·931	0·877	0·835	0·800	0·771
0·80	1·000	0·944	0·899	0·862	0·832	0·806
0·85	1·000	0·957	0·921	0·891	0·865	0·844
0·90	1·000	0·970	0·944	0·921	0·902	0·884
0·95	1·000	0·984	0·969	0·955	0·943	0·931
1·00	1·000	1·000	1·000	1·000	1·000	1·000

Note that the above table applies only for values of β_I greater than $\alpha_I/(2 - \alpha_I)$.

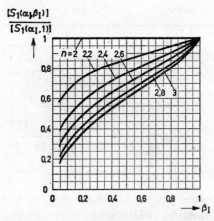

Figure 8.19. First harmonic of the elastance with arbitrary bias partial pumping

5.4. Higher harmonics of the elastance

Table 8.16. Full pumping. Value of $|S_k(1, 1)|/S_{\max}$

k	$n = 2\cdot 0$	$n = 2\cdot 2$	$n = 2\cdot 4$	$n = 2\cdot 6$	$n = 2\cdot 8$	$n = 3\cdot 0$
2	0·0000	0·0143	0·0248	0·0326	0·0383	0·0424
3	0·0000	0·0043	0·0086	0·0123	0·0156	0·0182
4	0·0000	0·0020	0·0042	0·0063	0·0083	0·0101
5	0·0000	0·0011	0·0024	0·0038	0·0052	0·0064

With partial pumping and minimum bias we saw earlier that $|S_k(\alpha_I, 1)|/|S_k(1, 1)|$ is the same as $|S_0(\alpha_I, 1)|/|S_0(1, 1)|$. All that is needed therefore is to refer back to Table 8.12 and Fig. 8.16.

The values of $|S_k|$ with k higher than 1 are used too rarely to justify a tabulation of partial pumping. However, if required, these values can be obtained rapidly from the values of $|S_0|$ and $|S_1|$, thanks to the recurrence formulae (8.130) and (8.131).

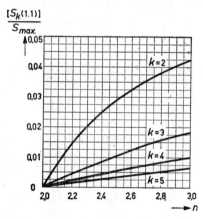

Figure 8.20. Higher harmonics of the elastance with full pumping

5.5. Polarization

Table 8.17. Full pumping

n	$\dfrac{1 + V_0(1, 1)/\phi}{1 + V_B/\phi}$	n	$\dfrac{1 + V_0(1, 1)/\phi}{1 + V_B/\phi}$
2·0	0·375	2·6	0·411
2·2	0·389	2·8	0·417
2·4	0·400	3·0	0·426

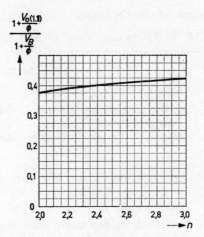

Figure 8.21. Polarization with full pumping

Table 8.18. Partial pumping with minimum bias
$[1 + V_0(\alpha_I, 1)/\phi]/[1 + V_0(1, 1)/\phi]$

α_I	$n = 2\cdot 0$	$n = 2\cdot 2$	$n = 2\cdot 4$	$n = 2\cdot 6$	$n = 2\cdot 8$	$n = 3\cdot 0$
0·0	0·000	0·000	0·000	0·000	0·000	0·000
0·1	0·010	0·015	0·019	0·024	0·028	0·032
0·2	0·040	0·052	0·063	0·073	0·082	0·090
0·3	0·090	0·112	0·128	0·142	0·154	0·164
0·4	0·160	0·186	0·208	0·226	0·241	0·253
0·5	0·250	0·280	0·305	0·324	0·340	0·354
0·6	0·360	0·392	0·417	0·436	0·452	0·465
0·7	0·490	0·520	0·543	0·560	0·574	0·586
0·8	0·640	0·664	0·682	0·696	0·707	0·715
0·9	0·810	0·824	0·835	0·843	0·849	0·854
1·0	1·000	1·000	1·000	1·000	1·000	1·000

Table 8.19. Arbitrary bias partial pumping
$[1 + V_0(\alpha_I, \beta_I)/\phi]/[1 + V_0(\alpha_I, 1)/\phi]$

β_I	$n = 2\cdot 0$	$n = 2\cdot 2$	$n = 2\cdot 4$	$n = 2\cdot 6$	$n = 2\cdot 8$	$n = 3\cdot 0$
0·00	∞	∞	∞	∞	∞	∞
0·05	267·000	175·071	129·268	102·850	86·033	74·544
0·10	67·000	49·268	39·485	33·409	29·316	26·392
0·15	29·963	23·539	19·780	17·341	15·644	14·400
0·20	17·000	13·983	12·144	10·914	10·038	9·384

PUMPING OF DIODES FOR MICROWAVE FREQUENCIES

Table 8.19. (*continued*)

β_I	$n = 2\cdot0$	$n = 2\cdot2$	$n = 2\cdot4$	$n = 2\cdot6$	$n = 2\cdot8$	$n = 3\cdot0$
0·25	11·000	9·367	8·340	7·638	7·128	6·743
0·30	7·741	6·775	6·152	5·718	5·400	5·156
0·35	5·776	5·168	4·769	4·487	4·278	4·116
0·40	4·500	4·102	3·836	3·646	3·503	3·392
0·45	3·626	3·356	3·174	3·042	2·943	2·865
0·50	3·000	2·813	2·686	2·593	2·523	2·468
0·55	2·537	2·406	2·316	2·250	2·200	2·160
0·60	2·185	2·092	2·027	1·980	1·944	1·916
0·65	1·911	1·845	1·798	1·765	1·739	1·718
0·70	1·694	1·646	1·614	1·589	1·571	1·557
0·75	1·519	1·485	1·462	1·445	1·432	1·422
0·80	1·375	1·352	1·336	1·325	1·316	1·309
0·85	1·256	1·241	1·231	1·223	1·218	1·213
0·90	1·156	1·148	1·141	1·137	1·134	1·131
0·95	1·072	1·068	1·065	1·063	1·062	1·061
1·00	1·000	1·000	1·000	1·000	1·000	1·000

Note that the above table applies only for values of β_I greater than $\alpha_I/(2 - \alpha_I)$.

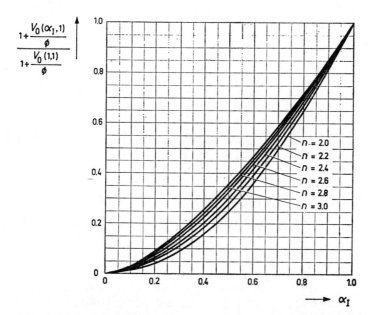

Figure 8.22. Polarization with partial pumping and minimum bias

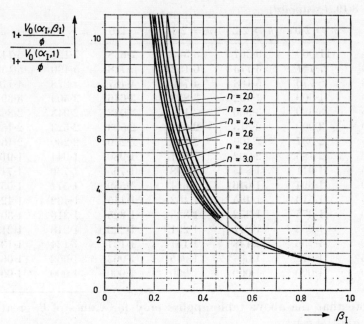

Figure 8.23. Polarization with arbitrary bias partial pumping

5.6. Pumping power

Table 8.20. Full pumping

n	$\dfrac{P(1, 1)}{P_{\text{norm}}(\omega_p/\omega_{c\max})^2}$	n	$\dfrac{P(1, 1)}{P_{\text{norm}}(\omega_p/\omega_{c\max})^2}$
2·0	0·500	2·6	0·330
2·2	0·420	2·8	0·301
2·4	0·366	3·0	0·281

The tabulation for partial pumping with minimum polarization is very simple since $P(\alpha_I, 1)/P(1, 1)$ does not depend on n.

Table 8.21. Partial pumping with minimum bias

α_I	$\dfrac{P(\alpha_I, 1)}{P(1, 1)}$	α_I	$\dfrac{P(\alpha_I, 1)}{P(1, 1)}$
0·0	0·000	0·6	0·360
0·1	0·010	0·7	0·490
0·2	0·040	0·8	0·640
0·3	0·090	0·9	0·810
0·4	0·160	1·0	1·000
0·5	0·250		

Finally, the case of arbitrary pumping presents no problems since P does not depend on β_I. We then still have:

$$\frac{P(\alpha_I, \beta_I)}{P(\alpha_I, 1)} = 1$$

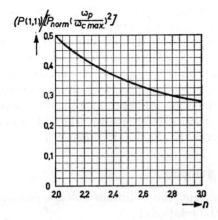

Figure 8.24. Pump power with full pumping

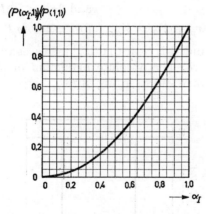

Figure 8.25. Pump power with partial pumping

6. Pumping optimization

6.1. General. Pumping of negative-resistance amplifiers and sum-frequency converters

The choice of the best conditions for pumping is specific to each parametric device. Depending on whether we are designing a multiplier, a frequency divider, a negative-resistance amplifier, a low-frequency amplifier or any other parametric system, it is essential to obtain optimum values for the $|C_k|$'s or the $|S_k|$'s or for more or less complex functions of these quantities. It is not possible here to envisage all the problems that may occur. For example, we may be considering the specially important case of three-frequency amplifiers and converters, degenerate or non-degenerate.

The performance of these devices improves as the dynamic cut-off frequency, $\omega_c' = |S_1|/R_s$, rises. For a given varactor, R_s is a constant and it is required to give $|S_1|$ a maximum value. We will compare the values of $|S_1|$ supplied by voltage and current pumping.

We will then study the influence of ν on the maximum value of $|S_1|$. For the purposes of reference we will first study square wave pumping which gives the maximum possible $|S_1|$.

6.2. Square-wave pumping

For a given amplitude, the optimum voltage pumping is a square wave defined by the relations:

$$v_p + V_0 = V_{\max} \quad \text{for} \quad 0 \leqslant \omega_p t \leqslant \pi/2$$
$$\text{and} \; 3\pi/2 \leqslant \omega_p t \leqslant 2\pi$$
$$v_p + V_0 = V_{\min} \quad \text{for} \quad \pi/2 \leqslant \omega_p t \leqslant 3\pi/2$$

On the other hand, it is clear that $|S_1|$ increases with the amplitude of the square wave. The absolute maximum of $|S_1|$ will then be supplied by the wave shown in Fig. 8.26.

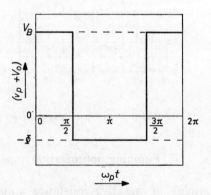

Figure 8.26. Optimum waveform of the pumping voltage

The elastance varies also as a square wave from $S_{\max}$ to 0 (Fig. 8.27).

PUMPING OF DIODES FOR MICROWAVE FREQUENCIES   239

We then at once obtain:

$$\frac{|S_1|}{S_{\max}} = \frac{1}{2\pi} \int_{-\pi/2}^{+\pi/2} \cos x \, dx = \frac{1}{\pi} \simeq 0.318 \tag{8.158}$$

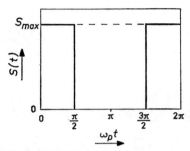

Figure 8.27. Shape of the elastance for square-wave pumping

6.3. Comparison between voltage and current pumping

6.3.1. VOLTAGE PUMPING

In the most general case (any given values of α_v and β_v) we have:

$$\frac{|S_1|}{S_{\max}} = \frac{\alpha_v^\nu}{4^\nu} \cdot \frac{\Gamma(1+2\nu)}{\Gamma(2+\nu)\Gamma(\nu)}$$

$$\times \beta_v^{1-\nu} \frac{\Gamma\left(1+\dfrac{\nu}{2}\right)\Gamma\left(\dfrac{3+\nu}{2}\right)}{\Gamma\left(\dfrac{1}{2}+\nu\right)} F\left(\frac{1-\nu}{2}, \frac{2-\nu}{2} \Big| 2|\beta_v^2\right) \tag{8.159}$$

We then have at once:

$$\frac{d}{d\beta_v}\left(\frac{|S_1|}{S_{\max}}\right) = \frac{\alpha_v^\nu}{4^\nu} \cdot \frac{\Gamma(1+2\nu)\Gamma\left(1+\dfrac{\nu}{2}\right)\Gamma\left(\dfrac{3+\nu}{2}\right)}{\Gamma(\nu)\Gamma(2+\nu)\Gamma\left(\dfrac{1}{2}+\nu\right)}$$

$$\times \left[(1-\nu)\beta_v^{-\nu} F\left(\frac{1-\nu}{2}, \frac{2-\nu}{2} \Big| 2|\beta_v^2\right)\right.$$

$$\left. + 2\beta_v^{2-\nu} \frac{(1-\nu)(2-\nu)}{8} F\left(\frac{3-\nu}{2}, \frac{4-\nu}{2} \Big| 3|\beta_v^2\right)\right] > 0 \tag{8.160}$$

$|S_1|$ will then be a maximum for a maximum value of β_v, namely, $\beta_v = 1$ (minimum bias). Further, we see at once that $|S_1|$ increases with α_v. The optimum pumping is therefore full pumping ($\alpha_v = \beta_v = 1$). We then have:

$$\frac{|S_1|}{S_{\max}} = \frac{1}{4^\nu} \left| \frac{\Gamma(1 + 2\nu)}{\Gamma(2 + \nu)\Gamma(\nu)} \right| \tag{8.161}$$

The pumping power required is:

$$\frac{P}{P_{\text{norm}}\left(\dfrac{\omega_p}{\omega_{c\,\max}}\right)^2} = \frac{1}{(1-\nu)16^{1-\nu}} \cdot \frac{\Gamma(3-4\nu)}{[\Gamma(2-2\nu)]^2} \tag{8.162}$$

6.3.2. CURRENT PUMPING

It can be shown that $|S_1|$ is a maximum with full pumping. We then have:

$$\frac{|S_1|}{S_{\max}} = \frac{1}{4^{\nu/(1-\nu)}} \left| \frac{\Gamma[1 + 2\nu/(1-\nu)]}{\Gamma[2 + \nu/(1-\nu)]\Gamma[\nu/(1-\nu)]} \right| \tag{8.163}$$

The pumping power required is:

$$\frac{P}{P_{\text{norm}}(\omega_p/\omega_{c\,\max})^2} = \frac{1}{8} \cdot \frac{1}{(1-\nu)^2} \tag{8.164}$$

6.3.3. COMPARISON BETWEEN METHODS OF PUMPING

We saw in section 1.1 that the number n lay between 2 and 3; ν then lies between $\frac{1}{2}$ and $\frac{1}{3}$. We will calculate the curves of $|S_1|$ and P as functions of n. The numerical values are given in Table 8.22. The corresponding curves are plotted in Figs. 8.28 and 8.29.

Table 8.22

Pumping		Voltage		Current	
n	ν	$\dfrac{\|S_1\|}{S_{\max}}$	$\dfrac{P}{P_{\text{norm}}(\omega_p/\omega_{c\max})^2}$	$\dfrac{\|S_1\|}{S_{\max}}$	$\dfrac{P}{P_{\text{norm}}(\omega_p/\omega_{c\max})^2}$
2·0	0·500	0·212	0·500	0·250	0·500
2·2	0·454	0·204	0·408	0·243	0·419
2·4	0·416	0·197	0·353	0·236	0·366
2·6	0·385	0·191	0·316	0·228	0·330
2·8	0·357	0·184	0·289	0·221	0·302
3·0	0·333	0·179	0·267	0·212	0·281

Examination of these curves shows that, whatever the varactor

PUMPING OF DIODES FOR MICROWAVE FREQUENCIES 241

used (value of n), current pumping gives a higher value of $|S_1|/S_{max}$. However, current pumping requires a slightly higher power when n is greater than 2. When, for practical reasons, the pump power is limited, we must compare the $|S_1|$ obtained for the same pump power. We have therefore calculated the curves for $|S_1|/S_{max}$ as a function of ν by fixing the power at the value required for full voltage pumping. The current pumping is therefore partial, and assumed to be with minimum bias. The numerical values are given in Table 8.23. The corresponding curves are plotted in Fig. 8.30.

Table 8.23

Pumping	Voltage		Current	
n	$\dfrac{|S_1|}{S_{max}}$	α_I^2	$\alpha_I \dfrac{\nu}{1-\nu}$	$\dfrac{|S_1|}{S_{max}}$
2·0	0·212	1·000	1·000	0·250
2·2	0·204	0·983	0·993	0·241
2·4	0·197	0·960	0·986	0·233
2·6	0·191	0·958	0·987	0·225
2·8	0·184	0·954	0·987	0·217
3·0	0·178	0·951	0·988	0·209

We may conclude that current pumping is preferable to voltage pumping when we wish to make $|S_1|$ a maximum.

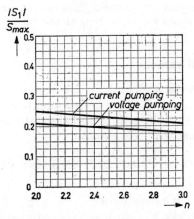

Figure 8.28. Comparison between the $|S_1|$ obtained for complete voltage and current pumping

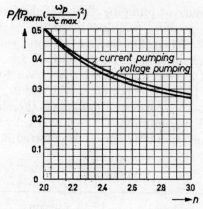

Figure 8.29. Power required for complete voltage and current pumping

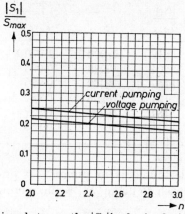

Figure 8.30. Comparison between the $|S_1|$'s obtained with voltage and current pumping when the pump power is fixed at the value required to give full voltage pumping.

6.4. Comparison of varactors

The following criteria are used for this comparison: the dynamic cut-off frequency ω'_c and the required pump power P. Since full current pumping is the best from all points of view, we will assume for this purpose that it is being used.

6.4.1. DYNAMIC CUT-OFF FREQUENCY WITH ANY VALUE OF P

When an unlimited pump power is available, all that is needed is to compare the ω'_c. We may write:

$$\omega'_c = \frac{|S_1|}{R_s} = \frac{|S_1|}{S_{\max}} \cdot \frac{S_{\max}}{R_s} = \frac{|S_1|}{S_{\max}} \omega_{c\,\max} \qquad (8.165)$$

$$\frac{\omega_c'}{\omega_{c\,\text{max}}} = \frac{|S_1|}{S_{\text{max}}} = \frac{1}{4^{\nu/(1-\nu)}} \left| \frac{\Gamma[1 + 2\nu/(1-\nu)]}{\Gamma[2 + \nu/(1-\nu)]\Gamma[\nu/(1-\nu)]} \right| \quad (8.166)$$

Table 8.24 and Fig. 8.31 show the value of $\omega_c'/\omega_{c\,\text{max}}$ for indices of n varying from 2 to 3.

Table 8.24

n	$\omega_c'/\omega_{c\text{max}}$	n	$\omega_c'/\omega_{c\text{max}}$
2·0	0·250	2·6	0·228
2·2	0·243	2·8	0·220
2·4	0·236	3·0	0·212

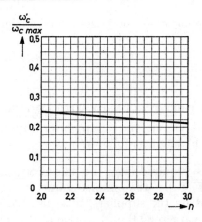

Figure 8.31. Maximum dynamic cut-off frequency as a function of n

We see that with a fixed $\omega_{c\,\text{max}}$, the square root varactors ($n = 2$) are the best. Designating by $\Omega_{c\,\text{max}}$ the static cut-off frequency of a square root varactor, a varactor with any value of n will enable us to obtain the same ω_c' if it possesses a static cut-off frequency $\omega_{c\text{max}}$ given by:

$$\omega_{c\,\text{max}} = k_1(n)\Omega_{c\,\text{max}} \quad (8.167)$$

with $k_1(n)$ given by Table 8.25 and Fig. 8.32.

Table 8.25

n	$k_1(n)$	n	$k_1(n)$
2·0	1·000	2·6	1·095
2·2	1·030	2·8	1·135
2·4	1·060	3·0	1·180

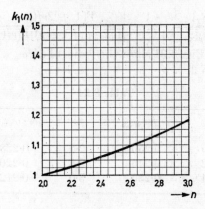

Figure 8.32. Relative importance of the static cut-off frequency as a function of n

6.4.2. DYNAMIC CUT-OFF FREQUENCY WITH LIMITED P

When, for practical reasons, the pump power is limited, the choice of the varactor may be made as follows:

The performance required from the parametric amplifier fixes the minimum value of ω_c' that the pumped varactor must possess. From the data furnished by the suppliers, it is possible to determine for all the available varactors the $\omega_{c\,\max}$ and (by means of the coefficient $k_1(n)$ defined in the previous paragraph) the ω_c' obtainable by means of full current pumping. The next step is to calculate the power required for full pumping of the varactors which give ω_c' the required minimum value. This power depends on the factors $\omega_{c\,\max}$, V_B, ϕ and R_s which are not directly tied to the type of varactor used (value of n). There is therefore no general rule which, for a given ω_c' enables one value of n to be preferred to another. The choice depends entirely on numerical calculation. If, among the varactors which give the required minimum value of ω_c', there is one for which the pumping does not require a power greater than the power available, the problem is solved. If this does not happen, the choice is made of a varactor of a higher quality than would be used in partial pumping. The coefficient β_I will be taken equal to 1 (minimum polarization) and the coefficient α_I such that $\alpha_I^{\nu/(1-\nu)}\omega_{c\,\max}'$ is equal to the required value of ω_c' (where $\omega_{c\,\max}'$ designates the ω_c' obtained by full pumping). It should be noted that if the varactor is characterized by $n = 2$, β_I can be arbitrarily fixed without affecting ω_c'.

References

[1] Morse, P. M. and Feshbach, H., *Methods of Theoretical Physics*, McGraw-Hill, 1953.

[2] Angot, A., 'Compléments de mathématiques à l'usage des ingénieurs de l'électrotechnique et des télécommunications', *Editions de la Revue d'Optique*, 1957.

[3] Jahnke, E. and Emde, F., *Tables of Functions*, Dover Publications, 1945.

[4] Dwight, H. B., *Tables of Integrals and other Mathematical Data*, Macmillan, 1957.

[5] Sensiper, S. and Weglein, R. D., 'Capacitance and charge coefficient for parametric diode devices', *Proc. I.R.E.*, pp. 1482–1483, August 1960.

[6] Penfield, P. and Rafuse, R., *Varactor Applications*, M.I.T. Press, 1962.

[7] Lienard, J. C., 'Le pompage des diodes paramétriques', *Revue M.B.L.E.*, vol. VII, No. 2, (8ᵉ année).

9

MEASUREMENT OF PARAMETRIC AMPLIFIERS

by J. C. LIENARD

Introduction

In this chapter we will give a brief outline of the measurements required for the design, adjustment and testing of parametric amplifiers; noise measurements will not be dealt with as they will be discussed in the next chapter.

The first section treats of 'cold' measurements, based on the special properties of the Smith chart. We will be dealing with only the simplest cases: sum-frequency converters, adjusted for either maximum gain, or minimum noise; inverting converters adjusted for minimum intrinsic noise and reflection amplifiers adjusted for minimum noise. Other cases can be treated in just the same way without difficulty except that lengthy calculations will be involved.

In the second section, we will discuss the adjustment of parametric amplifiers. The adjustment of the pump circuit, in which there are some special problems, will be dealt with first. Next, by way of an example, and because of its practical importance, we will discuss the adjustment of the other circuits concerned in the reflection amplifier. We will not be dealing with other parametric systems, since the principles pertaining to the reflection amplifier can be readily extended to them.

The third section is devoted to the measurement of varactors and their equivalent circuits, including the parasitic elements L_c and C_c.

Finally, in the fourth section we will summarize the properties and the methods of measurements for two-ports equivalent to the circuit elements.

1. The circle diagram and its interpretation

1.1. Properties of the Smith chart

The immitance of a one-port may be represented by a point on a graph whose co-ordinates respectively correspond to the real and imaginary parts of this immittance. This immittance graph is shown in Fig. 9.1. The unit chosen is the characteristic immittance of the transmission line to which the one-port is connected. A conformal mapping of this diagram can be effected by an inversion of radius 2 for which the pole is the coordinate point $r = -1$, $x = 0$ ($g = -1$, $b = 0$). In this way, we obtain the well-known Smith chart. All the

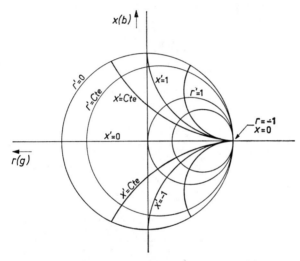

Figure 9.1. Relation between the graph (r, x) and the Smith chart

impedances (admittances) represented in the graph are designated by $z(y)$, and these same quantities represented on the Smith chart are designated by $z'(y')$.

This section will be devoted to the measurement of an impedance locus characterized by $r = $ const. through a passive, non-dissipative and reciprocal two-port. It will be seen later (in section 4 of this Chapter) that a two-port of this type may be represented by an equivalent circuit applicable to a single frequency and for which there are numerous variants. We will represent it by Fig. 9.2 which includes two sections of transmission line of unit impedance incorporating a series reactance jx_a. The point representing the impedance z_1 is a part of a locus characterized by the relation $r = r_1$.

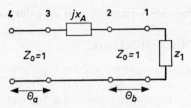

Figure 9.2. Equivalent circuit of a non-dissipative, passive and reciprocal two-port

Figure 9.3 shows the impedance loci measured at 1, 2, 3 and 4 in the plan (r, x) and on the Smith chart. In order to make the figure clearer the loci x have been shown by full lines, the loci x' by broken lines and the constructions by chain-dotted lines.

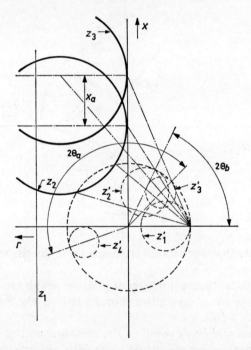

Figure 9.3. Transformation of a locus of constant resistance by a reciprocal non-dissipative and passive circuit

The representation of the loci z_1 and z'_1 is direct. The locus z'_2 is obtained from z'_1 by rotation through an angle $2\theta_b$ around the origin. The locus z_2 is derived from z'_2 by inversion. As the inversion is a conformal transform, the point of tangency between circles z'_2 and $r' = 0$ gives a tangency point between z_2 and the axis $r = 0$. The im-

MEASUREMENT OF PARAMETRIC AMPLIFIERS 249

pedance measured at 3 is obtained by adding jx_a to the impedance measured at 2. The locus z_3 is then obtained by translation of z_2 parallel to the axis $r = 0$. The locus of z'_3, inverse of the locus of z_3, possesses a point of tangency with the circle $r' = 0$. Finally, the locus z'_4 is obtained from z'_3 by rotation through an angle $2\theta_a$ around the origin.

From what has been stated above, we can formulate the following theorem: if an impedance locus is defined by the relation $r = $ const., the transformation of this locus by a reciprocal, non-dissipative and passive two-port produces a circle tangential to the circle $r' = 0$ in the Smith chart, that is, a locus which, after a suitable rotation, may be superimposed on a locus of constant resistance. It is clear that the placing of a resistance of any value r_p, in series with z_1, does not modify this conclusion since it is equivalent simply to replacing the relation $r = r_1$ by the relation $r = r_p + r_1$ for which the above theorem is applicable. Conversely, the placing in series of a resistance, r_p, at any point in the two-port considered above will usually cause the disappearance of the point of tangency between the measured locus and the circle $r' = 0$, as it corresponds to a translation parallel to the axis $x = 0$ in the graph (r, x).

This property has already been applied to the analysis of the circuits of several types of parametric amplifiers [1]. In the following paragraphs, we will consider this analysis in some detail for sum-frequency and difference-frequency converters and for the reflection amplifier. A description will be given of a method of measuring the pumping coefficient, α_I, and the maximum static cut-off frequency $\omega_{c\,\max}$.

It will be assumed that the parametric device is adjusted. In one of the circuits we will replace the external resistance, R, by an impedance-measuring system whose internal resistance is equal to R. The designation, 'circle diagram', will be given to the impedance locus of this circuit measured by varying the polarization of the varactor from $-\phi$ to V_B. As the varactor consists of a resistance, R_s, in series with a reactance, which is a function of the polarization, the above-mentioned properties may be applied to the circle diagrams of parametric amplifiers.

1.2. Analysis of pump circle diagram

1.2.1. IDEAL CASE

We will first consider the ideal case where the pump circuit has no losses and the pumping itself is strictly current pumping. An impedance-measuring set (for example a slotted line) is placed between the amplifier and the pump generator. The impedance is read as a function of the bias applied to the varactor while maintaining the level of the signal sufficiently low to avoid any pumping liable to modify the impedance. The locus of the impedance is shown on the Smith chart. We already know, from the theorem mentioned in the previous paragraph, that this locus is a circle (more precisely, an arc of a circle) tangential to the circle $r' = 0$. Further, there is a value of bias which gives to the reactive part of the impedance the value it has when pumping takes place. When this bias is applied, the non-pumped varactor is matched with the impedance of the generator,

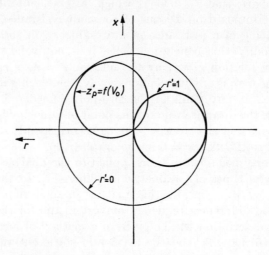

Figure 9.4. Pump circle diagram. Ideal case

that is, of the transmission line connecting it to the pump circuit. The locus of the impedance of the pump, z'_p, then passes through the centre of the Smith chart and may, if suitably rotated, coincide with the circle, $r' = 1$ (Fig. 9.4).

1.2.2. INFLUENCE OF THE PUMP HARMONICS

In theory it is possible to obtain pure current pumping by means of a generator if it has infinite internal impedance. This condition cannot be realized in practice, since the impedance of the generator is determined by the transmission lines used. In order to obtain current pumping, it is therefore necessary to use filters tuned to the main harmonic of the pump frequency to prevent currents of frequencies different from the fundamental flowing in the varactor. If these filters are badly adjusted, harmonic currents flow in the parasitic resistance, R_s, of the varactor, where they dissipate power. The apparent resistance of the pumped varactor seen from the pump generator

MEASUREMENT OF PARAMETRIC AMPLIFIERS 251

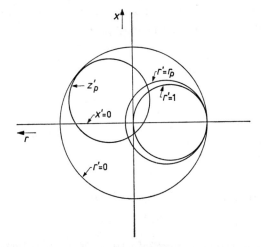

Figure 9.5. Pump circle diagram. Harmonics incorrectly terminated

becomes larger than R_s. The circle for the impedance, z'_p, of the non-pumped varactor will then no longer pass through the centre of the chart, but by rotation around this centre, will coincide with a circle of constant resistance, $r' = r_p < 1$ (Fig. 9.5).

1.2.3. INFLUENCE OF THE LOSSES IN THE PUMP CIRCUIT

The losses in the pump circuit usually cannot be ignored; good quality varactors have resistances, R_s, of the order of several ohms. On the other hand, the pump generator (the frequency of which is

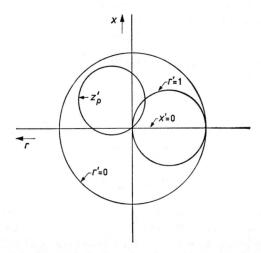

Figure 9.6. Pump circle diagram: the pump circuit has losses

very high) will most often be connected to the amplifier by means of a waveguide. We know that the matching of a low resistance to a waveguide requires a high-ratio transformation circuit which will inevitably introduce considerable losses. Even when the generator is connected by means of a coaxial line (the impedance of which is several tens of ohms), the impedance transformation required is too large to be effected without appreciable losses. The loss resistances of the matching circuit may even be greater than R_s. The pump circle diagram will usually be as shown in Fig. 9.6. There is no longer a point of tangency between the locus z'_p and circle $r' = 0$.

1.3. Circle diagrams of small-signal circuits

The following assumptions will be made in all that follows:

(a) current pumping is effected with $\beta_I = 1$ and arbitrary value of α_I (partial pumping with minimum bias)
(b) the transformation which converts the characteristic impedance of the line used to the impedance required to terminate the varactor is loss free.
(c) the locus of the varactor impedance is measured through the matching circuit by varying the d.c. bias from $-\phi$ to V_B; the level of the signal chosen is low enough to avoid any change of the measured impedance. We will discuss in detail the case of the input circuit.

The circuit, equivalent to the measured circuit, is shown in Fig. 9.7. At the terminals AA', the signal generator is connected. The impedance, Z_0, of this generator is that of the transmission line used.

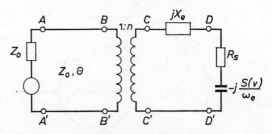

Figure 9.7. Equivalent circuit for the measurement of the input circuit

The two-port inserted between the terminals AA' and DD' represents the loss-free matching circuit. Finally, a varactor reduced to R_s and $S(V)$ in series is connected to terminals DD'. The parasitic elements L_c and C_c are considered as part of the matching circuit.

The impedance seen by the varactor (that is measured towards the left of DD' by replacing the varactor by a generator) is equal to $(R_e + jX_e)$ with:

$$R_e = n^2 Z_0 \tag{9.1}$$

in which R_e should have a value fixed by the type of amplifier and the chosen performance.

The impedances measured towards the right at DD', CC', and BB' successively when V varies are:

$$Z_{DD'}(V) = R_s - j\frac{S(V)}{\omega_e} \tag{9.2}$$

$$Z_{CC'}(V) = R_s + j\left(X_e - \frac{S(V)}{\omega_e}\right) \tag{9.3}$$

$$Z_{BB'}(V) = \frac{1}{n^2}\left[R_s + j\left(X_e - \frac{S(V)}{\omega_e}\right)\right] \tag{9.4}$$

The impedance reduced to Z and measured at BB' is equal, if we take account of (9.1), to:

$$\frac{Z_{BB'}}{Z_0} = \frac{n^2 Z_{BB}}{R_e} = \frac{R_s}{R_e} + j\frac{X_e - S(V)/\omega_e}{R_e} \tag{9.5}$$

Reactance jX_e should be chosen so as to cancel the reactance $-j[S_0(\alpha_I, 1)/\omega_e]$ of the pumped varactor. We then have:

$$X_e = \frac{S_0(\alpha_I, 1)}{\omega_e} = \alpha_I^{1/(n-1)}\frac{S_0(1, 1)}{\omega_e} \tag{9.6}$$

The coefficient $k_0(n)$ will be defined as the ratio of $S_0(1, 1)/S_{max}$ when $n = 2$ to $S_0(1, 1)/S_{max}$ when n has any value. Table I shows the value of $k_0(n)$ as a function of n.

Table 9.1. Values of $k_0(n)$

n	$S_0(1, 1)/S_{max}$	$k_0(n)$	n	$S_0(1, 1)/S_{max}$	$k_0(n)$
2·0	0·500	1·000	2·6	0·592	0·845
2·2	0·535	0·935	2·8	0·618	0·809
2·4	0·566	0·884	3·0	0·637	0·785

We then have:

$$X_e = \frac{\alpha_I^{1/(n-1)}}{2k_0(n)}\cdot\frac{S_{max}}{\omega_e} = R_s\frac{\alpha_I^{1/(n-1)}}{2k_0(n)}\cdot\frac{\omega_{c\,max}}{\omega_e} \tag{9.7}$$

$$\frac{Z_{BB'}}{Z_0} = \frac{R_s}{R_e}\left[1 + j\left(\frac{\alpha_I^{1/(n-1)}}{2k_0(n)}\cdot\frac{\omega_{c\,max}}{\omega_e} - \frac{S(V)}{R_s\omega_e}\right)\right] \tag{9.8}$$

In particular, we have:

for $V = -\phi$,
$$z'_1 = \frac{Z_{BB}(-\phi)}{Z_0} = \frac{R_s}{R_e}\left[1 + j\frac{\alpha_I^{1/(n-1)}}{2k_0(n)} \cdot \frac{\omega_{c\,max}}{\omega_e}\right] \quad (9.9)$$

for $V = V_B$,
$$z'_2 = \frac{Z_{BB}(V_B)}{Z_0} = \frac{R_s}{R_e}\left[1 - j\frac{\omega_{c\,max}}{\omega_e}\left(1 - \frac{\alpha_I^{1/(n-1)}}{2k_0(n)}\right)\right] \quad (9.10)$$

The locus of z'_{BB} (representation of Z_{BB}/Z_0 on the Smith chart) is then an arc of a constant resistance circle (Fig. 9.8).

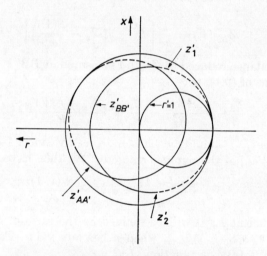

Figure 9.8. Circle diagram of the input circuit

We then have at once:

$$z'_1 = r'_1 + jx'_1 \quad (9.11)$$

$$x'_1 = \frac{R_s}{R_e} \cdot \frac{\alpha_I^{1/(n-1)}}{2k_0(n)} \frac{\omega_{c\,max}}{\omega_e} \quad (9.12)$$

$$z'_2 = r'_2 + jx'_2 \quad (9.13)$$

$$x'_2 = -\frac{R_s}{R_e} \cdot \frac{\omega_{c\,max}}{\omega_e}\left(1 - \frac{\alpha_I^{1/(n-1)}}{2k_0(n)}\right) \quad (9.14)$$

$$r'_1 = r'_2 = \frac{R_s}{R_e} \quad (9.15)$$

MEASUREMENT OF PARAMETRIC AMPLIFIERS   255

The following points can be checked by measurement of the locus $z'_{BB'}$:

(a) the absence of parasitic losses: the circle extended beyond z'_1 and z'_2 should be tangential to the external circle of the Smith chart;

(b) the value of R_e/R_s, determined by

$$r'_1 = r'_2 = \frac{R_s}{R_e} \qquad (9.16)$$

(c) the value of $\omega_{c\ max}$, determined by the relation

$$x'_1 - x'_2 = \frac{R_s}{R_e} \cdot \frac{\omega_{c\ max}}{\omega_e} \qquad (9.17)$$

(d) the value of α_I, determined by the relation (9.12).

Measurement of the circuit does not produce the locus $z'_{BB'}$ but the locus $z'_{AA'}$, which is the same as $z'_{BB'}$, rotated through an angle 2θ around the centre of the chart. All that is needed to obtain the locus of $z'_{BB'}$, therefore is to rotate $z'_{AA'}$ until it coincides with a circle of constant resistance.

The above paragraphs concerning the input circuit, of course, also apply to other circuits. All we need to do is to replace the index e in the formulae (9.1) to (9.15) by the index $(-)$ for the difference-frequency circuit and by $(+)$ for the sum-frequency circuit.

In the following paragraphs, we will apply this result to the main types of amplifiers studied in the previous chapters. We will limit ourselves to approximate formulae, which are applicable to high-quality varactors.

1.4. Sum-frequency converter adjusted for maximum gain.

1.4.1. INPUT CIRCUIT

The value of R_e is given by:

$$\frac{R_e}{R_s} = \sqrt{1 + \frac{q^2}{n_{(+)}}} \simeq \frac{q}{\sqrt{n_{(+)}}} \qquad (9.18)$$

This may be written:

$$\frac{R_e}{R_s} = \frac{\omega'_c}{\sqrt{\omega_e \omega_{(+)}}} = \frac{|S_1|}{R_s} \cdot \frac{1}{\sqrt{\omega_e \omega_{(+)}}}$$

$$= \frac{\alpha_I^{1/(n-1)} S_{max}}{4k_1(n) R_s \sqrt{\omega_e \omega_{(+)}}} = \frac{\alpha_I^{1/(n-1)}}{4k_1(n)} \cdot \frac{\omega_{c\ max}}{\sqrt{\omega_e \omega_{(+)}}} \qquad (9.19)$$

We then have:

$$r'_1 = r'_2 = \frac{4k_1(n)}{\alpha_I^{1/(n-1)}} \cdot \frac{\sqrt{\omega_e \omega_{(+)}}}{\omega_{c\,\text{max}}} \quad (9.20)$$

$$x'_1 = \frac{2k_1(n)}{k_0(n)} \cdot \sqrt{\frac{\omega_{(+)}}{\omega_e}} \quad (9.21)$$

$$x'_2 = -2k_1(n)\sqrt{\frac{\omega_{(+)}}{\omega_e}} \left(\frac{2}{\alpha_I^{1/(n-1)}} - \frac{1}{k_0(n)} \right) \quad (9.22)$$

1.4.2. OUTPUT CIRCUIT (SUM-FREQUENCY)

Since R_L is equal to R_e, we at once obtain:

$$r'_1 = r'_2 = \frac{4k_1(n)}{\alpha_I^{1/(n-1)}} \cdot \frac{\sqrt{\omega_e \omega_{(+)}}}{\omega_{c\,\text{max}}} \quad (9.23)$$

$$x'_1 = \frac{2k_1(n)}{k_0(n)} \sqrt{\frac{\omega_e}{\omega_{(+)}}} \quad (9.24)$$

$$x'_2 = -2k_1(n)\sqrt{\frac{\omega_e}{\omega_{(+)}}} \left(\frac{2}{\alpha_I^{1/(n-1)}} - \frac{1}{k_0(n)} \right) \quad (9.25)$$

1.5. Sum-frequency converter adjusted for minimum intrinsic noise.

1.5.1. INPUT CIRCUIT

The value of R_e is given by:

$$\frac{R_e}{R_s} = \sqrt{1 + q^2} \simeq q \quad (9.26)$$

This may be written:

$$\frac{R_e}{R_s} = \frac{\omega'_c}{\omega_e} = \frac{\alpha_I^{1/(n-1)}}{4k_1(n)} \cdot \frac{\omega_{c\,\text{max}}}{\omega_e} \quad (9.27)$$

We then have:

$$r'_1 = r'_2 = \frac{4k_1(n)}{\alpha_I^{1/(n-1)}} \cdot \frac{\omega_e}{\omega_{c\,\text{max}}} \quad (9.28)$$

$$x'_1 = \frac{2k_1(n)}{k_0(n)} \quad (9.29)$$

$$x'_2 = -2k_1(n)\left(\frac{2}{\alpha_I^{1/(n-1)}} - \frac{1}{k_0(n)} \right) \quad (9.30)$$

1.5.2. OUTPUT CIRCUIT

The value of R_L is given by:

$$\frac{R_L}{R_s} = 1 + \frac{q^2/n_{(+)}}{1 + \sqrt{1 + q^2}} \simeq 1 + \frac{q}{n_{(+)}} \quad (9.31)$$

This may be written:

$$\frac{R_L}{R_s} = 1 + \frac{\omega_c'}{\omega_{(+)}} = 1 + \frac{\alpha_I^{1/(n-1)}}{4k_1(n)} \cdot \frac{\omega_{c\,\max}}{\omega_{(+)}} \tag{9.32}$$

We then have:

$$r_1' = r_2' = \frac{1}{1 + \frac{\alpha_I^{1/(n-1)}}{4k_1(n)} \cdot \frac{\omega_{c\,\max}}{\omega_{(+)}}} \tag{9.33}$$

$$x_1' = \frac{2k_1(n)}{k_0(n)} \cdot \frac{1}{1 + \frac{4k_1(n)}{\alpha_I^{1/(n-1)}} \cdot \frac{\omega_{(+)}}{\omega_{c\,\max}}} \tag{9.34}$$

$$x_2' = \frac{-2k_1(n)\left(\frac{2}{\alpha_I^{1/(n-1)}} - \frac{1}{k_0(n)}\right)}{1 + \frac{4k_1(n)}{\alpha_I^{1/(n-1)}} \cdot \frac{\omega_{(+)}}{\omega_{c\,\max}}} \tag{9.35}$$

1.6. The inverting converter adjusted for minimum intrinsic noise

1.6.1. INPUT CIRCUIT

Resistance R_e is fixed by the instability factor chosen $S_{(-)}^e$. We then have:

$$\frac{R_e}{R_s} = \frac{q(S_{(-)}^e + 1)}{1 + \frac{n_{(-)}}{q}(S_{e(-)} - 1)} \simeq \frac{q^2}{n_{(-)}} \cdot \frac{S_{e(-)} + 1}{S_{e(-)} - 1} \tag{9.36}$$

This may be written:

$$\frac{R_e}{R_s} = \frac{\omega_c'^2}{\omega_e \omega_{(-)}} \cdot \frac{S_{e(-)} + 1}{S_{e(-)} - 1} = \frac{\alpha_I^{2/(n-1)}}{16 k_1^2(n)} \cdot \frac{\omega_{c\,\max}^2}{\omega_e \omega_{(-)}} \cdot \frac{S_{e(-)} + 1}{S_{e(-)} - 1} \tag{9.37}$$

We then have:

$$r_1' = r_2' = \frac{16 k_1^2(n)}{\alpha_I^{2/(n-1)}} \cdot \frac{\omega_e \omega_{(-)}}{\omega_{c\,\max}^2} \cdot \frac{S_{e(-)} - 1}{S_{e(-)} + 1} \tag{9.38}$$

$$x_1' = \frac{8 k_1^2(n)}{k_0(n)} \cdot \frac{1}{\alpha_I^{1/(n-1)}} \cdot \frac{\omega_{(-)}}{\omega_{c\,\max}} \cdot \frac{S_{e(-)} - 1}{S_{e(-)} + 1} \tag{9.39}$$

$$x_2' = -\frac{8 k_1^2(n)}{k_0(n)} \cdot \frac{\omega_{(-)}}{\omega_{c\,\max}} \left(\frac{2}{\alpha_I^{1/(n-1)}} - \frac{1}{k_0(n)}\right) \frac{S_{e(-)} - 1}{S_{e(-)} + 1} \tag{9.40}$$

1.6.2. OUTPUT CIRCUIT

Resistance R_L is given by:

$$\frac{R_L}{R_s} = \frac{q}{n_{(-)}(S_{(-)}^e - 1)} \tag{9.41}$$

In normal conditions $S^e_{(-)}$ is large by comparison with unity and the output circle will therefore be a very high resistance circle, the measurement of which is not very precise and without practical interest.

1.7. The reflection amplifier adjusted for minimum noise

1.7.1. INPUT CIRCUIT

Resistance R_e is fixed by the gain G_{ee} chosen. We then have:

$$\frac{R_g}{R_s} = \frac{q^2}{n_{(-)}} \frac{\sqrt{G_{ee}} + 1}{\sqrt{G_{ee}} - 1} \tag{9.42}$$

The expressions for r'_1, r'_2, x'_1 and x'_2 are therefore those found for the inverting converter provided $S_{e(-)}$ is replaced by $\sqrt{G_{ee}}$.

1.7.2. IDLER CIRCUIT

In the absence of an external load ($R_{(-)} = 0$) it is not possible to measure the circle.

2. Adjustment of a parametric amplifier

2.1. Adjustment of the pump circuit

2.1.1. GENERAL

This adjustment may be effected by means of the circuit shown in Fig. 9.9.

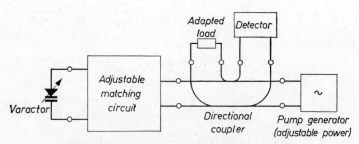

Figure 9.9. Adjustment of the pump circuit

The directional coupler, with its detector, measures the power of the wave reflected from the input of the matching circuit. This is adjusted so as a negligible reflection is obtained for a couple of values determined by the d.c.bias V_0 applied to the varactor, and by the power absorbed by the varactor. If one of these values is changed,

MEASUREMENT OF PARAMETRIC AMPLIFIERS 259

S_0 changes and the circuit is mismatched. It is for this reason that matching should be adjusted for the normal pumping conditions.

2.1.2. ARBITRARY BIAS PARTIAL PUMPING

The relations studied in the chapter on pumping enable us to calculate the bias voltage, V_0, to be applied to the terminals of the varactor. The required pumping cannot be obtained by adjusting the output power of the generator to the value given by the pumping equations and by matching. In fact, we saw, earlier that the loss resistances of the circuits absorb an appreciable fraction of the pump power. It is therefore necessary to experimentally determine the power required to pump the varactor up to near the potential $-\phi$, which can be checked by observing that a slight increase in the pump power involves the appearance of a direct current rectified by the varactor operating as a peak limiting device.

2.1.3. PARTIAL PUMPING WITH ANY GIVEN TYPE OF BIASING

This time, the pump voltage no longer reaches potential $-\phi$ and it is impossible to observe the flowing of the small current that indicates when the varactor is pumped. The procedure is therefore as follows: Firstly, the values of V_0 and P, corresponding to the required pumping, are calculated, and then the bias voltage, V_0', that must be applied to the varactor so that, when subjected to the pump power P, it will be pumped with minimum biasing. The matching and the pump power are adjusted by the small passing-current method described above and by applying this bias voltage, V_0'. The correct bias voltage, V_0, is then applied and the matching is again adjusted without changing the power of the generator. The pumping thus obtained will be of the required value, on condition that the losses of the matching circuit are the same for the two adjustments, which can be taken as correct provided these adjustments do not differ too widely.

2.1.4. BIAS SWEEPING METHOD

The presence of appreciable losses in the pump matching circuits may cause trouble. It happens sometimes that the generator is matched not with R_s but with the loss resistance of the circuit. This trouble can be obviated by adopting the bias sweeping method. A low-frequency alternating potential (50 Hz for example) is superimposed on the normal d.c. biasing. The output detector (Fig. 9.9) energizes the vertical deviation system of a cathode-ray oscillograph. The horizontal deviation is obtained by means of a voltage synchronized with the low-frequency voltage.

If the varactor absorbs an appreciable fraction of the power of the pump generator, its reactance will affect the matching with this generator and the reflected wave will depend on the instantaneous value of the low-frequency voltage applied to the varactor.

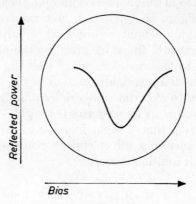

Figure 9.10. Matching of the pump circuit by bias scanning

It is thus possible during the matching to ensure that the varactor has pumping current flowing through it. Figure 9.10 shows a typical oscillograph trace recorded during the process of matching.

2.2. Adjustment of a reflection amplifier

The experimental set-up is illustrated in Fig. 9.11.

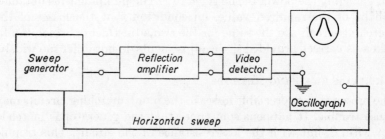

Figure 9.11. Adjustment of a reflection amplifier

We will assume at the start that the idler circuit is not exactly tuned to the correct frequency. This circuit reflects into the input circuit a negative resistance lower than expected and also a reactance. Input matching is used in order to obtain a response curve centred on a good frequency. When this is done, the input circuit has been detuned so as to compensate for the reactance introduced by the detuning of the idler.

If we modify the tuning of the idler, the response curve is modified and the peak amplitude changes to become a maximum when the sum of the tuning frequencies of the input and idler circuits becomes equal to f_p. Figure 9.12 shows several curves corresponding to successive positions of the idler matching transformer as well as the envelope of their maxima. The frequency f_m at which the maximum maximorum occurs is such that the idler is tuned to $f_p - f_m$. The input and idler matching transformers are adjusted so as to cause f_m

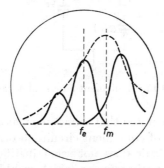

Figure 9.12. Response curves of a reflection amplifier

and f_e to coincide. At this instant, the idler is tuned to the correct frequency and the required gain can readily be obtained by an adjustment of the input circuit.

3. Measurement of varactors

3.1. Equivalent circuit of a varactor

The essential elements of a varactor are the non-linear capacitance $C(V)$ and the series parasitic resistance R_s. A knowledge of them enables us to calculate the limiting performance of a parametric amplifier in which a varactor is incorporated. The calculation of the impedance matching and the bandwidth, the evaluation of the influence of the circuit losses, and in short, the general conception of the amplifier requires, in addition, a knowledge of the parasitic reactive elements. These are the capacitance C_c of the cartridge and the inductance L_c of the cartridge. The complete circuit is shown in Fig. 9.13.

Blackwell and Kotzebue [2] have adopted the equivalent circuit in Fig. 9.13 after having verified its soundness by numerous tests.

We have tested the behaviour of numerous varactors of different manufacturers, and the results are in full accord with the circuit of

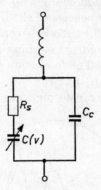

Figure 9.13. Equivalent circuit of a varactor

Fig. 9.13. It should be noted that numerous manufacturers and users employ another type of equivalent circuit, shown in Fig. 9.14.

The use of this circuit does not seem justified, as the varactor does not possess the antiresonance which would be expected when the orders of magnitude of L_c and C_c are taken into account. The

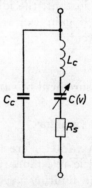

Figure 9.14. Another type of equivalent circuit (most probably inaccurate)

use of an equivalent circuit consisting entirely of lumped elements would seem to be dangerous in the microwave field. Nevertheless, the results of experience confirm the calculations performed on the basis of this circuit for frequencies which extend up to at least the X band (8.2–12.4 GHz).

MEASUREMENT OF PARAMETRIC AMPLIFIERS 263

3.2. Characteristics specified by the manufacturer

In order to predict the behaviour of a varactor, we must know the following parameters:

- ϕ Contact potential.
- V_B Zener voltage.
- R_s Series parasitic resistance.
- $C_{\min}$ Differential capacitance of the junction biased to the Zener voltage.
- n Exponent of the root defining $C(V)$.
- C_c Capacitance of the cartridge.
- L_c Inductance of the cartridge.

The manufacturer usually supplies the values ϕ, V_B, r, C_c and L_c. The values of C_c and L_c may be considerably influenced by the system adopted for mounting the varactor. The value given should therefore be considered as an order of magnitude and checked by measurement.

Catalogues often give the total capacitance for a given biasing. This capacitance enables $C(V)$ to be determined by subtracting C_c from it (provided that the value given for C_c is confirmed by measurement). Finally, from our knowledge of n and ϕ, we can determine $C_{\min}$ by means of the relations:

$$C_{\min} = \frac{C_{0v}}{\sqrt[n]{1 + V_B/\phi}} = C(V)\sqrt[n]{\frac{1 + V/\phi}{1 + V_B/\phi}} \tag{9.43}$$

The value of R_s is seldom supplied. Catalogues tend rather to give the quality factor measured under standardized conditions and defined by the relation:

$$Q(V,f) = \frac{1}{2\pi R_s C(V) f} \tag{9.44}$$

Also, the static cut-off frequency f_c corresponding to a given bias is often given:

$$f_c(V) = \frac{1}{2\pi R_s C(V)} \tag{9.45}$$

We see, from (9.44) and (9.45), that we may write:

$$f_c(V) = fQ(V,f) \tag{9.46}$$

Finally, it should be noted that several manufacturers adopt a sign convention different from ours for the voltage applied to the varactor: the voltage in the conducting direction is considered as positive.

3.3. Measurement of the total capacitance

This measurement is made on a low-frequency bridge (100 kHz to 1 MHz), with fixed biasing. The accuracy of this measurement could be suspected since the results are used for a frequency range very different from the test frequency. However, it is confirmed by experience that the values of the capacitances are accurate even for very high frequencies.

3.4. Measurement of the quality factor Q

We saw in section 3.2 that catalogues often give the quality factor $Q(V, f)$. The principle of the measurement is as follows:

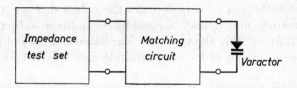

Figure 9.15. Measurement of the quality factor of a varactor

Let f be the test frequency. The matching circuit is adjusted until for a given bias the varactor is matched with the transmission line. The impedance is then measured by varying the bias from $-\phi$ to the voltage V at which we wish to measure Q. This measurement is analogous to those described in the first section of this chapter. The locus of the measured impedance is, on the Smith chart, the circle of unit resistance rotated through a certain angle around the centre (Fig. 9.16).

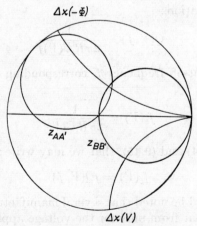

Figure 9.16. Measurement of the impedance for the determination of Q

The circuit equivalent to the matching system (in which the parasitic elements L_c and C_c are introduced) is, for example, that in Fig. 9.17.

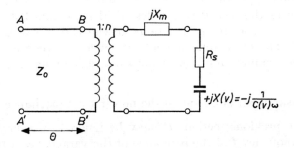

Figure 9.17 Equivalent circuit of the matching system

By rotation of the circle $z_{AA'}$ through an angle 2θ around the centre of the chart we obtain the locus $z_{BB'}$ (Fig. 9.16). We then at once have:

$$\frac{n^2 R_s}{Z_0} = 1 \tag{9.47}$$

$$\frac{n^2 X_m}{Z_0} = \Delta(x-\phi) \tag{9.48}$$

$$\frac{n^2[X_m + X_v(V)]}{Z_0} = +\Delta x(V) \tag{9.49}$$

$$X_v(V) = -\frac{1}{2\pi f C(V)} \tag{9.50}$$

$$Q(f, V) = \frac{-X_v(V)}{R_s} = \frac{\dfrac{n^2 X_m}{Z_0} - \dfrac{n^2}{Z_0}[X_m + X_v(V)]}{\dfrac{n^2 R_s}{Z_0}} \Bigg\} \tag{9.51}$$

$$= \Delta x(-\phi) - \Delta x(V)$$

The quality factor Q is then given by the range of the reduced reactance read on the Smith chart. However, it should be noted that the circuit losses are liable to cause the measurement to be incorrect, as any resistance R' placed in series with R_s, decreases the measured value of Q in the ratio $R_s/(R_s + R')$. This ratio may differ considerably from unity even for low values of R', since R_s is itself of the order of several ohms. The matching circuit should therefore be studied with very great care.

3.5. Measurement of L_c

Let us refer back to the circuit in Fig. 9.13. As R_s is low compared with the reactances of the circuit capacitances, it may be ignored with a good approximation. As the value of $C_T(V) = C(V) + C_c$ is known from the measurement described in section 3.3, all that is needed is to measure the reactance of the varactor at a given frequency f. The measurement made at several frequencies supplies a second determination of $C_T(V)$ and some check points.

3.6. Measurement of C_c and of the non-linear capacitance

As in the previous section, R_s may be ignored. Measurement at a known frequency f of the reactance of the varactor as a function of the bias enables us, with a known L_c, to calculate the curve for the total capacitance $C_T(V)$ as a function of V. For the purposes of verification, the same curve may be calculated from measurements made at other frequencies. It will be seen that from the curve $C_T = f(V)$ we can derive the quantities C_c, $C_{\min}$, n and ϕ. We may write:

$$C_T(V) = C_c + \frac{C_{0v}}{\sqrt[n]{1 + V/\phi}} \tag{9.52}$$

By graphical differentiation we obtain the curve $dC_T/dV = f'(V)$ which, furthermore, conforms with the equation:

$$\frac{dC_T}{dV} = -\frac{C_{0v}}{n\phi}\left(1 + \frac{V}{\phi}\right)^{-(n+1)/n} \tag{9.53}$$

The value of dC_T/dV at $V = 0V$ then supplies the value of $C_{0v}/n\phi$. We may then write:

$$\log\left(-\frac{n\phi}{C_{0v}}\cdot\frac{dC_T}{dV}\right) = -\frac{n+1}{n}\log\left(1 + \frac{V}{\phi}\right) \tag{9.54}$$

Let

$$\log\left(-\frac{n\phi}{C_{0v}}\cdot\frac{dC_T}{dV}\right) = g(V) \tag{9.55}$$

The function $g(V)$ is known numerically. If we consider two values V' and V'' of V, we may write:

$$\frac{\log\left(1 + \dfrac{V'}{\phi}\right)}{\log\left(1 + \dfrac{V''}{\phi}\right)} = \frac{g(V')}{g(V'')} = A \tag{9.56}$$

The equation in ϕ

$$\log\left(1 + \frac{V'}{\phi}\right) - A \log\left(1 + \frac{V''}{\phi}\right) = 0 \tag{9.57}$$

will be solved numerically or graphically in order to find ϕ. The parameter $g(V)$ is then put in diagram as a function of $\log(1 + V/\phi)$. The resulting locus is a straight line with a slope $-(n+1)/n$. It is thus easy to determine n.

A knowledge of n and ϕ allows us to calculate C_{0v}, as the value of $C_{0v}/n\phi$ has already been found. C_c is then obtained from the relation:

$$C_c - C_T(ov) - C_{ov} \tag{9.58}$$

Finally, we find $C_{\min}$:

$$C_{\min} = \frac{C_{ov}}{\sqrt[n]{1 + V_B/\phi}} \tag{9.59}$$

For verification purposes we may plot the curve of $\log(C_T - C)$ as a function of $\log(1 + V/\phi)$. We should obtain a straight line with slope $-1/n$.

4. Equivalent circuits of passive two-ports

4.1. Representation of the circuit elements

This section deals with the representation at a fixed frequency of reciprocal and purely reactive two-ports. The behaviour of a two-port of this type is completely determined by any one of its matrices (impedance, admittance, chain or scattering) that is specified by three parameters. The operation could therefore be examined by means of an equivalent two-port of three elements. It should be noted that this equivalent circuit applies only to the frequency considered and cannot therefore serve for the calculation of the frequency response nor for the checking of the stability conditions of a system of which this circuit forms part.

A circuit element for microwave frequencies is always connected by means of transmission lines. We will therefore use equivalent circuits applicable to impedances reduced with respect to the characteristic impedances of these lines. Equivalent circuits can be classified into three groups:

(1) Circuits with lumped parameters.
(2) Circuits consisting of one or several lines of unit characteristic impedance (reduced characteristic impedance).

(3) Circuits consisting of one or several lines of arbitrary characteristic impedance.

All the above applies to a passive and purely reactive reciprocal two-port. If the circuit is symmetrical, the number of independent parameters is reduced to two and the same applies to the number of independent elements of the equivalent circuit.

4.2. Chain matrix of elementary two-ports

All three-element equivalent circuits may be constituted by the chain formation of three elementary two-ports (with a single ele-

Table 9.2

Quadripole	Circuit	Chain matrix
Series impedance	jX	$a = 1$ $b = jX$ $c = 0$ $d = 1$
Parallel admittance	jB	$a = 1$ $b = 0$ $c = jb$ $d = 1$
Ideal transformer	$n:1$	$a = n$ $b = 0$ $c = 0$ $d = 1/n$
Line section	Z_0, θ	$a = \cos\theta$ $b = jZ_0 \sin\theta$ $c = j(\sin\theta/Z_0)$ $d = \cos\theta$

ment). There are four elementary two-ports: series impedance, parallel admittance, ideal transformer, and a line section of characteristic impedance Z_0 and of electrical length θ. Table 9.2 shows the chain matrices of these two-ports.

MEASUREMENT OF PARAMETRIC AMPLIFIERS 269

4.3. Discussion: practical importance of the various equivalent circuits

The usefulness of an equivalent circuit is assessed in terms of the facility with which calculations can be made on the Smith chart. The operations which the use of equivalent circuits involve are:
 (a) addition of a reactance to an impedance;
 (b) addition of a susceptance to an admittance;
 (c) inversion of an immittance;
 (d) multiplication of an immittance by a constant (the effect of an ideal transformer or of a change in the characteristic impedance);
 (e) transformation of the immittance by a section of line.

It is difficult to perform operation (d) with accuracy. On the other hand, it is useful to reduce the number of operations to be performed. It is preferable, therefore, to use the equivalent circuits containing only sections of line of unit impedance and immittances of a single type (reactances or susceptances).

4.4. Study of the usual equivalent circuits

There are four circuits which satisfy the conditions laid down in the previous paragraph. Two important circuits may be added to these. The first was used in section 1.1. The second will be used in section 4.5. The chain matrices of these circuits are obtained by multiplication of the chain matrices of the elementary two-ports of which they are composed. The results are given in Table 9.3, which contains also the formulae for the calculation of the elements by means of the chain matrix for the first four circuits (the only ones for which this calculation is of interest).

4.5. Measurement of equivalent circuits by continuous displacement of a short circuit

This method of measurement is well known [3]. The results will be given here as a reminder.

1) General theory

Amongst all the possible forms of the equivalent circuit of a two-port, we will use that consisting of an ideal transformer (with $n \geqslant 1$) connected between two sections of line of unit characteristic

270 PARAMETRIC AMPLIFIERS

Table 9.3. The usual equivalent circuits

Circuit	Chain matrix	Calculation of the elements
jX_1 — jX_2 series, $Z_0=1$, length θ	$a = \cos\theta - X_1 \sin\theta$ $b = j[X_2(\cos\theta - X_1 \sin\theta)$ $\quad + \sin\theta + X_1 \cos\theta]$ $c = j\sin\theta$ $d = \cos\theta - X_2 \sin\theta$	$\theta = \arcsin(-jc)$ $X_1 = \cotan\theta - j\dfrac{a}{c}$ $X_2 = \cotan\theta - j\dfrac{d}{c}$
jB_1, jB_2 shunt, $Z_0=1$, length θ	$a = \cos\theta - B_2 \sin\theta$ $b = j\sin\theta$ $c = j[B_2(\cos\theta - B_1 \sin\theta)$ $\quad + \sin\theta + B_1 \cos\theta]$ $d = \cos\theta - B_1 \sin\theta$	$\theta = \arcsin(-jb)$ $B_1 = \cotan\theta - j\dfrac{d}{b}$ $B_2 = \cotan\theta - j\dfrac{a}{b}$
jX between two lines $Z_0=1$ of lengths θ_1, θ_2	$a = \cos(\theta_1 + \theta_2) - X\cos\theta_1 \sin\theta_2$ $b = j[\sin(\theta_1 + \theta_2) + X\cos\theta_1 \cos\theta_2]$ $c = j[\sin(\theta_1 + \theta_2) - X\sin\theta_1 \sin\theta_2]$ $d = \cos(\theta_1 + \theta_2) - X\sin\theta_1 \cos\theta_2$	$X = \pm\sqrt{(a-d)^2 - (b-c)^2}$ $\theta_1 - \theta_2 = \arctan\left(j\dfrac{a-d}{b-c}\right)$ $\theta_1 + \theta_2$ $= \arctan\left[\dfrac{-(a+d)X + j2(b+c)}{2(a+d) - j(b+c)X}\right]$

Table 9.3. (continued)

Circuit	Parameters	
$Z_0=1$ — jB — $Z_0=1$, with θ_1, θ_2	$a = \cos(\theta_1 + \theta_2) - B \sin\theta_1 \cos\theta_2$ $b = j[\sin(\theta_1 + \theta_2) - B \sin\theta_1 \sin\theta_2]$ $c = j[\sin(\theta_1 + \theta_2) + B \cos\theta_1 \cos\theta_2]$ $d = \cos(\theta_1 + \theta_2) - B \sin\theta_2 \cos\theta_1$	$B = \pm\sqrt{(a-d)^2 - (b-c)^2}$ $\theta_2 - \theta_1 = \arctan\left(j\dfrac{a-d}{b-c}\right)$ $\theta_1 + \theta_2$ $= \arctan\left[-\dfrac{(a+d)B + j2(b+c)}{2(a+d) - j(b+c)B}\right]$
$n{:}1$ transformer with jX, $Z_0=1$, θ	$a = n\cos\theta - \dfrac{X}{n}\sin\theta$ $b = j\!\left(n\sin\theta + \dfrac{X}{n}\cos\theta\right)$ $c = j\,\dfrac{1}{n}\sin\theta$ $d = \dfrac{1}{n}\cos\theta$	
$Z_0=1$ — $n{:}1$ transformer — $Z_0=1$, with θ_1, θ_2	$a = n\cos\theta_1 \cos\theta_2 - \dfrac{1}{n}\sin\theta_1 \sin\theta_2$ $b = j\!\left[n\cos\theta_1 \sin\theta_2 + \dfrac{1}{n}\sin\theta_1 \cos\theta_2\right]$ $c = j\!\left[n\sin\theta_1 \cos\theta_2 + \dfrac{1}{n}\cos\theta_1 \sin\theta_2\right]$ $d = \dfrac{1}{n}\cos\theta_1 \cos\theta_2 - n\sin\theta_1 \sin\theta_2$	

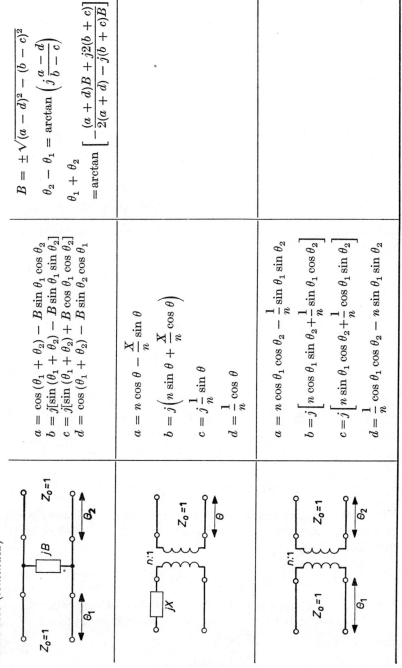

impedance. A mobile piston produces a short circuit at an electrical distance θ_{sc} (expressed in radians) from the output of the two-port. A measurement is made of the electrical distance θ_{min} separating the

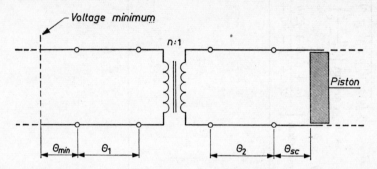

Figure 9.18. Measurement of a two-port by displacement of a short circuit

position of the minimum voltage from the input of the two-port Fig. (9.18). We then at once have:

$$n^2 \tan(\theta_2 + \theta_{sc}) = -\tan(\theta_1 + \theta_{min}) \tag{9.60}$$

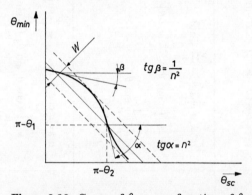

Figure 9.19. Curve of θ_{min} as a function of θ_{sc}

The curve of θ_{min} as a function of θ_{sc} then oscillates with a periodicity 2π around a straight line of slope -1 (Fig. 9.19). The slope of this curve is given by:

$$\frac{d\theta_{min}}{d\theta_{sc}} = -\frac{n^2}{1 + (n^4 - 1)\sin^2(\theta_2 + \theta_{sc})} \tag{9.61}$$

This slope is a maximum for $\theta_{sc} = \pi - \theta_2$ and the value $\theta_{min} = \pi - \theta_1$ corresponds to it. The maximum slope is equal to:

$$\left(\frac{d\theta_{min}}{d\theta_{sc}}\right)_{max} = -n^2 \tag{9.62}$$

MEASUREMENT OF PARAMETRIC AMPLIFIERS 273

The point of maximum slope is that at which the curve $\theta_{\min} = f(\theta_{sc})$ cuts the mean straight line of slope -1. It can also be shown that the minimum slope is:

$$\left(\frac{d\theta_{\min}}{d\theta_{sc}}\right)_{\min} = -\frac{1}{n^2} \tag{9.63}$$

The elements of the equivalent circuit may then be measured on the curve of Fig. 9.19. The following relation may also be proved:

$$n^2 = \cotan^2 2\pi\left(\frac{1}{8} - \frac{\sqrt{2}}{4}w\right) \tag{9.64}$$

which enables a more precise determination of n^2 to be made.

(2) *Determination of the elements*

The point of maximum slope is difficult to localize precisely. Let us then plot approximately the mean straight line of slope -1. The values n', θ'_1 and θ'_2 are thus found by a first approximation. The curve $\theta'_{\min} = f(\theta_{sc})$ corresponding to these values is then recalculated from (9.60). If the divergence between the measured and calculated curves is a random one, the values calculated are correct. If, on the other hand, there is a certain regularity in the error, the first-approximation values are incorrect and we may write:

$$n^2 = n'^2 + \Delta n^2 \tag{9.65}$$

$$\theta_1 = \theta'_1 + \Delta\theta_1 \tag{9.66}$$

$$\theta_2 = \theta'_2 + \Delta\theta_2 \tag{9.67}$$

$$\theta_{\min} = \theta'_{\min} + \Delta\theta_{\min} \tag{9.68}$$

If the errors are small we may write after (9.60)

$$\Delta\theta_{\min} \cong -\Delta\theta_1 - \frac{\tan(\theta'_2 + \theta_{sc})}{1 + n'^4 \tan^2(\theta'_2 + \theta_{sc})} \cdot \frac{\Delta n^2}{2\pi}$$

$$- \frac{n'^2}{1 + (n'^4 - 1)\sin^2(\theta'_2 + \theta_{sc})} \Delta\theta_2 \tag{9.69}$$

Let Δ_a, Δ_b and Δ_c represent respectively the errors $\Delta\theta_{\min}$ corresponding to the values, 0, $\pi/4$ and $\pi/2$ of $(\theta'_2 + \theta_{sc})$. We then have

$$\Delta_a = -\Delta\theta_1 - n'^2\Delta\theta_2 \tag{9.70}$$

$$\Delta_b = -\Delta\theta_1 - \frac{1}{1 + n'^4}\Delta n^2 - \frac{2n'^2}{1 + n'^4}\Delta\theta_2 \tag{9.71}$$

$$\Delta_c = -\Delta\theta_1 - \frac{1}{n'^2}\Delta\theta_2 \tag{9.72}$$

Again, let:

$$\Delta\theta_1 = \frac{1}{n'^4 - 1}(\Delta_a - n'^4 \Delta_c) \quad (9.73)$$

$$\Delta\theta_2 = \frac{n'^2}{n'^4 - 1}(\Delta_c - \Delta_a) \quad (9.74)$$

$$\Delta n^2 = 2\pi[-(1 + n'^4)\Delta_b + \Delta_a + n'^4 \Delta_c] \quad (9.75)$$

The second-approximation values are then:

$$n''^2 = n'^2 + \Delta n^2 \quad (9.76)$$

$$\theta_1'' = \theta_1' + \Delta\theta_1 \quad (9.77)$$

$$\theta_2'' = \theta_2' + \Delta\theta_2 \quad (9.78)$$

The curve $\theta''_{\min} = f(\theta_{sc})$ can be recalculated from (9.60) and, if necessary we can calculate the approximations of a higher order.

(3) *Special case*: $n^2 \simeq 1$

The measured curve $\theta_{\min} = f(\theta_{sc})$ departs very little from the mean straight line of slope -1. The determination of n^2, θ_1 and θ_2 by the method described above is then not very accurate. Let:

$$n^2 = 1 + \epsilon \quad (0 < \epsilon \ll 1) \quad (9.79)$$

(9.60) may be written:

$$\pi - (\theta_1 + \theta_{\min}) = \theta_2 + \theta_{sc} + \frac{\epsilon}{4\pi}\sin 2(\theta_2 + \theta_{sc}) \quad (9.80)$$

$$\theta_{\min} + \theta_{sc} = \pi - \left[(\theta_1 + \theta_2) + \frac{\epsilon}{4\pi}\sin 2(\theta_2 + \theta_{sc})\right] \quad (9.81)$$

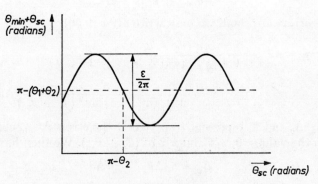

Figure 9.20. Case in which $n^2 \simeq 1$

The curve $\theta_{\min} + \theta_{sc}$ as a function of θ_{sc} is shown in Fig. 9.20. It is then easy to derive ϵ, $(\theta_1 + \theta_2)$ and $(\pi - \theta_2)$, from which we obtain θ_1, θ_2 and $n^2 = 1 + \epsilon$.

(4) *Calculation of the equivalent circuits*

Since we know, from the measurements just described, the elements of which the equivalent circuit in Fig. 9.18 consists, we may, by means of the formulae in Table 9.3, determine the corresponding chain matrix. We may then, again with the aid of Table 9.3, calculate the values of the elements of the equivalent circuit chosen.

References

[1] Kurokawa, K., 'On the use of passive circuit measurements for the adjustment of variable capacitance amplifiers', *Bell System Technical Journal*, pp. 361–381, January 1962.
[2] Blackwell, L. A. and Kotzebue, K. L., *Semiconductor-Diode Parametric Amplifier*, Prentice-Hall, Inc., Englewood Cliffs, N.J., 1961.
[3] Marcuwitz, *Waveguide Handbook*, McGraw-Hill, Radiation Laboratory Series, Chapter 3, pp. 101–167.

10

MEASUREMENT OF THE NOISE TEMPERATURE OF AN AMPLIFIER

by L. LAURENT

1. Equipment common to all methods of measurement

All methods of measurement of noise temperature involve a direct or comparative measurement of the output power of the amplifier energized either by a noise generator or by a resistance brought to a given temperature. In these conditions the output power of the amplifier is very low and in order to measure it we must use a post-amplifier preceded usually by a frequency changer (Fig. 10.1).

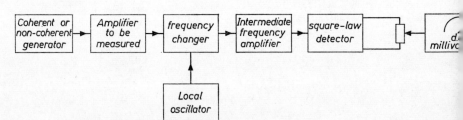

Figure 10.1. Equipment common to all the methods of measurement

The intermediate-frequency amplifier should have a noise bandwidth much narrower than that of the amplifier to be measured for the formula (1.125) to be applicable and the calculation facilitated. The value thus determined is a spectral measurement of the noise. The noise temperature T_1 of the amplifier to be measured is equal to:

$$T_1 = T_T - \frac{T_2}{G_1} \qquad (10.1)$$

where T_T is the noise temperature of the whole system,
   T_2 is the noise temperature of the frequency changer followed by the I.F. amplifier,
   G_1 is the gain of the amplifier to be measured.

It may be difficult to measure the value of the gain G_1 accurately, especially for parametric converters in which the input and output frequencies are different. Hence, it is always of advantage to keep the noise temperature of the combination frequency changer-intermediate-frequency amplifier to a low value. The influence of the error of the measured amplifier gain on the precision of the measurement is then minimized.

As a detector we may use a bolometer or an R.F. diode operating in its square-law region followed by a millivoltmeter. This last method is the most convenient. We may assume that any R.F. diode follows a square-law sufficiently closely up to 5 mV D.C. output.

It is useful to connect a potentiometer between the output of the diode and the millivoltmeter since, most of the measurements, as we shall see, being made by comparison, it is then possible to make the two readings on the same scale and in this way to obtain the maximum accuracy and speed.

2. Direct measurement of the noise temperature

This method will only be mentioned in passing, since it is not very accurate. The available noise power at the output of an amplifier of which the input is energized by a resistance equal to the internal resistance of the signal generator is equal to

$$N_1 = kBG_T(T_0 + T_T) \tag{10.2}$$

where B is the equivalent noise bandwidth of the I.F. amplifier,
   G_T is the transducer gain of the whole system,
   T_0 is the temperature of the input resistance,
   T_T is the noise temperature of the whole system.

Hence:

$$T_T = \frac{N_1}{kBG_T} - T_0 \tag{10.3}$$

Since this is a direct rather than a comparative method of measurement, it has the disadvantage that it requires exact values of B, N_1 and G_T.

3. Measurement by comparison with a 'monochromatic' signal generator

The resistive input termination used in the previous method is replaced by a calibrated adjustable generator of the same source resistance (Fig. 10.2).

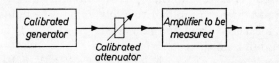

Figure 10.2. Measurement of noise temperature by comparison with a generator

Two successive measurements are made. During the first measurement the generator is switched off and the reading on the millivoltmeter is l_1, proportional to N_1 (see (10.2)).

During the second measurement the generator is switched on and the attenuator is adjusted until the reading is doubled: $l_2 = 2l_1$. Let P_G be the available power corresponding to the output of the attenuator; we then have

$$T_1 = \frac{P_G}{kB} - T_0 \qquad (10.4)$$

The accuracy depends chiefly on the calibration of the combination generator–attenuator. As before, however, we must know exactly the equivalent noise bandwidth.

4. Hot and cold resistance method

This is an absolute calibration method. It has the advantage that it may be used at any frequency, but has the disadvantage that it is complicated to set up. It is particularly applicable to the measurement of low noise temperatures.

The noise source consists of a matched load connected to a low-loss section of line or waveguide, made of low heat conductivity material.

The termination is brought successively to two temperatures, T' and T'':

the first reading l_1 is proportional to

$$N_1 = kBG_T(T_T + T') \qquad (10.5)$$

NOISE TEMPERATURE OF AN AMPLIFIER

the second reading, l_2, to

$$N_2 = kBG_T(T_T + T'') \qquad (10.6)$$

whence

$$\frac{l_1}{l_2} = \frac{T_T + T'}{T_T + T''} \qquad (10.7)$$

or, if we make

$$l_1/l_2 = l \qquad (10.8)$$

$$T_T = \frac{lT'' - T'}{1 - l} \qquad (10.9)$$

The accuracy of the measurement can be improved by choosing three temperatures T', T'', T''' instead of two.

The great advantage of this method compared to the previous ones is that it requires neither a knowledge of the bandwidth nor of the total gain of the chain. The gain of the amplifier itself need only be known because it enters into the calculation of the noise contribution of the extra amplifier.

Further, the measuring set constitutes an absolute standard: there is no necessity to know any empirical factors, since the measurement is equivalent to that of a temperature.

The only disadvantage of this method is that it is dependent on the gain remaining constant between the two measurements. It therefore becomes less accurate when the gain of the amplifier to be measured fluctuates, as happens with high-gain negative-resistance amplifiers.

PRACTICAL DETAILS

(1) *Choice of temperatures*

If high accuracy is required, temperatures T' and T'' should be made to differ considerably and should be of the order of magnitude of the noise temperature to be measured. On the other hand, these temperatures should be easily reproducible.

The simplest procedure is, of course, to use temperatures at which pure bodies change their state, for example the freezing or boiling points of water, the boiling point of liquid nitrogen or the temperature of sublimation of carbon dioxide.

(2) *Choice of termination*

The value of the resistance should not be affected by temperature changes. In order to avoid having to wait for the stabilization of the temperature, which involves the risk that the gain may drift from the time of one measurement to the next, it is preferable to use two different loads, one at T' and the other at T'', and to allow these temperatures to become stable before connecting the loads successively to the input of the amplifier.

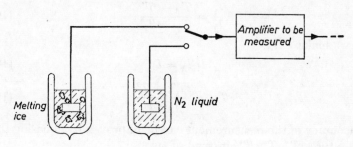

Figure 10.3. Method of hot and cold resistances

This procedure has the additional advantage that it decreases the number of thermal shocks suffered by the load, especially during a series of measurements.

Further, the load and the waveguide or the line need only to be matched at a single temperature.

(3) *Construction of the line or waveguide connecting the resistance to the amplifier*

In order to avoid heat exchanges with the exterior, this section of line or waveguide should be a bad conductor of heat and at the same time have very low losses at the working frequency.

Thin silvered monel metal fulfils this requirement. Further, monel tubes can easily be obtained, as they are commonly used in the chemical industry.

In order to avoid condensation during use at low temperatures, the section of line or waveguide should have a current of dry gas of lower liquefaction temperature flowing through it.

The line-resistance combination should be matched. This matching can be effected by means of a matching transformer with very low losses or at ambient temperature, or placed in the Dewar flask and brought to the temperature of the resistance. A convenient procedure when working in a narrow band of frequencies is to make a local constriction in the external monel tube and to control the matching by means of a directional coupler. The elasticity of the very thin walls makes it easy to determine the spot at which to make the constriction. Figure 10.4 shows the schema of the equipment for a coaxial line.

NOISE TEMPERATURE OF AN AMPLIFIER

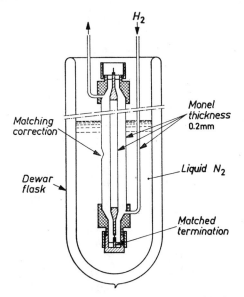

Figure 10.4. Cooled termination and its insulating vessel

(4) *Choice of measurement scales*

If we insert an adjustable attenuator between the post amplifier and the detector, or a potentiometer between the detector and the millivoltmeter, it is possible to read l directly on the voltmeter. If we arrange for the higher of the two readings to correspond to the marking 1 at the bottom of the scale, the second reading gives l directly.

Table 10.1 and Fig. 10.5 show the noise temperature as a function of l where ($l < 1$, $T'' > T'$) when using boiling liquid nitrogen (77°K), and melting ice (273°K) or boiling water (373°K).

5. Use of a thermionic diode

We know that the effective noise current of a vacuum tube diode operating in the saturated state is given by the Schottky equation:

$$I_b = \sqrt{2eI_0 B} \qquad (10.10)$$

where e is the charge of the electron: $1 \cdot 603 \times 10^{-19}$ Cb,

I_0 is the direct current flowing through the diode, depending on the voltage applied to the heater and the construction of the tube.

B is the equivalent noise bandwidth in which the noise current is measured.

Table 10.1. Hot and cold resistance method

l	Boiling nitrogen Melting ice	Boiling nitrogen Boiling water	l	Boiling nitrogen Melting ice	Boiling nitrogen Boiling water
0·21	—	1	0·60	217	367
0·29	3	43·7	0·62	243	406
0·30	7	50	0·64	271	449
0·32	15	62	0·66	303	497
0·34	24	75	0·68	339	552
0·36	33	89	0·70	380	613
0·38	43	104	0·72	427	684
0·40	53	120	0·74	481	765
0·42	65	137	0·76	543	860
0·44	77	155	0·78	618	972
0·46	90	175	0·80	707	1110
0·48	104	196	0·82	816	1270
0·50	119	219	0·84	952	1480
0·52	135	243	0·86	1130	1740
0·54	153	270	0·88	1360	2090
0·56	172	300	0·90	1690	2590
0·58	193	332	1·00	∞	∞

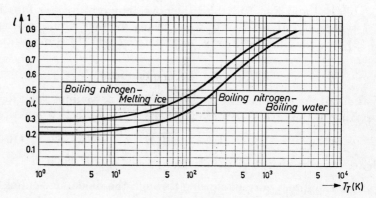

Figure 10.5. Noise temperature as a function of l

NOISE TEMPERATURE OF AN AMPLIFIER

The area of applicability of the formula is limited towards the lower extreme by the flicker effect (negligible below several tens of Hz for a tungsten cathode) and towards the upper extreme by the transit time of the electrons. Care must be taken also to see that the diode is fully saturated, and this assumes a sufficiently high anode voltage (in practice higher than about a hundred volts). We will assume that the diode is connected to a coaxial line† terminated by a resistance R_0 equal to its characteristic impedance.

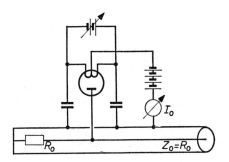

Figure 10.6. Noise generator with a saturated thermionic diode

The available noise power N is equal to:

$$N_s = \tfrac{1}{4} R_0 I_b^2 + kT_0 B \qquad (10.11)$$

or, if we replace I_b by its value derived from (10.10):

$$N_s = B\left(\frac{eR_0 I_0}{2} + kT_0\right) \qquad (10.12)$$

where T_0 is the temperature of R_0. This gives us the excess noise temperature of the source:

$$T_s = \frac{N_s}{kB} - T_0 = \frac{1}{2}\frac{e}{k} R_0 I_0 \qquad (10.13)$$

If we now connect the system to the input of the amplifier to be measured, the reading on the output millivoltmeter will be l, proportional to the output power N of the post amplifier. When the heater voltage of the noise diode is zero: $I_0 = 0$ and the output power is given by (10.2). If we adjust the heater voltage so as to obtain a

† The usual precautions applicable to these frequencies must of course be taken; very short connections, good decouplings and correction for the discontinuity introduced into the line.

doubled reading $l_2 = 2l_1$, the direct current of the diode reaches a value I'_0 and

$$N_2 = 2N_1 = kBG_T(T_0 + T_s + T_T) \tag{10.14}$$

hence

$$T_T = T_s - T_0 \tag{10.15}$$

If we replace e and k by their values, we obtain:

$$T_T = 5833 R_0 I_0 - T_0 \tag{10.16}$$

For $R_0 = 50\Omega$ and $T_0 = 290°K$, (10.16) can be expressed in a convenient manner by:

$$F_T \approx I_0 \text{ (mA)} \tag{10.17}$$

It should be noted that the diode current corresponding to $T_T = 0$ is not zero but is equal to:

$$I_{0\,(T_T=0)} = \frac{T_0}{5833 R_0} \tag{10.18}$$

This method of measurement has the same advantage as the previous one: it constitutes a primary standard, since the measurement is equivalent to that of the intensity of a direct current.

The saturated diode can no longer be used as a noise source at frequencies above 1 GHz because of the transit time of the electrons.

5.1. Use of a cooled load

Formula (10.16) shows that the measurement of low noise temperatures is not very accurate when T_0 is the ambient temperature: T_T is the difference between two nearly equal numbers. It is therefore of advantage to cool R_0, for example by immersing it in liquid nitrogen (Fig. 10.7).

If the apparatus has a 50 Ω termination, we have

$$T_T = 292 I_0 - T'_0 \tag{10.19}$$

or

$$T_T = (292 I_0 - 77)°K \tag{10.20}$$

when liquid nitrogen is used (see Table 3 in Chapter 12) and

$$T_T = (292 I_0 - 195)°K \tag{10.21}$$

when solid carbon dioxide is used.

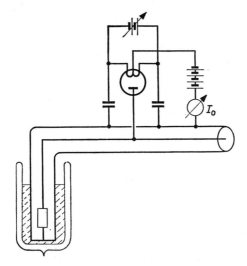

Figure 10.7. Noise generator with a saturated diode and cooled termination

6. Use of a gas-discharge tube

A gas-discharge tube with a hot or cold cathode constitutes a source of noise. The positive column of a plasma emits a random electromagnetic radiation resulting from the slowing down of the electrons due to collision (Bremsstrahlung). Since the mean energy of the other particles (ions or non-ionized atoms) is negligible compared with that of the electrons, the electronic temperature is much higher than that of the temperature of the gas. This makes available a noise source of an equivalent temperature of the order of $10^4\,°K$ although the tube itself gives out only a small amount of heat.

The gas tube is coupled to the measuring circuit by means of a helix when a coaxial line is used, or by placing the tube at an angle in a waveguide, for the higher frequencies (Fig. 10.8), so as to suppress the reflections between the plasma and the line or waveguide.

Unlike the saturated diode, the gas-discharge does not constitute a primary standard as its noise temperature depends on numerous factors. However, the stability and the uniformity of the noise temperature are satisfactory throughout a range of tubes, which makes this type of source reliable once they are calibrated.

(1) *Equivalent noise temperature of a gas-discharge tube*

The gas-discharge tube is assumed to be perfectly matched with the waveguide or coaxial line to which it is connected. Let α_p represent the attenuation due to the plasma; the combination terminal load

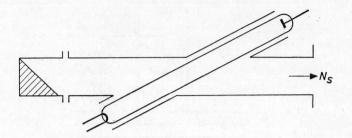

Figure 10.8a. Coupling of a gas-discharge tube with a waveguide

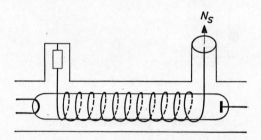

Figure 10.8b. Coupling of a gas-discharge tube with a coaxial line

gas-discharge tube behaves as a load at the temperature T_L followed by an attenuator at the temperature T_{el}. From which (1.117), the equivalent noise temperature of the source, is given by:

$$T_{eq} = \alpha_p T_L + (1 - \alpha_p) T_{eq} \tag{10.22}$$

and the excess noise temperature by:

$$T_s = T_{eq} - T_0 \tag{10.23}$$

The excess noise factor of this source:

$$F_s = \frac{T_s}{T_0} \tag{10.24}$$

is usually given for $T_L = T_0 = 290°K$ and is equal to 15 to 18 dB depending on the type of tube.

(2) *First method of measurement: the attenuator is inserted between the noise source and the amplifier*

The equivalent noise temperature, which varies only slightly with the current, cannot be adjusted, and consequently different methods of measurement from those used for saturated diodes have to be employed. If we insert an adjustable attenuator between the noise source and the amplifier to be measured (Fig. 10.9):

(a) when the source is extinguished, the reading l_1 is proportional to N_1 where N_1 is given by (10.2);
(b) when the source is illuminated, l_2 is proportional to:

$$N_2 = kBG_T[\alpha(T_s + T_0) + (1 - \alpha)T_0 + T_T] \qquad (10.25)$$

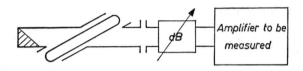

Figure 10.9. First method of measurement with gas-discharge tube

If the attenuator is adjusted to a value α such that $l_2 = 2l_1$, then:

$$T_T = \alpha T_s - T_0 \qquad (10.26)$$

or using the noise figure:

$$F_{T(\text{dB})} = F_{s(\text{dB})} - |\alpha_{(\text{dB})}| \qquad (10.27)$$

(3) *Second method of measurement: the attenuator is located after the amplifier*

A variable attenuator is inserted between the post amplifier and the detector (Fig. 10.10).

Figure 10.10. Second method of measurement with gas-discharge tube

The attenuator is adjusted to zero, the noise source being extinguished: the reading on the millivoltmeter is proportional to N_1 with N_1 given by (10.2). The noise source is illuminated and the attenuator is adjusted to a value of α such that the same reading is obtained at the output.

Hence
$$N_2 = N_1 = \alpha KBG_T(T_s + T_0 + T_T) \tag{10.28}$$
or
$$T_T = \frac{\alpha}{1-\alpha} T_s - T_0 \tag{10.29}$$

$$F_T = \frac{\alpha}{1-\alpha} F_s \tag{10.30}$$

This method has the advantage over the previous one in that the accuracy of the measurement does not depend on the law governing the detector. The accuracy depends solely on the accuracy of the attenuator and of the source.

(4) *Third method of measurement: measurement by direct comparison without an attenuator*

The gas-discharge tube is connected directly to the amplifier to be measured.

(a) When the source is disconnected, the reading l_1 is proportional to N_1 given by (10.2).
(b) When the source is illuminated, the reading l_2 is proportional to

$$N_2 = kBG_T(T_0 + T_s + T_T) \tag{10.31}$$

hence we obtain:
$$T_T = \frac{T_s}{(N_2/N_1) - 1} - T_0 \tag{10.32}$$

Making $l_1/l_2 = l$ as before and adjusting the potentiometer preceding the millivoltmeter to make $l_2 = 1$ corresponding to the full scale, we have

$$T_T = \frac{T_s}{(1/l) - 1} - T_0 \tag{10.33}$$

Table 10.2 and Fig. 10.11 give T_T as a function of l for the gas discharge tube with $T_s = 9600°K$. We see that this method is not very suitable for the measurement of low noise temperatures.

7. Automatic measurement of noise temperature

All the above methods require two successive measurements, on the assumption that the total gain remains absolutely constant. These

Table 10.2. Noise temperature as a function of l using a gas-discharge tube with $T_0 = 9600°K$

l	T	l	T	l	T
0·04	110	0·20	2110	0·36	5134
0·06	321	0·22	2407	0·38	5595
0·08	543	0·24	2734	0·40	6103
0·10	776	0·26	3079	0·42	6660
0·12	1025	0·28	3444	0·44	7265
0·14	1284	0·30	3828	0·46	7918
0·16	1534	0·32	4241	0·48	8600
0·18	1812	0·34	4654	0·50	9310

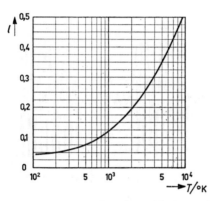

Figure 10.11. Noise temperature as a function of l when using a gas-discharge tube with $T_s = 9600°K$

methods are therefore not very accurate for the determination of the high-gain noise temperature of negative-resistance converters and amplifiers; further, they are very laborious when the noise temperature is optimized by adjusting several parameters.

Automatic measurement is also based on two successive measurements, but they are performed several hundred times a second so as to give continuous readings, after integration, of the noise temperature or noise figure. The block diagram of an apparatus of this nature is shown in Fig. 10.12.

The method of operation is as follows: the noise source (a saturated diode or a gas-discharge tube) is pulsed. The output signal of the intermediate-frequency preamplifier consists of modulated noise in which the troughs correspond to (10.2) (Fig. 10.13), and the peaks to (10.31).

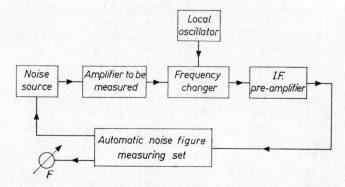

Figure 10.12. Block diagram of an automatic noise figure measurement system

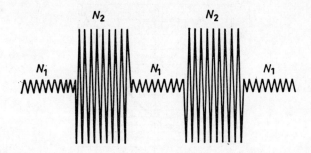

Figure 10.13. Crenellated noise-wave at the output of the IF amplifier

The apparatus is provided with an intermediate amplifier with automatic gain control (A.G.C.), a square-law detector and a video amplifier. The A.G.C. is controlled by a circuit which is synchronized with the pulses of the noise source. This circuit samples the video signal when the source is illuminated, integrates it, and applies it to the A.G.C. If the gain of the pre-amplifier is adequate, we have:

$$N_2 = \text{const.} \qquad (10.34)$$

whatever the value of N_1.

The meter is connected to the video amplifier during the source extinction by means of a switching circuit which is also synchronized with the pulses. After integration, the reading is then proportional to N_1, and T_T is given by (10.32).

The graduation of the scale is established as follows: let l be the meter reading, assuming that it is graduated linearly from zero to one, and let $T_{\max}$ be the noise temperature corresponding to the full scale $l = 1$: since N_2 is constant and l proportional to N_1 we have:

$$l = \frac{C_{st}}{N_2/N_1} \tag{10.35}$$

whence from (10.32)

$$l = \frac{C_{st}}{1 + [T_s/(T_0 + T_T)]} \tag{10.36}$$

and since $l_{max} = 1$

$$l = \frac{1 + T_s/(T_0 + T_{max})}{1 + T_s/(T_0 + T_T)} \tag{10.37}$$

Industrial sets have a scale graduated in F, with $F_{max} = \infty$, and this simplifies the calibration (calibration is performed simply by allowing the source to remain illuminated during the measurement of N_1: from (10.32), when $N_1 = N_2$, $T_T = \infty$). For $T_{max} = \infty$, T_T is given by (10.33).

For very low noise temperatures this method of measurement is of course no more accurate than that described in section 6 (4). We will see in the next section how to resolve this difficulty and nevertheless to obtain a high degree of accuracy.

8. Use of a directional coupler and a cooled termination

(1) *Principle*

This is a general method and does not apply only to automatic measurement. Instead of connecting the noise source directly to the amplifier to be measured, we insert between them a directional coupler provided with two terminations, of which one is cooled, as shown in Fig. 10.14; let T_r be the temperature of the cooled load and α_c be the coupling coefficient of the directional coupler.

The available noise power at the output of the coupler is equal to:

(a) when the source is extinguished

$$N_1 = kB[T_r(1 - \alpha_c) + T_0\alpha_c] \tag{10.38}$$

(b) when the source is illuminated:

$$N_2 = kB[T_r(1 - \alpha_c) + (T_0 + T_s)\alpha_c] \tag{10.39}$$

hence the new reference temperature:

$$T_0' = T_r(1 - \alpha_c) + T_0\alpha_c \tag{10.40}$$

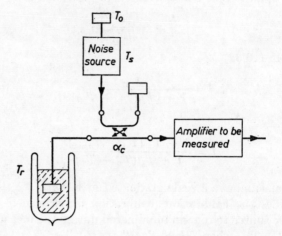

Figure 10.14. Use of a directional coupler and a cooled termination

and the excess temperature of the new source with respect to T'_0:

$$T'_s = T_s \alpha_c \tag{10.41}$$

(2) *Application to the case of section 5 (saturated diode, non-automatic measurement)*

Equations (10.14), (10.17), (10.40) and (10.41) give us:

$$T'_s = 5833 R_0 I_0 \alpha_c \tag{10.42}$$

and:

$$T_T = (5833 R_0 I_0 - T_0 + T_r)\alpha_c - T_r \tag{10.43}$$

The minimum diode current, corresponding to $T_T = 0$ is equal to

$$I_{0(T_T=0)} = \frac{T_r[(1/\alpha_c) - 1] + T_0}{5833 R_0} \tag{10.44}$$

For $R_0 = 50\,\Omega$, $T_0 = 290°K$, $T_r = 77°K$:

$$T_T = 290(I_{0(\text{mA})} - 0.734)\alpha_c - 77 \tag{10.45}$$

Table 10.3 and Fig. 10.15 show the comparison between the three methods of use of the saturated diode for $T_0 = 290°K$, $T_r = 77°K$, $\alpha_c = 0.5$ and $R_0 = 50\,\Omega$.

Table 10.3. Use of a saturated diode (non-automatic measurement). Comparison of readings for the three methods of measurement

I_0(mA)	With termination at 290°K	With termination cooled to 77°K	With 3 dB directional coupler and termination cooled to 77°K
	T	T	T
0·3	—	9·5	—
0·4	—	39	—
0·5	—	68	—
0·6	—	97	—
0·7	—	126	—
0·8	—	155	—
0·9	—	185	—
1·0	2	214	—
1·1	31	243	—
1·2	60	272	—
1·3	89	301	—
1·4	118	330	19
1·5	146	359	34
1·6	177	389	48
1·7	206	418	63
1·8	235	447	77
1·9	264	476	92
2·0	293	505	106
2·1	323		121
2·2	352		135
2·3	381		150
2·4	410		164
2·5	439		179
2·6	468		193
2·7	498		208
2·8	527		222
2·9	556		237
3·0	585		251
3·1	614		266
3·2	644		280

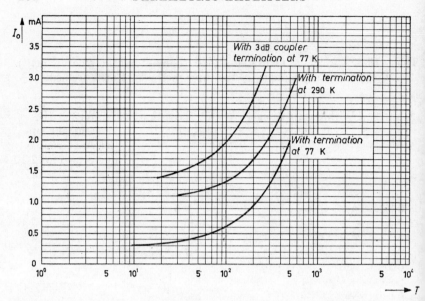

Figure 10.15. Use of a saturated diode. Comparison of readings for the three methods of measurement

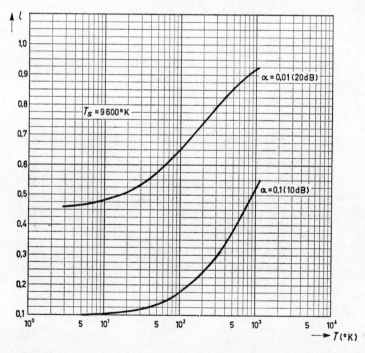

Figure 10.16. Automatic measurement of T_T. Use of a directional coupler and a termination cooled to 77°K

NOISE TEMPERATURE OF AN AMPLIFIER

Table 10.4. Automatic measurement of T. Use of a directional coupler and a termination cooled to 77°K ($T_s = 9600°K$)

l	T for $\alpha = 0{\cdot}01$	T for $\alpha = 0{\cdot}1$	l	T for $\alpha = 0{\cdot}01$	T for $\alpha = 0{\cdot}1$
0·10	—	8	0·52	25	941
0·12	—	33	0·54	36	1051
0.14	—	59	0·56	43	—
0·16	—	84	0·58	54	—
0·18	—	112	0·60	67	—
0·20	—	142	0·62	78	—
0·22	—	172	0·64	93	—
0·24	—	204	0·66	105	—
0·26	—	239	0·68	125	—
0·28	—	275	0·70	144	—
0·30	—	314	0·72	167	—
0·32	—	355	0·74	196	—
0·34	—	396	0·76	221	—
0·36	—	444	0·78	264	—
0·38	—	491	0·80	305	—
0·40	—	541	0·82	358	—
0·42	—	597	0·84	426	—
0·44	—	657	0·86	521	—
0·46	3	723	0·88	607	—
0·48	10	791	0·90	781	—
0·50	16	862			

Table 10.5. Automatic measurement of the noise figure. Use of a 20 dB directional coupler and a termination cooled to 77°K. Conversion of F_{read} into F.

F_{read}(dB)	F(dB)	F_{read}(dB)	F(dB)	F_{read}(dB)	F(dB)
15	0·19	18·5	1·6	24	5·1
15·5	0·25	19	1·8	25	5·9
16	0·50	19·5	2·0	26	6·7
16·5	0·70	20	2·4	27	7·6
17	0·90	21	3·0	28	8·5
17·5	1·1	22	3·7	29	9·4
18	1·3	23	4·4	30	10·3

(3) *Application to case of section 7 (automatic measurement)*

The apparatus operates with the new reference temperature T'_0 and the new excess noise temperature T'_s, whence, from (10.33), (10.40) and (10.41) we obtain:

$$T_T = \left[\frac{T_s}{(1/l) - 1} - T_0\right]\alpha_c - T_r(1 - \alpha_c) \qquad (10.46)$$

Fig. 10.16 and Table IV give T_T as a function of l for $T_0 = 290°K$, $T_r = 77°K$, $\alpha_c = 0\cdot1$ or $0\cdot01$ and $T_s = 9600°K$.

If T_{read} is the noise temperature reading of the apparatus when graduated for use with a cooled load:

$$T_T = T_{\text{read}}\alpha_c - T_r(1 - \alpha_c) \qquad (10.47)$$

or, if we use the noise figures:

$$F_T = F_{\text{read}}\alpha_c + [1 - (T_r/T_0)](1 - \alpha_c) \qquad (10\cdot48)$$

These results are, of course, applicable to the method of section 6 (4)

Table 10.5 gives the conversion for $T_0 = 290°K$, $T_r = 77°K$ and $\alpha_c = 0\cdot01$.

References

[1] Sorger, G. U. and Weinschel, B. D., 'Comparison of deviations from square law for RF crystal diodes and barretters', *I.R.E. Trans. on Instrumentation*, December 1959.
[2] Hart, P. A. H., 'Les sources de bruit Etalons', *Revue Technique Philips*, vol. 23 (No. 9), 1961–62.
[3] *Automatic Noise Figure Meter 113 Manual*, Magnetic A.B., Stockholm
[4] Poulter, H. C., 'An automatic noise figure meter for improving microwave device performance', *Hewlett Packard Journal*, vol. 9 (No. 5), January 1958.
[5] Negrete, M. R., 'Additional conveniences for noise figure measurements', *Hewlett Packard Journal*, vol. 10 (Nos. 6 and 7), February and March 1959.

11

TECHNICAL ASPECTS

by L. LAURENT

It is not proposed in this chapter to discuss in detail the construction of parametric amplifiers, as this would require a review of all the technical aspects of microwave frequencies. We will assume that the reader is familiar with them and we will therefore limit ourselves to a discussion of the chief problems peculiar to parametric amplifiers and illustrate it by two examples.

1. Reactive networks including a varactor

The hypotheses which have served as a basis for the establishment of the small-signal theory imply that the filters should satisfy two conditions: they should allow the loss-free passage of the signal for which they are designed, and they should present an infinite impedance at the terminals of the junction to all the other signals. We also saw that in the equivalent circuit selected, each branch should include a reactive element tuning the varactor to the corresponding frequency and an impedance transformer providing the terminal impedances from the impedances of the transmission lines used (except of course when the branch is an idler).

In practice the three components: filter, tuning reactance and impedance transformer of a branch do not necessarily exist in the form of separate units in accordance with the configuration of the equivalent circuit. They will be included in a single reactive network M and we will see later how this network can be produced.

Designating by M_e, $M_{(+)}$, $M_{(-)}$ and M_p the networks corresponding to the main frequencies, the general block diagram of a parametric amplifier or converter is given in Fig. 11.1.

In this figure we must not forget that the parasitic inductance and capacitance of the varactor produce impedance transformations

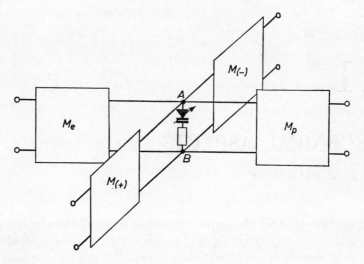

Figure 11.1. Block diagram of a parametric amplifier or converter

which cannot be ignored, and they must therefore be included in the networks M.

Our next problem is to examine the conditions which the networks must fulfil:

(1) At their corresponding frequency they must produce the required impedance transformation whilst at the same time having the lowest possible losses.
(2) Their bandwidth should be wide enough for the information to be transmitted, but narrow enough to appear as a pure reactance at the main frequencies other than those corresponding to them. This reactance should be compensated for by the transformation circuits of the other networks.
(3) The resultant reactances of the four networks between point A and B should be infinite for the main unused frequencies $f_{(+)}$ or $f_{(-)}$. The network corresponding to the unused branch is reduced to a simple reactance.

The ideal solution is of course that in which each network presents an infinite impedance at terminals A and B to all frequencies other than that corresponding to it. For a single frequency, this condition can always be realized by adjusting the electrical distance between the network and the junction.

In practice, condition (3) is rarely satisfied: a compromise is to check that the reactance at the terminals of the junction is large enough for the power consumption at the unused frequency to be negligible.

The first condition requires the use of high-quality materials: silver or gold plated metals and low-loss dielectrics. We will not enlarge on the technical aspects involved, since they are standard in the microwave field.

The second and third conditions lead to the production of filters in which the bandwidth and the reactance outside it should be determined with great accuracy. Some examples of the design of filters will be given after the discussion of the different possible configurations of the networks M.

2. Construction of reactive networks corresponding to the different branches

We saw that the reactive network corresponding to a branch should fulfil the double function of filtering and impedance transformation. It is usually simpler to assign these two functions separately to independently adjustable units. We saw also that, except for degenerate amplifiers, the only amplifiers which are of interest from the point of view of noise temperature are those in which the pumping frequency is considerably in excess of the input frequency. In these conditions, the frequencies of the pump, sum or difference branches are relatively close, and the same types of circuits may be used for them. The filters may be either of the band-pass or band-stop type, those of the former being especially suitable for waveguides, and the latter for coaxial lines.

The filter of the input circuit, on the other hand, is almost always of the band-stop type because of the difficulty in controlling the behaviour of a band-pass filter at frequencies f_p, $f_{(+)}$ and $f_{(-)}$, which are much higher than that of the passing frequency f_e. In order to simplify the process of adjustment, the filters are usually adjusted between matched terminations before insertion into the circuit. The impedance-transformation circuits are then adjusted by the methods described in Chapter 9.

When the input frequency is not too high (a few GHz), the input circuit is almost always in the form of a coaxial line, whilst either waveguides or coaxial lines can be used for the pump, idler or sum-frequency circuits. Amplifiers have been produced with 'stripline' but this technique appears to be used only with relatively low frequencies. Some manufacturers have also used a cavity resonator simultaneously operating on two modes, thus making possible

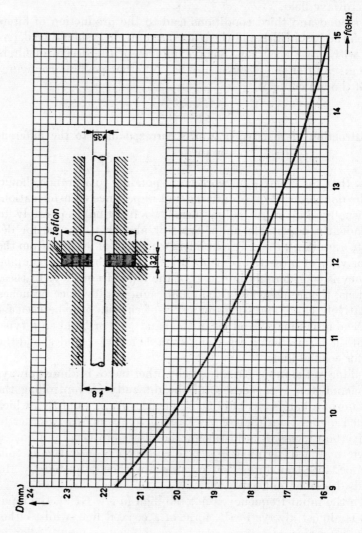

Figure 11.2. Coaxial band-stop filter

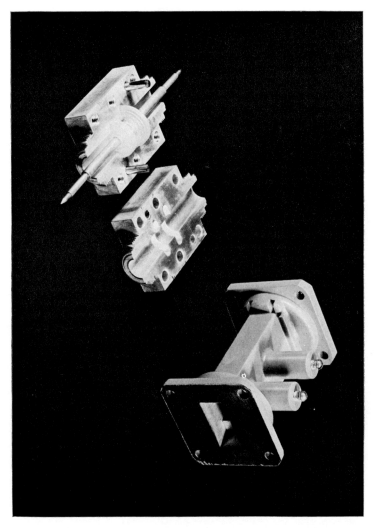

Plate 1. Examples of X band filters. (*By permission of M.B.L.E. 55154*)

Plate 2. Sum-frequency converter 408–9480 MHz. (*By permission of M.B.L.E.* 55156)

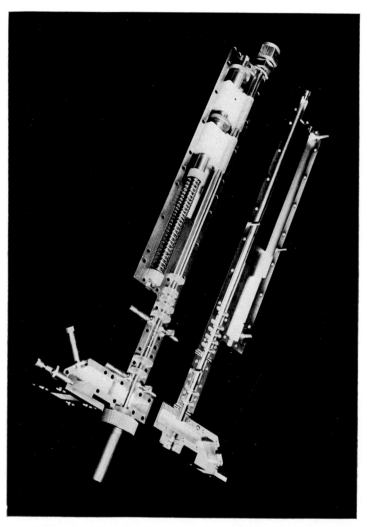

Plate 3. 1535 MHz parametric amplifier opened up. (*By permission of M.B.L.E.* 55153)

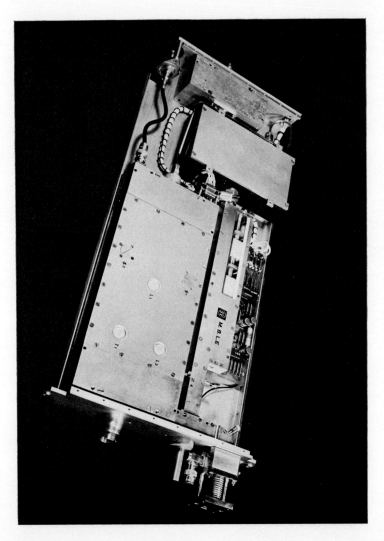

Plate 4. 1535 MHz parametric amplifier complete with its pump amplitude stabilization circuit, frequency changer and I.F. preamplifier. (*By permission of M.B.L.E. 55155*)

the construction of the input and pump filters of a degenerate amplifier with the same cavity.

3. Filters

3.1. Types of filters most frequently used

The theory of filters for microwave frequencies [1], [2], will not be dealt with here, but several examples of the most commonly used types will be given. Band-pass waveguide filters are most often produced by means of a succession of cavities spaced in the waveguide by means of inductive-irises. Each cavity is fitted with an adjustable screw made either of metal or of a material having a high dielectric constant; this makes it possible to avoid contact problems. A filter with two cavities for the X band is represented in the top part of Plate 1. The lower part shows an example of a coaxial line band-stop filter, also in the X band, produced with the aid of grooves filled with a dielectric [3]. The rejection bands are relatively narrow, and this involves the use of several filters when it is required to block signals corresponding to several frequency bands, even when they are close together, as in the case of f_p, $f_{(+)}$ and $f_{(-)}$. The advantage of this type of filter is that it constitutes a mechanical support for the central conductor of the coaxial line, but it has the drawback that it cannot be adjusted. It is essential, therefore, that it should be constructed with great precision. Figure 11.2 shows the dimensions obtained experimentally of a filter of this type using a teflon cylindrical disc entirely filling the groove, for frequencies between 9 and 15 GHz.

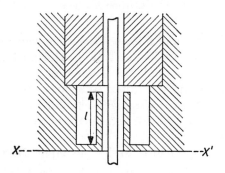

Figure 11.3. Band-stop filter of the $\lambda/4$ re-entrant type.

A second type of coaxial filter often used is the reentrant type shown in Fig. 11.3. At the rejection frequency the filter presents a short circuit in the plane XX'. Due to the parasitic capacitance of the end, the length l is always less than $\lambda/4$ and should be determined experimentally.

3.2. Measurement of response curves and short-circuit planes

The response curve of a filter is measured by the standard method with a frequency-sweep generator and a detector, the filter being inserted between two matched attenuators in order to minimize the effect of parasitic reflections. The measurement of the equivalent short-circuit planes at the frequencies to be 'blocked' is made with a

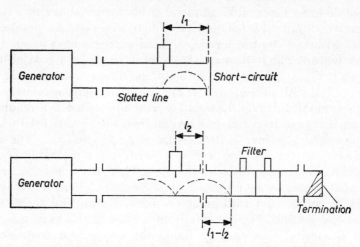

Figure 11.4. Measurement of the short-circuit plane of a filter by the direct-substitution method

slotted line. If the line and filter use the same waveguide the direct-substitution method can be used (Fig. 11.4); the slotted line is terminated by a short circuit, the position of the zero is noted, the short circuit is replaced by the filter to be measured and the probe is moved until a new zero is obtained. The difference in the readings gives the distance of the short-circuit plane of the filter from the face of the assembly.

If the connection of the filter to the line requires the insertion of an intermediate two-port (connector or waveguide-to-coaxial transition), this method may give incorrect results if the Weissfloch curve of the two-port has not been determined beforehand. It is then better to use a mobile short circuit: the filter is connected to the line by means of the two-port, the probe is placed at zero and the filter is replaced by a calibrated piston using the same type of transmission line. All we need then is to adjust the piston so as to obtain a zero reading at the slotted line: the position of the short-circuit of the piston coincides with that of the short-circuit of the filter. For waveguide bandpass filters the plane of the short-circuit usually remains near the first iris. For the dielectric-disc rejectors described earlier, it is located at approximately a quarter wavelength from the rear face of the disc.

4. Location of impedance transformers

The adjustable tuning circuits or impedance transformers between a filter and the varactor also affect the signals corresponding to the other branches by displacing the position of the short-circuit plane of the filter for these signals. Now, it is essential that the adjustments of the different branches should, as far as possible, have no coupling in order to avoid difficulties in construction caused by the difficulty of interpreting the measurements.

This trouble may be avoided by placing impedance transformers in positions where they have the least disturbing effect on the other branches, and this can easily be done for a three-frequency circuit, but becomes almost impossible for a four-frequency circuit. It is for this reason that the latter have been practically abandoned. The optimum positioning, which depends on the configuration used, is on each occasion a problem on its own. The following descriptions of a sum-frequency converter and a reflection amplifier will give an idea of the systems used.

5. Description of a sum-frequency converter

The apparatus described below (Figure 11.5 and Plate 2) is a sum-frequency converter in which the input frequency is 408 MHz and the sum-frequency is 9480 MHz. The ratio of the input to the output frequency is thus 23, which would enable a gain of 13·6 dB to be obtained with a perfect varactor.

This quite low input frequency converter was built for experimental purposes. The input circuit is in coaxial line. The impedance matching transformer has no contacts and consists of a parallel susceptance and a series capacitor. The adjustable susceptance is obtained by means of a section of short-circuited line containing a mobile dielectric cylinder. The short circuit consists of a capacitance allowing the introduction of the varactor bias. The rejector blocking the signals at frequencies f_p, $f_{(+)}$ and $f_{(-)}$ is of the $\lambda/4$ reentrant type described earlier. The varactor is located in a waveguide at the output of the rejector. An adjustable coaxial piston located beyond the varactor enables the varactor to be tuned at the pumping frequency so as to reduce the transformation to be made by the waveguide adapters. The piston, which operates also on the other branches, is

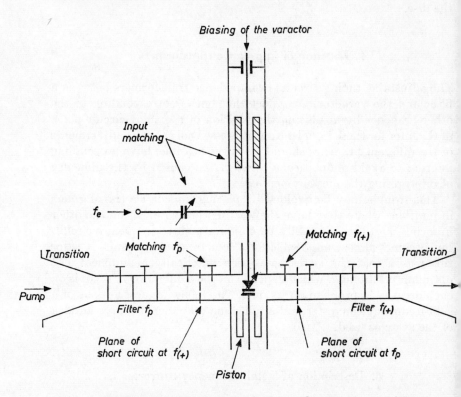

Figure 11.5. Basic diagram of the sum-frequency converter

adjusted initially by the method described in Chapter 9 and then blocked. The problem of the decoupling of the adjustments is then solved quite simply: as the signal at f_e is not propagated in branches f_p and $f_{(+)}$, for which it is below the cut-off frequency, its circuit is not affected by the waveguide matchings. On the other hand, the input matching circuit does not react on the other circuits because of the presence of the rejector. All that is needed therefore is to prevent interference between the adjustments of the pump and sum-frequency circuits. The matching in each branch is effected by

TECHNICAL ASPECTS 305

means of two adjustable susceptances with $\lambda/8$ spacing and arranged symmetrically on each side of the short circuit of the filter for the other frequency. It can be shown that in these conditions these susceptances have a negligible effect on the position of the plane of the short circuit.

It was, of course, necessary to check:

(1) that the impedance transformations could be effected by susceptances located in this way; if not, a fixed impedance transformer should have been inserted between the filter and the adjustable susceptances, and the filters displaced correspondingly;
(2) that the circuit was sufficiently detuned at frequency $f_{(-)}$ for the power consumed at this frequency to be negligible.

The waveguide impedance matching devices pose a second problem: in a high-quality varactor, R_s is of the order of a few ohms and for the frequencies chosen the output impedance at $f_{(+)}$ of the order of a few tens of ohms. Now, the impedance seen from the varactor, when the waveguide is terminated by a short circuit at $\lambda/4$ on one side and by a matched load on the other, is approximately equal to:

$$R = 754\left(1 - \frac{\lambda^2}{4\alpha^2}\right)^{-1/2} \frac{b}{\alpha} \sin^2 \frac{\pi\alpha}{\alpha}$$

where d is the distance between the varactor and the wall of the waveguide (Fig. 11.6) and λ the wavelength in air.†

If the calculation is performed for the pumping frequency of 9·072 MHz, and for a normal WR 90 waveguide taking $\alpha = a/2$, we find a value of 470 Ω for R. The impedance transformation required to match the pump would be enormous and would inevitably involve excessive losses and selectivity. The trouble was solved by the use of waveguide of reduced height (3 mm) and by staggering the varactor with respect to the centre of the waveguide. A combination of these two methods enables the impedance to be decreased tenfold, non-selectively and practically without losses. Taking into account the impedance transformation due to the parasitic capacitance of the cartridge, the varactor is then almost exactly matched with the waveguide for the output circuit, whilst the impedance transformation on the pump side is still reasonable and may be performed by means of dielectric screws.

† This formula was established in [4] for $b \ll \lambda/4$ and for an infinitely thin wire.

The filters also consist of waveguides of reduced height and are terminated in progressive transitions, allowing connection of the apparatus to standard waveguides.

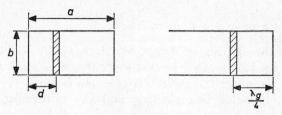

Figure 11.6

6. Description of a reflection amplifier

The reflection amplifier described is one in which the input frequency is 1535 MHz and the pumping frequency 12·4 MHz. The amplifier is followed by a frequency changer which converts the output frequency to 30 MHz and by an I.F. preamplifier.

6.1. The parametric amplifier

A photograph of the amplifier open is shown in Plate 3. The circuit at f_e is obtained by means of a coaxial line and uses an impedance transformer without contacts, fitted with mobile dielectric cylinders (double slug). A circulator with five ports (not shown in the figure) is connected to it. This circulator has a decoupled output connected to the central conductor of the coaxial line, giving access to the bias of the varactor. A group of rejectors containing dielectric discs of the types described earlier suppresses the signals at $f_p, f_{(+)}$, and $f_{(-)}$, preventing them from passing into the input circuit. These rejectors, whose frequency cannot be adjusted, are machined to a precision of a few microns.

Unlike the previous apparatus, the varactor is not located in a waveguide, but is placed in series in the coaxial line. The idler circuit is the section of the line containing the varactor and is enclosed between two rejectors at $f_{(-)}$. The idler is tuned by means of two screws of dielectric material placed in the neighbourhood of the short-circuit plane of the pump rejector, which prevents any reaction on the adjustment of the latter. At the output of the idler the coaxial line is coupled to the waveguide through which the pumping signal is fed in. The pump is matched by means of four screws of dielectric material which, placed after the rejectors, do not affect the behaviour of the circuit at the idler and sum frequencies.

Two small radial rejectors visible on the central conductor on both sides of the varactor, prevent the formation of parasitic idlers corresponding to $2f_p$ and $3f_p$. These filters consist of grooves filled with dielectric rings, the dielectric constants of which are adjusted in order to obtain the correct rejection frequencies.

6.2. Auxiliary circuits of the reflection amplifier

In a reflection amplifier the gain is considerably influenced by the pumping amplitude which must be strictly stabilized if the gain is to be maintained at a constant value. Due to the selectivity of the circuits, the pump frequency has also to be very stable. The two stabilization circuits used are very effective.

Stabilization of the pump oscillator frequency

The stability of the frequency of a pump source (klystron or solid-state oscillator) may be ensured by coupling to it a resonant cavity possessing a very high quality factor. In the case of a solid state oscillator, the cavity also improves the spectrum purity.

The type of cavity employed here uses the TE_{011} mode in a cylindrical resonator. The quality factor obtained is 18,000. A reduction in the temperature coefficient is obtained by the use of gold-plated invar. Details of the calculations for these cavities can be found under reference [5].

The choice of the coupling between the cavity and the oscillator is the result of a compromise. The higher the ratio Q without load/Q under load of the cavity, the better the stabilization, but at the expense of insertion loss. In practice, a loss of at least 3 dB should be accepted for a good stability. The loss is here 5 dB for a tenfold increase of the frequency stability of the klystron source.

Stabilization of the pumping amplitude

The bias circuit is of the 'self bias' type: the varactor is used as a detector and there is no need for a bias source. If the rectified current is limited to a few microamps, the performance of the amplifier is not degraded, whilst the d.c. voltage at the terminals of the varactor is a function of the pumping power and can be used for stabilizing it.

This detected voltage is applied via the bias output of the circulator to the input of a differential amplifier which compares it with a reference voltage. The error signal thus obtained is amplified and controls a P.I.N. diode attenuator connected in the pumping circuit (Fig. 11.7). The only critical part in this system is the differential amplifier which must have a small drift and high input impedance (10 MΩ).

Plate 4 shows the complete amplifier, fitted with its power stabilization system.

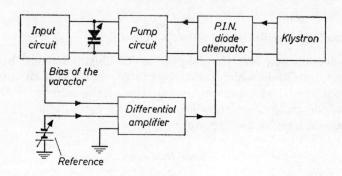

Figure 11.7. Stabilization circuit of the pumping amplitude

References

[1] Ragan, *Microwave Transmission Circuits*, McGraw-Hill, New York, 1947.
[2] Matthaei, G. L., Young, L. and Jones, E. M. T., *Microwave Filters, Impedance Matching Networks and Coupling Structures*, McGraw-Hill, New York, 1947.
[3] De Loach, B. C., 'Radial line coaxial filters in the microwave region', *I.E.E.E. Trans.*, January 1963.
[4] Schelkunoff, S. A., *Electromagnetic Waves*, Van Nostrand Company, Princeton, 1956.
[5] Montgomery, *Technique of Microwave Measurements*, McGraw-Hill, New York, 1948.

Part III

LOW-FREQUENCY PARAMETRIC AMPLIFIERS

12

PARAMETRIC AMPLIFICATION OF LOW-FREQUENCY SIGNALS

by J. C. DECROLY

1. Introduction

We will deal in this chapter with the possibility of using a parametric device for the amplification of signals whose spectrum extends from 0 to several hundreds of Hz. The type of parametric amplifier used for this purpose is based on the operating principles of four-frequency amplifiers for which the equivalent circuits, following the nomenclature adopted at the beginning of this book, are as shown in Figs. 12.1 and 12.2. In these types of converters the signal to be amplified enters at frequency $f_e (f_e < f_p)$ and is received at the output at frequencies $f_+ = f_p + f_e$ and/or $f_- = f_p - f_e$.

With the low frequencies being considered the frequencies f_p, f_+ and f_- are always very close together (when $f_e = 0$ they are equal) and the two circuits tuned to the sum and difference frequencies can be replaced by a single circuit tuned to the pumping frequency f_p. However, it is no longer possible to decouple the pumping and output circuits (tuned to the same frequency) by a simple filtering of the components of a different frequency and it is necessary to use a differential structure with two diodes. The circuit used in this study is as shown in Fig. 12.3 [1].

In this circuit a voltage source e_p of frequency ω_p feeds the primary of a transformer T_1, the secondary of which is closed by a circuit consisting of two semiconductor diodes which are inverse biased. The centre point of the circuit is connected to the generator of the low-frequency signal e_e by means of an anti-resonant G.L.C. circuit tuned to the frequency ω_p of the pump. The capacitor C_e is a decoupling capacitance. As will be seen later, the output signal (received at the terminals of the anti-resonant circuit) is by a first

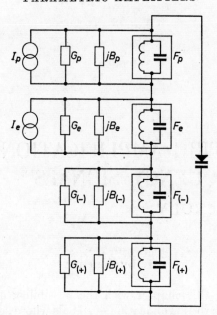

Figure 12.1. Series configuration

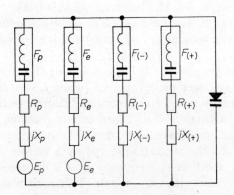

Figure 12.2. Parallel configuration

approximation a wave of frequency ω_p amplitude-modulated by the input signal of frequency ω_e. The parametric amplifier will then be followed by a detector. A conventional amplifier stage (for example, transistorized) will usually be connected between the parametric device and the detector. The basic diagram is then as shown in Fig. 12.4.

AMPLIFICATION OF LOW-FREQUENCY SIGNALS

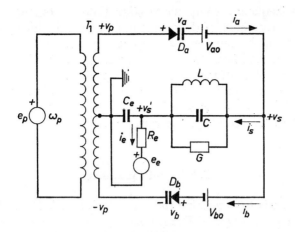

Figure 12.3

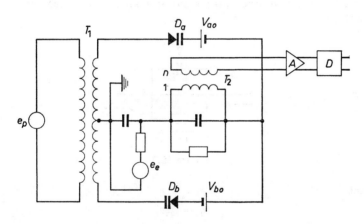

Figure 12.4

The following assumptions will be made when studying the theory of operation with the help of Fig. 12.3.

(1) The voltage v_s resulting from the injection of the signal e_e is low by comparison with voltages $v_p - V_{ao}$ and $v_p - V_{bo}$ (small-signal hypothesis).

(2) The frequency ω_e of the input signal remains at a very low value with respect to the frequency ω_p of the pump.

(3) The frequency components *other* than ω_p, $\omega_p + \omega_e$, $\omega_p - \omega_e$ of the voltage $v_s - v_s'$ at the terminals of the anti-resonant circuit are negligible by comparison with the components of frequency ω_p, $\omega_p + \omega_e$ and $\omega_p - \omega_e$.

(4) The frequency components *other* than ω_e of the voltage v'_s at the terminals of the capacitor C_e are negligible by comparison with the components of frequency ω_e.
(5) The internal impedance of the pump generator is negligible.
(6) The coupling of transformer T_1 is perfect.
(7) The input admittance G_i and the voltage gain A of the conventional amplifier terminated by the detector are constant and real in the frequency range in question.

2. General equations

The first step is to determine the currents i_a and i_b (Fig. 12.3) flowing through diodes D_a and D_b respectively. As will be seen in Chapter 13, the semiconductor junction (for low frequencies) may be represented by a capacitance and a non-linear conductance in parallel (Fig. 12.5).

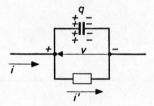

Figure 12.5

Let
$$q = Q(v)$$
$$i = I(v)$$

be the laws governing the charge q and the current i at the applied voltage v. By virtue of assumption (1) and taking account of the notations used in Fig. 12.3, we have for diode D_a:

$$\left. \begin{array}{l} q_a = Q_a(v'_a - v_s) = Q_a(v'_a) - C_a(v'_a)v_s \\ i'_a = I_a(v'_a - v_s) = I_a(v'_a) - G_a(v'_a)v_s \end{array} \right\} \quad (12.1)$$

with $v'_a = v_p - V_{ao}$. For diode D_b:

$$\left. \begin{array}{l} q_b = Q_b(v'_b + v_s) = Q_b(v'_b) + C_b(v'_b)v_s \\ i'_b = I_b(v'_b + v_s) = I_b(v'_b) + G_b(v'_b)v_s \end{array} \right\} \quad (12.2)$$

with $v'_b = v_p - V_{bo}$.

AMPLIFICATION OF LOW-FREQUENCY SIGNALS

In expressions (12.1) and (12.2) we put

$$C(v) = \frac{dQ}{dv} \quad \text{Differential capacitance}$$

$$G(v) = \frac{dI}{dv} \quad \text{Differential conductance}$$

(12.1) and (12.2) give us:

The current i_a flowing through diode D_a:

$$i_a = I_a(v'_a) + C_a(v'_a)\frac{dv'_a}{dt} - G_a(v'_a)v_s - \frac{d}{dt}C_a(v'_a)v_s \quad (12.3)$$

The current i_b flowing through diode D_b:

$$i_b = I_b(v'_b) + C_b(v'_b)\frac{dv'_b}{dt} + G_b(v'_b)v_s + \frac{d}{dt}C_b(v'_b)v_s \quad (12.4)$$

The current i_s flowing through the centre branch is equal to:

$$i_s = i_a - i_b$$

whence, by virtue of (12.3) and (12.4)

$$i_s = I_a(v'_a) - I_b(v'_b) + C_a(v'_a)\frac{dv'_a}{dt} - C_b(v'_b)\frac{dv'_b}{dt}$$

$$-[G_a(v'_a) + G_b(v'_b)]v_s - \frac{d}{dt}[C_a(v'_a) + C_b(v'_b)]v_s \quad (12.5)$$

We will consider the case in which the pumping voltage v_p is periodic, with a fundamental frequency ω_p, and is symmetrical with respect to the centre of the interval $0, 2\pi/\omega_p$ (see Chapters 8 and 14). In these conditions we have:

$$v_p = V_{p_1}[e^{j\omega_p t} + e^{j\omega_p t}] + \sum_{k=2}^{\infty} V_{p_k}[e^{jk\omega_p t} + e^{-jk\omega_p t}] \quad (12.6)$$

Developing the functions $I(v')$, $C(v')$ and $G(v')$ into a Fourier series, we obtain (see Chapter 14):

$$\left.\begin{array}{l} I_m(v'_m) = I_m(t) = I_{m_0} + \sum_l I_{m,n}\, e^{jn\omega_p t} + \sum_l I_{m,n}\, e^{-jn\omega_p t} \\[4pt] C_m(v'_m) = C_m(t) = C_{m_0} + \sum_l C_{m,n}\, e^{jn\omega_p t} + \sum_l C_{m,n}\, e^{-jn\omega_p t} \\[4pt] G_m(v'_m) = G_m(t) = G_{m_0} + \sum_l G_{m,n}\, e^{jn\omega_p t} + \sum_l G_{m,n}\, e^{-jn\omega_p t} \\[4pt] m = a, b \end{array}\right\} \quad (12.7)$$

It should be noted that a symmetrical waveform (in the sense defined above) gives Fourier series developments of $C_m(t)$ and $G_m(t)$ for which, by a suitable choice of the time origin, the initial *phase of all the harmonics* is zero. Any given type of waveform can give rise to developments of $C_m(t)$ and $G_m(t)$ containing the complex amplitudes $C_{m,n} = |C_{m,n}|^{j\phi_n}$, $G_{m,n} = |G_{m,n}|^{j\psi_n}$.

We will return in section 3.41 to the influence that an initial phase difference between the first and second harmonics of $C(t)$ may have on the performance of the circuit. On the other hand, by virtue of assumptions (3) and (4), the voltage v_s may be written in the form:

$$v_s = V_e\,e^{j\omega_e t} + V_e^*\,e^{-j\omega_e t} + V_+\,e^{j\omega_+ t} + V_+^*\,e^{-j\omega_+ t}$$
$$+ V_-\,e^{j\omega_- t} + V_-^*\,e^{-j\omega_- t} + V_{sp}\,e^{j\omega_p t} + V_{sp}^*\,e^{-j\omega_p t} + v_{sc} \quad (12.8)$$

in which we have made

$$\omega_+ = \omega_p + \omega_e$$
$$\omega_- = \omega_p - \omega_e$$

The relations (12.5) being satisfied for each of the components, we obtain, when (12.6), (12.7) and (12.8) are taken into account:

Frequency

$\omega_e \quad I_{se} = -[G_0' + j\omega_e C_0']V_e - [G_1' + j\omega_e C_1'](V_+ + V_-^*)$

$\omega_+ \quad I_{s+} = -[G_0' + j\omega_+ C_0']V_+$
$\qquad\quad\;\; -[G_1' + j\omega_+ C_1']V_e - [G_2' + j\omega_+ C_2']V_-^*$

$-\omega_- \quad I_{s-}^* = -[G_0' - j\omega_- C_0']V_-^*$
$\qquad\quad\;\; -[G_1' - j\omega_- C_1']V_e - [G_2' - j\omega_- C_2']V_+$

$\omega_p \quad I_{sp} = -[G_0' + j\omega_p C_0']V_{sp} - [G_1' + j\omega_p C_1']V_{sc}$
$\qquad\quad\;\; -[G_2' + j\omega_p C_2']V_{sp}^* + I_{a_1} - I_{b_1}$
$\qquad\quad\;\; + \sum_{k=1} jk\omega_p[C_{a_{k-1}} - C_{b_{k-1}} - C_{a_{k+1}} + C_{b_{k+1}}]V_{p_k}$

$0 \quad I_{sc} = -G_0' V_{sc} - G_1' \cdot (V_{sp} + V_{sp}^*) + I_{a0} - I_{b0} \quad (12.9)$

with,

$$G_n' = G_{a_n} + G_{b_n}$$
$$C_n' = C_{a_n} + C_{b_n} \qquad n = 0, 1, 2$$

On the other hand we have:

$$I_{se} = j\omega_e C_e V_e + \frac{V_e}{R_e} - \frac{E_e}{R_e}$$

$$I_{s+} = \left[G_A + G + j\left(\omega_+ C - \frac{1}{\omega_+ L}\right)\right]V_+$$

$$I_{s-}^* = \left[G_A + G - j\left(\omega_- C - \frac{1}{\omega_- L}\right)\right]V_-^*$$

AMPLIFICATION OF LOW-FREQUENCY SIGNALS

$$I_{sp} = \left[G_A + G + j\left(\omega_p C - \frac{1}{\omega_p L}\right)\right] V_{sp}$$

$$I_{sc} = \frac{V_{sc}}{R_{e_2}}$$

where $G_A = n^2 G_i$ is the conductance of the load of the parametric amplifier (input conductance G_i of the conventional amplifier seen by the anti-resonant circuit);
G is the conductance of the anti-resonant circuit.

This gives:

$$\left. \begin{aligned} \frac{E_e}{R_e} &= Y_e V_e + (G_1' + j\omega_e C_1')(V_+ + V_-^*) \\ 0 &= (G_1' + j\omega_+ C_1') V_e + Y_+ V_+ + (G_2' + j\omega_+ C_2') V_-^* \\ 0 &= (G_1' - j\omega_- C_1') V_e + (G_2' - j\omega_- C_2') V_+ + Y_-^* V_-^* \end{aligned} \right\} \quad (12.10)$$

$$\left. \begin{aligned} I_{a_1} - I_{b_1} + \sum_k jk\omega_p [C_{a_{k-1}} - C_{b_{k-1}} - C_{a_{k+1}} + C_{b_{k+1}}] V_{p_k} \\ = (G_1' + j\omega_p C_1') V_{sc} + Y_p V_{sp} + (G_2' + j\omega_p C_2') V_{sp}^* \\ I_{a_0} - I_{b_0} = \left(G_0' + \frac{1}{R_e}\right) V_{sc} + G_1'(V_{sp} + V_{sp}^*) \end{aligned} \right\} \quad (12.11)$$

in which we made:

$$Y_e = \frac{1}{R_e} + G_0' + j\omega_e(C_0' + C_e) \quad (12.12)$$

$$Y_p = G_A + G + G_0' = G_T \quad (12.13)$$

$$\left. \begin{aligned} Y_+ &= G_T \left[1 + j\omega_p \frac{(C_0' + C)}{G_T} \frac{(2\omega_p + \omega_e)}{(\omega_p + \omega_e)}\right] \\ Y_-^* &= G_T \left[1 + j\omega_p \frac{(C_0' + C)}{G_T} \frac{(2\omega_p - \omega_e)}{(\omega_p - \omega_e)\omega_p}\right] \end{aligned} \right\} \quad (12.14)$$

with:

$$(C_0' + C)L = \frac{1}{\omega_p^2}$$

The operation of the circuit is governed by the systems of equations (12.10) and (12.11). Since the system (12.10) is linear with respect to V_e, V_+ and V_-, we obtain, for $E_e = 0$, the solution $V_e = V_+ = V_- = 0$, if we assume that the conditions for the initiation of the oscillations are not satisfied (zero determinant for this system).

Further, as equations (12.11) show, it is possible, even for $E_e = V_e = V_+ = V_- = 0$ to receive at the output a certain signal of frequency ω_p, which is solely a function of the pumping conditions $(I_a, I_b, \ldots)$. Since this signal cannot be suppressed by filtering, the first members of equations (12.11) must be zero, that is, we have:

$$I_{a_0} - I_{b_0} = 0$$
$$I_{a_1} - I_{b_1} = 0$$
$$[C_{a_{k-1}} - C_{b_{k-1}} - C_{a_{k+1}} + C_{b_{k+1}}]V_{p_k} = 0$$

These are the conditions for the balancing of the circuit. These conditions and the methods of adjustment enabling us to satisfy them will be studied in greater detail in Chapter 13. It should be noted that for cissoidal pumping, the last condition reduces to

$$C_{a_0} - C_{b_0} - C_{a_2} + C_{b_2} = 0$$

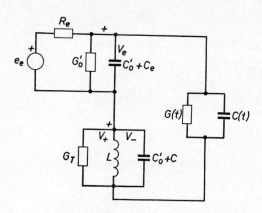

Figure 12.6

Assuming that the conditions for balance are satisfied, equations (12.10) lead to the equivalent circuit of Fig. 12.6. In this circuit, $C(t)$ and $G(t)$ are periodic functions of time expressed by:

$$C(t) = C'_1[e^{j\omega_p t} + e^{-j\omega_p t}] + C'_2[e^{2j\omega_p t} + e^{-2j\omega_p t}]$$
$$G(t) = G'_1[e^{j\omega_p t} + e^{-j\omega_p t}] + G'_2[e^{2j\omega_p t} + e^{-2j\omega_p t}]$$

NOTES:

(1) The introduction of the terms G'_0, G'_1 and G'_2, characterizing the losses of the diodes, calls for some comment. As we see from equations (12.10), even if these terms can be ignored in the components of frequency ω_p, ω_+ or ω_-, we cannot, *a priori*, ignore their effects for the components of frequency ω_e.

(2) In the theory of parametric amplifiers developed in Chapter 3,

the small-signal hypotheses and those appertaining to the filtering of the components of different frequencies, make it possible to substitute *for the varactor and its pumping circuit* a capacitance (elastance) varying periodically with time.

In the case considered, this substitution, that is the transition from the circuit of Fig. 12.3 to that of Fig. 12.6, can only take place if the circuit is balanced. Before undertaking the solution of the system of equations (12.10), it is appropriate to make some notes concerning the special case in which we are dealing with a d.c. input signal ($\omega_e = 0$). This may be considered as a limiting case of the previous one ($\omega_e \neq 0$), that is that in the expressions obtained after the solution of (12.10), we make ω_e approach zero. It can, however, always be treated directly.

For $\omega_e = 0$, the voltages V_+ and V_- combine into a single component V of frequency ω_p. By virtue of assumptions (3) and (4), the voltage v_s may then be written in the form:

$$v_s = V_{e_0} + V e^{j\omega_p t} + V^* e^{-j\omega_p t} \qquad (12.15)$$

By reasoning similar to that which led to the system of equations (12.10) and assuming that the conditions for balance are satisfied, we obtain, when (12.12) and (12.13) are taken into account:

$$\frac{E_{e_0}}{R_e} = \left(\frac{1}{R_e} + G_0'\right) V_{e_0} + G_1'(V + V^*) \qquad (12.16)$$

$$0 = (G_1' + j\omega_p C_1') V_{e_0} + G_T V + (G_2' + j\omega_p C_2') V^* \qquad (12.17)$$

It will be seen that the two last equations (12.10) reduce for $\omega_e = 0$ to the single relation (12.17) *causing both V and V* to be included*. We are therefore in the presence of a degeneracy similar to that which occurred in parametric amplifiers for which the pumping frequency f_p was equal to double that of the signal to be amplified. We will discuss this question later.

3. Analysis of the operation for $\omega_e = 0$

The analysis of equations (12.10) usually presents considerable difficulties. In order to justify certain approximations enabling a considerable simplification of these equations to be obtained, we will deal first with the analysis of the operation when we are dealing with a d.c. input signal.

3.1. Phase and amplitude of the output signal of the parametric amplifier

The direct method will be used and equations (12.16) and (12.17) will be written in the form:

$$\frac{E_{e0}}{R_e} = \left(\frac{1}{R_e} + G'_0\right)V_{e0} + G'_1(V + V^*) \tag{12.16}$$

$$0 = \beta[\gamma_1 + j]V_{e0} + V + \alpha(\gamma_2 + j)V^* \tag{12.18}$$

with

$$\left.\begin{aligned}\beta &= \frac{\omega_p C'_1}{G_T} \\ \alpha &= \frac{\omega_p C'_2}{G_T}\end{aligned}\right\} \tag{12.19}$$

$$\left.\begin{aligned}\gamma_1 &= \frac{G'_1}{\omega_p C'_1} \\ \gamma_2 &= \frac{G'_2}{\omega_p C'_2}\end{aligned}\right\} \tag{12.20}$$

Let

$$V = |V|\,e^{-j\theta} \tag{12.21}$$

Substituting this expression for V in (12.18) we find, after having separated the real and imaginary parts,

$$-\beta\gamma_1 V_{e0} = |V|[(1 + \alpha\gamma_2)\cos\theta - \alpha\sin\theta] \tag{12.22}$$

$$-\beta V_{e0} = |V|[\alpha\cos\theta - (1 - \alpha\gamma_2)\sin\theta] \tag{12.23}$$

(12.22) and (12.23) give us:

$$\cos\theta = \frac{\alpha(1 + \gamma_1\gamma_2) - \gamma_1}{\sqrt{[\alpha(1 + \gamma_1\gamma_2) - \gamma_1]^2 + [1 - \alpha(\gamma_1 - \gamma_2)]^2}} \tag{12.24}$$

$$\sin\theta = \frac{1 - \alpha(\gamma_1 - \gamma_2)}{\sqrt{[\alpha(1 + \gamma_1\gamma_2) - \gamma_1]^2 + [1 - \alpha(\gamma_1 - \gamma_2)]^2}} \tag{12.25}$$

$$|V|^2 = \frac{[\alpha(1 + \gamma_1\gamma_2) - \gamma_1]^2 + [1 - \alpha(\gamma_1 - \gamma_2)]^2}{1 - (\gamma_2^2 + 1)\alpha^2}\beta^2\,V_{e0}^2 \tag{12.26}$$

Replacing $V + V^*$ in equation (12.16) by $2|V|\cos\theta$, we obtain, when (12.25) and (12.26) are taken into account,

$$\frac{E_{e0}}{R_e} = \left[\frac{1}{R_e} + G'_0 + 2G'_1 \cdot \frac{[\alpha(1 + \gamma_1\gamma_2) - \gamma_1]\beta}{1 - (\gamma_2^2 + 1)\alpha^2}\right]V_{e0} \tag{12.27}$$

AMPLIFICATION OF LOW-FREQUENCY SIGNALS   321

or

$$\frac{E_{e_0}}{R_e} = \left(\frac{1}{R_e} + Y_{\text{in}_0}\right) V_{e_0} \qquad (12.28)$$

with

$$Y_{\text{in}_0} = G_0'' + 2G_1'' \frac{[\alpha(1 + \gamma_1 \gamma_2) - \gamma_1]\beta}{1 - (\gamma_2^2 + 1)\alpha^2} \qquad (12.29)$$

By virtue of (12.26) and (12.28) we have:

$$|V| = \frac{\beta \sqrt{[\alpha(1 + \gamma_1 \gamma_2) - \gamma_1]^2 + [-1 + \alpha(\gamma_1 - \gamma_2)]^2}}{[1 - (\gamma_2^2 + 1)\alpha^2][1 + R_e Y_{\text{in}_0}]} |E_{e_0}| \qquad (12.30)$$

3.2. Admittance Y_{in_0}

Where I_{e_0} is the direct current supplied by the signal source E_{e_0}, we have:

$$I_{e_0} = \frac{E_{e_0} - V_{e_0}}{R_e}$$

whence, by virtue of (12.28):

$$\frac{I_{e_0}}{V_{e_0}} = Y_{\text{in}_0}$$

We thus see that Y_{in_0} is the input admittance of the parametric amplifier.

3.3. Discussion of the relations (12.29) and (12.30)

Relation (12.30) shows that the components of the second harmonic of the Fourier series development of the conductance and differential capacitance give rise to a negative-resistance effect. However, as shown by (12.29), these same parameters may give rise to a very high input admittance. The result is that the negative-resistance effect appearing in (12.30) may be partially compensated for by the accompanying increase in the term $R_e Y_{\text{in}}$.

On the other hand, as (12.13) shows, the conductance G_0' due to the diodes constitutes a supplementary load for the anti-resonant circuit, resulting in a decrease in the reduced parameter α. It is, however, easy to show from the preceding equations that the ratio $|V|/|E_{e_0}|$ cannot exceed unity in the absence of the parametric effect ($C_1 = C_2 = 0$) because $G_0' > G_1' > G_2'$ (see Chapter 14). These considerations tell us that *in order to obtain a high gain and a low input admittance*, it is necessary to reduce as far as possible the influence of the terms G_0' and G_1'. As will be seen later, it is possible, by an appropriate choice of diodes and the pumping conditions, to bring these terms to a low enough value such that

$$\gamma_1 \leqslant 10^{-4} \ll 1$$
$$\gamma_2 \leqslant 10^{-4} \ll 1$$

In these conditions we may ignore the terms γ_1 and γ_2 in expressions (12.24), (12.25), (12.26) and (12.27) and write

$$\cos \theta = \frac{\alpha}{\sqrt{1 + \alpha^2}} \qquad (12.31)$$

$$\sin \theta = \frac{1}{\sqrt{1 + \alpha^2}} \qquad (12.32)$$

$$|V| = \frac{\sqrt{1 + \alpha^2}}{(1 - \alpha)^2(1 + R_e Y_{\text{in}_0})} \beta |E_{e_0}| \qquad (12.33)$$

$$Y_{\text{in}_0} = G'_0 + 2G'_1 \frac{\alpha\beta}{1 - \alpha^2} \qquad (12.34)$$

3.4. Operation of the parametric amplifier when $R_e Y_{\text{in}_0} \ll 1$

As shown by the relation (12.33), this circuit is of interest only when the admittance $1/R_e$ of the source is much higher than the input admittance Y_{in_0}. Let us revert to equations (12.16) and (12.17), which, if we take the permissible approximations into account, may be written:

$$E_{e_0} = (1 + R_e Y_{\text{in}_0}) V_{e_0} \qquad (12.35)$$

$$-j\omega_p C'_1 V_{e_0} = G_T V + j\omega_p C'_2 V^* \qquad (12.36)$$

Assuming that, for a given value of α, the impedance R_e of the source should be such that we have:

$$R_e Y_{\text{in}} \ll 1 \qquad (12.37)$$

or:

$$R_e \left(G'_0 + \frac{2G'\alpha\beta}{1 - \alpha^2} \right) \ll 1$$

we may write approximately

$$E_{e_0} = V_{e_0}$$

and:

$$-j\omega_p C'_1 E_{e_0} = G_T V + j\omega_p C'_2 V^* \qquad (12.38)$$

or:

$$-j\beta E_{e_0} = V + j\alpha V^*$$

AMPLIFICATION OF LOW-FREQUENCY SIGNALS

It can easily be shown that equation (12.38) governs the operation of a coherent-phase degenerate amplifier of which the equivalent circuit is shown in Fig. 12.7, where:

$$i = -j\omega_p C_1' E_{e_0} e^{j\omega_p t} + j\omega_p C_1' E_{e_0} e^{-j\omega_p t} \quad (12.39)$$

$$C_2(t) = C_2'[e^{2j\omega_p t} + e^{-2j\omega_p t}] \quad (12.40)$$

We now see that the voltage E_{e_0}, which unbalances the circuit (Fig. 12.3), gives rise, in the load G_T, to a current of amplitude proportional to E_{e_0} and frequency ω_p (classic tuned-load modulation). This current is itself amplified by the effect of the negative resistance due to the second harmonic $C_2(t)$.

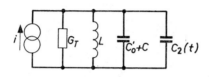

Figure 12.7

3.4.1. INFLUENCE OF A RELATIVE PHASE CHANGE BETWEEN THE FIRST AND SECOND HARMONIC OF $C(t)$

In the circuit of Fig. 12.7 and in accordance with (12.39) and (12.40), the initial phase of the current i (frequency ω_p), with relation to the parametric excitation $C_2(t)$ (frequency $2\omega_p$), depends only on the relative initial phase which may exist between the first and second harmonics of the Fourier series development of the differential capacitance. This relative phase plays a role similar to that which, in degenerate parametric amplifiers, is played by the initial phase of the input signal (frequency $\omega_e = \omega_p/2$) with respect to the parametric excitation (frequency ω_p) [2], [3]. To explain this role let us assume that by some means we can obtain a development of $C(t)$ in the form:

$$C(t) = C_0' + C_1' e^{j\omega_p t} + C_1'^* e^{-j\omega_p t} + C_2'(e^{2j\omega_p t} + e^{-2j\omega_p t})$$

$$+ \sum_{n=3} C_n' e^{jn\omega_p t} + \sum_{n=3} C_n'^* e^{-jn\omega_p t} \quad (12.41)$$

with $C_1' = |C_1'| e^{j\theta_p}$.

Since the time origin is arbitrary we may always choose it in such a way as to equate to zero the initial phase of the second harmonic. Taking (12.41) into account, we obtain after the solution of (12.38):

$$|V| = \frac{\sqrt{1 + \alpha^2 - 2\alpha \sin 2\theta_p}}{1 - \alpha^2} \beta |E_{e_0}| \quad (12.42)$$

$$\tan \theta = \frac{1 - \alpha \tan \theta_p}{\tan \theta_p - \alpha} \tag{12.43}$$

In the case of symmetrical pumping, $\theta_p = 0$ and (12.33) and (12.34) apply once more. Relations (12.42) and (12.43) as well as Fig. 12.8, which give $|V|/(\beta E_{e_0})$ as a function of θ_p, are to be compared with the

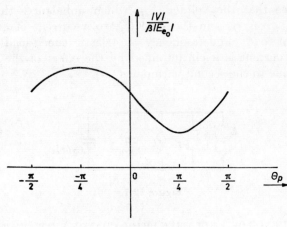

Figure 12.8

results obtained in references [2] and [3] for degenerate parametric amplifiers. For a given value of α, (12.42) gives us:

$$|V|_{\max} = \frac{\beta |E_{e_0}|}{1 - \alpha} \quad \text{for} \quad \theta_p = -\frac{\pi}{4} + k\pi$$

$$|V|_{\min} = \frac{\beta |E_{e_0}|}{1 + \alpha} \quad \text{for} \quad \theta_p = \frac{\pi}{4} + k\pi$$

The remainder of this section will be on the basis of $\theta_p = 0$.

3.5. Synchronous detection of the output signal

By virtue of (12.33) and (12.21) and of assumption (7), the instantaneous voltage $v'(t)$ at the detector is of the form:

$$v'(t) = \frac{2nA\beta\sqrt{1 + \alpha^2}}{(1 - \alpha^2)(1 + R_e Y_{\text{in}_0})} E_{e_0} \cos(\omega_p t - \theta)$$

and this, when $R_e Y_{\text{in}_0} \ll 1$, becomes:

$$v'(t) = \frac{2nA\beta\sqrt{1 + \alpha^2}}{(1 - \alpha^2)} E_{e_0} \cos(\omega_p t - \theta)$$

where A is the voltage gain of the conventional amplifier, and n the transformation ratio of the transformer T_2 inserted between the two amplifiers (Fig. 12.4).

As these relations show, the phase of this voltage swings from the value $-\theta$ to the value $-\theta + \pi$ when the d.c. input voltage changes sign. In order to restore a d.c. signal at the output, reproducing the input signal *with its sign*, it is essential to have synchronous detection.

If we multiply $v'(t)$ by a wave of the form:

$$A' \cos(\omega_p t - \phi)$$

we obtain, after elimination by filtering of the components of frequency $2\omega_p$ resulting from this multiplication,

$$V_{s_0} = \frac{2nK\beta\sqrt{1 + \alpha^2}}{1 - \alpha^2} \cos(\phi - \theta) E_{e_0}$$

in which we made

$$K = \frac{A \cdot A'}{2}$$

If the phase ϕ of the demodulation wave is adjusted so as to make $\phi = \theta$, the output voltage after synchronous detection is

$$V_{s_0} = \frac{2nK\beta\sqrt{1 + \alpha^2}}{1 - \alpha^2} E_{e_0}$$

The conversion efficiency η_0 of the parametric amplifier will be designated by the ratio defined by:

$$\eta_0 = \frac{V_{s_0}}{nKE_{e_0}} = \frac{2\beta\sqrt{1 + \alpha^2}}{1 - \alpha^2} \qquad (12.44)$$

It should be noted that for a d.c. input signal the conversion efficiency η_0 thus defined is equal to the ratio between the peak value of the output voltage of the parametric amplifier and the d.c. input voltage.

Further, the frequency of the demodulation wave is the same as that of the pump generator. Synchronous detection can therefore take the place of a signal from this same generator; the relative phase between the excitation of the pump and the demodulation wave is unaffected by variations in the phase of the local oscillator.

By virtue of (12.19), expression (12.44) may be written in the form:

$$\eta_0 = 2 \frac{C_1'}{C_0'} \cdot \frac{Q}{k_1} \cdot \frac{\sqrt{1 + \left(\frac{C_2'}{C_0'}\right)^2 \frac{Q^2}{k_1^2}}}{1 - \left(\frac{C_2'}{C_0'}\right)^2} \qquad (12.45)$$

in which we put:

$$k_1 = \frac{C_0' + C}{C_0'}$$

$$Q = \frac{\omega_p(C_0' + C)}{G_T} \qquad (12.46)$$

For well-matched diodes, we have almost exactly:

$$C_{a_n} + C_{b_n} = C_n$$

hence:

$$\frac{C_n'}{C_0'} \approx \frac{C_n}{C_0}$$

As will be seen in Chapter 14, the ratios C_n/C_0 are connected by simple relations to the pumping parameters and to the characteristics of the varactors.

4. Analysis of the operation when $\omega_e \neq 0$

It was shown in the previous section that it was justifiable to ignore the effect of the losses in the diodes for the components of frequency near to that of the pump. In these conditions the systems of equations (12.10) may be written in the form:

$$\left.\begin{aligned}\frac{E_e}{R_e} &= Y_e V_e + (G_1' + j\omega_e C_1')(V_+ + V_-^*) = \left(\frac{1}{R_e} + Y_{\text{in}}\right) V_e \\ 0 &= j\omega_+ C_1' V_e + Y_+ V_+ + j\omega_+ C_2' V_-^* \\ 0 &= -j\omega_- C_1' V_e - j\omega_- C_2' V_+ + Y_-^* V_-^* \end{aligned}\right\} \qquad (12.47)$$

If we assume that the admittance $1/R_e$ of the source in the frequency range f_e is much larger than the input admittance Y_{in} of the amplifier ($R_e Y_{\text{in}} \ll 1$), then E_e is almost exactly equal to V_e and the system of equations (12.47) can be reduced to two equations:

$$\left.\begin{aligned}-j\omega_+ C_1' E_e &= Y_+ V_+ + j\omega_+ C_2' V_-^* \\ j\omega_- C_1' E_e &= -j\omega_- C_2' V_+ + Y_-^* V_-^* \end{aligned}\right\} \qquad (12.48)$$

It can easily be shown that these relations govern the operation of the parametric amplifier the equivalent circuit of which is as shown in Fig. 12.9.

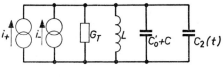

Figure 12.9

in which we put:

$$i_+ = -j\omega_+ C'_1 E_e \, e^{j\omega_+ t} + j\omega_+ C'_1 E_e^* \, e^{-j\omega_+ t}$$
$$i_- = -j\omega_- C'_1 E_e^* \, e^{j\omega_- t} + j\omega_- C'_1 E_e \, e^{-j\omega_- t}$$
$$C_2(t) = C'_2(e^{2j\omega_p t} + e^{-2j\omega_p t})$$

4.1. Phase and amplitude of the components of frequency f_+ and f_- of the voltage at the output of the parametric amplifier

It is first necessary to define the admittances Y_+ and Y_-^*. By virtue of hypothesis 2 and (12.14), we have almost exactly:

$$Y_+ = Y_-^* = G_T(1 + 2jQx) \qquad (12.49)$$

with:

$$Q = \frac{\omega_p(C + C'_0)}{G_T} \qquad (12.46)$$

and:

$$x = \frac{\omega_e}{\omega_p} \qquad (12.50)$$

Taking these notations and (12.19) into account, the system of equations (12.48) may be written in the form:

$$\left. \begin{array}{l} -j(1+x)\beta E_e = (1+2jQx)V_+ + j(1+x)\alpha V_-^* \\ j(1-x)\beta E_e = -j(1-x)\alpha V_+ + (1+2jQx)V_-^* \end{array} \right\} \quad (12.51)$$

$$\left. \begin{array}{l} V_+ = \dfrac{-j + \alpha + 2Qx\left[1 - \dfrac{\alpha}{2Q}\right]}{1 - \alpha^2 - 4Q^2 x^2 \left(1 - \dfrac{\alpha^2}{4Q^2}\right) + 4jQx}(1+x)\beta E_e \\[2ex] V_-^* = \dfrac{j + \alpha - 2Qx\left[1 - \dfrac{\alpha}{2Q}\right]}{1 - \alpha^2 - 4Q^2 x^2 \left(1 - \dfrac{\alpha^2}{4Q^2}\right) + 4jQx}(1-x)\beta E_e \end{array} \right\} \quad (12.52)$$

In these relations x is assumed to be much smaller than 1; on the other hand, by virtue of (12.19) and (12.46) we have:

$$\frac{\alpha}{Q} = \frac{C'_2}{C'_0 + C} \ll 1$$

We may therefore write approximately:

$$\left. \begin{aligned} V_+ &= \frac{-j + \alpha + 2Qx}{1 - \alpha^2 - 4Q^2x^2 + 4jQx} \beta E_e \\ V_-^* &= \frac{j + \alpha - 2Qx}{1 - \alpha^2 - 4Q^2x^2 + 4jQx} \beta E_e \end{aligned} \right\} \quad (12.53)$$

Let θ_e represent the initial phase of the input signal e_e with relation to the harmonics of $C(t)$ and put:

$$\left. \begin{aligned} V_+ &= \frac{A_+}{2} e^{-j(\theta_+ + \theta' + \theta_e)} \\ V_- &= \frac{A_-}{2} e^{-j(\theta_- - \theta' - \theta_e)} \end{aligned} \right\} \quad (12.54)$$

with

$$\left. \begin{aligned} A_+ &= 2\beta \frac{\sqrt{1 + (\alpha + 2Qx)^2}}{D} |E_e| \\ A_- &= 2\beta \frac{\sqrt{1 + (\alpha - 2Qx)^2}}{D} |E_e| \end{aligned} \right\} \quad (12.55)$$

$$\left. \begin{aligned} \sin \theta_+ &= \frac{1}{\sqrt{1 + (\alpha + 2Qx)^2}}, & \cos \theta_+ &= \frac{\alpha + 2Qx}{\sqrt{1 + (\alpha + 2Qx)^2}} \\ \sin \theta_- &= \frac{1}{\sqrt{1 + (\alpha - 2Qx)^2}}, & \cos \theta_- &= \frac{\alpha - 2Qx}{\sqrt{1 + (\alpha - 2Qx)^2}} \end{aligned} \right\} \quad (12.56)$$

$$\sin \theta' = \frac{4Qx}{D}, \quad \cos \theta' = \frac{1 - \alpha^2 - 4Q^2x^2}{D} \quad (12.57)$$

$$D = \sqrt{(1 - \alpha^2 - 4Q^2x^2)^2 + 16Q^2x^2} \quad (12.58)$$

Note: If we make x tend to zero in the expressions (12.55), (12.56), (12.57) and (12.58), the voltages V_+ and V_- approach the same value (modulus and phase), and we again derive the expressions obtained in section 3.

AMPLIFICATION OF LOW-FREQUENCY SIGNALS   329

4.2. Instantaneous voltage energizing the detector

By virtue of (12.8) and (12.54), the voltage energizing the detector is of the form:

$$v'(t) = nA[A_+ \cos[\omega_+ t - (\theta_+ + \theta_e + \theta')]$$
$$+ A_- \cos[\omega_- t - (\theta_- - \theta_e - \theta')]]$$

$$= nA(A_+ + A_-) \cos\left(\omega_e t - \frac{\theta_+ - \theta_- + 2\theta_e + 2\theta'}{2}\right)$$

$$\times \cos\left(\omega_p t - \frac{\theta_+ + \theta_-}{2}\right) - nA(A_+ - A_-)$$

$$\times \sin\left(\omega_e t - \frac{\theta_+ - \theta_- + 2\theta_e + 2\theta'}{2}\right)$$

$$\times \sin\left(\omega_p t - \frac{\theta_+ + \theta_-}{2}\right) \qquad (12.59)$$

For $x = 0$, (12.57), (12.58), (12.59) give us:

$$v'(t) = nA \frac{2\beta\sqrt{1+\alpha^2}}{1-\alpha^2} E_{e_0} \cos(\omega_p t - \theta)$$

Relation (12.59) shows that for $\alpha \neq 0$, the voltage energizing the detector (equal to a factor close to the output voltage of the parametric amplifier), is not a wave of frequency ω_p amplitude modulated by a signal of frequency ω_e. As the importance of the second beat component increases as the difference $A_+ - A_-$ increases, this difference is found to increase with ω_e. If it is desired to make this component negligible, α must be made near to zero, that is we must abandon parametric amplification by the second harmonic and operate the circuit as a standard tuned-load modulator. This disadvantage can be avoided by the use of synchronous detection.

4.3. Synchronous detection of the output signal

If we multiply the voltage $v'(t)$ by a wave of the form:

$$A' \cos(\omega_p t - \phi)$$

and eliminate by filtering the components of frequency $2\omega_p + \omega_e$ and $2\omega_p - \omega_e$ resulting from this operation, we obtain:

$$v_{se}(t) = |V_{se}| \cos\left(\omega_e t - \frac{\theta_+ - \theta_- + 2\theta'}{2} - \theta_e + \theta''\right) \qquad (12.60)$$

with

$$|V_{se}|^2 = n^2 K^2 [A_+^2 + A_-^2 + 2A_+ A_- \cos(\theta_+ + \theta_- - 2\phi)] \quad (12.61)$$

$$\tan \theta'' = \frac{A_+ - A_-}{A_+ + A_-} \tan\left(\frac{\theta_+ + \theta_-}{2} - 2\phi\right) \quad (12.62)$$

Note. It has been assumed that the transfer function of the LF filter after the synchronous detector was unity for the expected frequencies fe of the input signal. (Bandwidth of the filter wider than that of the parametric amplifier.)

The phase angle ϕ will be chosen so as to have for $\omega_e = 0$ ($x = 0$)

$$2\phi = \theta_+ + \theta_- = 2\theta \quad (12.63)$$

a value which makes the expression (12.61) a maximum.

If we take (12.61) and (12.63) into account we obtain:

$$\cos \phi = \frac{\alpha}{\sqrt{1 + \alpha^2}}, \quad \sin \phi = \frac{1}{\sqrt{1 + \alpha^2}} \quad (12.64)$$

By virtue of (12.55), (12.56), (12.58), (12.61) and (12.64) we find:

$$|V_{se}|^2 = \frac{n^2 K^2 \beta^2 [(1 + \alpha^2)^2 + 4Q^2 x^2]}{(1 + \alpha^2)\{[1 - \alpha^2 - 4Q^2 x^2]^2 + 16Q^2 x^2\}} 16|E_e|^2$$

hence

$$|V_{se}| = \frac{4nK\beta \sqrt{(1 + \alpha^2)^2 + 4Q^2 x^2}}{\sqrt{1 + \alpha^2} \sqrt{[1 - \alpha^2 - 4Q^2 x^2]^2 + 16Q^2 x^2}} |E_e| \quad (12.65)$$

If the input signal is

$$e_e(t) = E_e e^{j\omega_e t} + E_e^* e^{-j\omega_e t} = 2|E_e| \cos(\omega_e t - \theta_e)$$

the conversion efficiency will be designated as before by η, the ratio:

$$\eta = \frac{|V_{se}|}{nK2|E_e|} = \frac{2\beta \sqrt{(1 + \alpha^2)^2 + 4Q^2 x^2}}{\sqrt{1 + \alpha^2} \sqrt{[1 - \alpha^2 - 4Q^2 x^2] + 16Q^2 x^2}} \quad (12.66)$$

For $x = 0$ ($\omega_e = 0$, $2|E_e| = |E_{e_0}|$) we obtain

$$\eta_0 = \frac{2\beta \sqrt{1 + \alpha^2}}{1 - \alpha^2}$$

The result obtained in section 3.5 appears here again.

As shown by relations (12.59) and (12.66) there is no simple relation between the conversion efficiency η and the ratio of the peak

AMPLIFICATION OF LOW-FREQUENCY SIGNALS 331

values of the voltage $v'(t)$ to the input voltage. It should be noted, however, that for $\alpha = 0$, (12.55), (12.56) and (12.64) give us:

$$A_+ = A_-, \qquad \theta_+ + \theta_- = \pi = 2\phi$$

hence:

$$|V_{se}|^2 = n^2 K^2 (A_+ + A_-)^2$$

and in this case the conversion efficiency as defined by (12.66) is equal to the ratio of the peak values of the output and input voltages of the parametric amplifier.

4.4. Bandwidth at 3 dB

The following ratio can be used for the calculation of the bandwidth:

$$\frac{\eta_x^2}{\eta_0^2} = \frac{(1-\alpha^2)^2[(1+\alpha^2)^2 + 4Q^2 x^2]}{(1+\alpha^2)^2[(1-\alpha^2)^2 + 8(1+\alpha^2)Q^2 x^2 + 16Q^4 x^4]}$$

$$= \frac{(1-\alpha^2)^2[(1+\alpha^2)^2 + y^2]}{(1+\alpha^2)^2[(1-\alpha^2)^2 + 2(1+\alpha^2)y^2 + y^4]} \qquad (12.67)$$

in which we made:

$$y = 2Qx \qquad (12.68)$$

Next, find the value of y corresponding to:

$$\frac{\eta_y^2}{\eta_0^2} = \frac{1}{2}$$

(12.67) and (12.68) give us

$$(1+\alpha^2)^2 y^4 + 2[(1+\alpha^2)^3 - (1-\alpha^2)^2]y^2 - (1+\alpha^2)^2(1-\alpha^2)^2 = 0$$

hence

$$y_{3\,\text{dB}}^2 = \frac{-[(1+\alpha^2)^3 - (1-\alpha^2)^2]}{(1+\alpha^2)^2}$$

$$+ \frac{\sqrt{[(1+\alpha^2)^3 - (1-\alpha^2)^2]^2 + (1+\alpha^2)^4(1-\alpha^2)^2}}{(1+\alpha^2)^2} \qquad (12.69)$$

The curve $y_{3\,\text{dB}} = f(\alpha)$ is shown in Fig. 12.10. It will be seen that as a first approximation we can make:

$$y_{3\,\text{dB}} = 1 - \alpha$$

This, when (12.50) and (12.68) are taken into account, enables us to write:

$$\frac{\omega_e\,3\,\text{dB}}{\omega_p} = \frac{1-\alpha}{2Q} \qquad (12.70)$$

or

$$B_{3\,\text{dB}} = \frac{1-\alpha}{2Q} f_p \qquad (12.71)$$

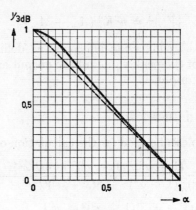

Figure 12.10

4.5. Input admittance

By virtue of relations (12.47) and designating by I_e the component of the frequency current ω_e delivered by the generator e_e, we obtain:

$$Y_{\text{in}} = \frac{I_e}{V_e} = G_0' + j\omega_p(C_0' + C_e)x$$
$$+ 2(G_1' + j\omega_p C_1' x)\beta \frac{[-jx(1 + 2jQx) + \alpha]}{(1 + 2jQx)^2 - \alpha^2} \qquad (12.72)$$

For $\omega_e = 0$ $(x = 0)$ we once more obtain the expression for the input admittance Y_{in_0} which we found when studying the operation for $\omega_e = 0$:

$$Y_{\text{in}_0} = G_0' + \frac{2G_0'\alpha\beta}{1 - \alpha^2}$$

In order that the circuit being studied may operate effectively it is necessary that $R_e Y_{\text{in}} \ll 1$.

This condition was discussed earlier for $\omega_e = 0$. From (12.72) we see that the admittance Y_{in} increases rapidly with the frequency ω_e of the input signal, and the condition $R_e Y_{\text{in}} \ll 1$, which we have assumed was satisfied, will limit, for a given bandwidth, the value of the internal impedance R_e compatible with this assumption. In order to determine the permissible value of R_e, the admittance Y_{in}

AMPLIFICATION OF LOW-FREQUENCY SIGNALS

for frequency $\omega_e = \omega_{e\,3\,\mathrm{dB}}$ must be calculated. (12.70) and (12.72) give us:

$$Y_{\mathrm{in\,3\,dB}} = G_0' + \frac{\omega_p C_1' \beta}{2Q(\alpha^2 + 1)}$$

$$\times \left[\left(\alpha(1 - \alpha) + \frac{2\alpha^2 Q}{1 - \alpha} - 1 \right) \gamma_1 + \frac{1 - \alpha}{2Q} + \alpha \right]$$

$$+ j\omega_p (C_0 + C_e)$$

$$\times \left[1 + \frac{C_1'^2}{2(\alpha^2 + 1)(C_0' + C_e)(C_0' + C)} (\alpha(\alpha - 1) - 1) \right]$$

$$\times \left(\frac{1 - \alpha}{2Q} \right) + \frac{j\omega_p C_1' \beta}{2Q(\alpha^2 + 1)} \left[\alpha^2 - \gamma_1 \left(1 + \frac{2Q\alpha}{1 - \alpha} \right) \right]$$

For $\alpha = 0$ (negligible effect of the second harmonic)

$$Y_{\mathrm{in\,3\,dB}} = G_0' + \frac{\omega_p C_1'}{2Q} \left[\frac{1}{2Q} - \gamma_1 \right]$$

$$+ j\omega_p \left(\frac{C_0 + C_e}{2Q} \right) \left[1 - \frac{C_1'^2}{2(C_0' + C_e)(C_0' + C)} \right] \cdot \frac{j\omega_p C_1' \beta \gamma_1}{2Q}$$

$$= \frac{\omega_p (C_0' + C_e)}{2Q} \left[\frac{2G_0'}{G_T} \frac{C + C_0'}{C_0' + C_e} + \frac{C_1'}{2(C + C_0')} \frac{C_1'}{C_0' + C_e} (1 - j) \right.$$

$$\left. - \frac{C_1' G_1'}{(C_0' + C_e) G_T} (1 + j) + j \right]$$

Under normal operating conditions we have:

$$\frac{G_0'}{G_T} \ll 1, \quad \frac{G_1'}{G_T} \ll 1, \quad \frac{C_1'}{C + C_0'} \frac{C_1'}{C_0' + C_e} \ll 1$$

and as a first approximation we can write:

$$Y_{\mathrm{in\,3\,dB}} = \frac{j\omega_p (C_0' + C_e)}{2Q}$$

The condition $R_e Y_{\mathrm{in}} \ll 1$ gives:

$$R_e(C_e + C_0') \frac{\omega_p}{2Q} = R_e G_T \frac{(C_0' + C_e)}{2(C_0' + C)} \ll 1$$

For values of α lying between 0 and 1, we have almost exactly

$$Y_{\mathrm{in\,3\,dB}} = G_0' + \frac{\omega_p C_1' \beta \alpha}{2Q(\alpha^2 + 1)} \left[\frac{2Q\gamma_1 \alpha}{1 - \alpha} + 1 \right]$$

$$+ j\omega_p \left[\frac{(C_0' + C_e)(1 - \alpha)}{2Q} + \frac{j\omega_p C_1' \beta \alpha}{2Q(\alpha^2 + 1)} \left[\alpha - \frac{2Q\gamma_1}{1 - \alpha} \right] \right]$$

and the following criterion will be adopted $R_e^2|Y_{\text{in}}|^2 \ll 1$, which gives:

$$R_e^2\left\{\left[G_0' + \frac{\omega_p C_1'}{2Q(\alpha^2+1)}\left(\frac{2Q\alpha\gamma}{1-\alpha}+1\right)\right]^2 + \left[(C_0'+C_e)(1-\alpha)+\frac{C_1'\beta\alpha}{\alpha^2+1}\left[\alpha-\frac{2Q\gamma_1}{1-\alpha}\right]\right]^2 \cdot \frac{\omega_p^2}{4Q^2}\right\} \ll 1 \quad (12.74)$$

The condition (12.74) determines as a first approximation the value of the impedance R_e as a function of the circuit parameters and vice versa. It is clear that for very high source impedances it will be necessary, in order to obtain a wide bandwidth and an admittance, $Y_{\text{in}} \ll 1/R_e$, to

(1) Decrease the value of α.
(2) Use a more complex input circuit.

5. Background noise

5.1. Internal noise sources in parametric amplifiers

The main internal source of noise in the parametric amplifier under consideration is the thermal-agitation noise generated by the conductance G of the anti-resonant circuit. For low frequencies the series resistance R_s of the diodes plays only a negligible role. On the other hand, the Shottky noise produced in the semi-conductor junctions (noise which depends on the currents i_a' and i_b' flowing through the diodes D_a and D_b (Fig. 12.5)) may be reduced to a very low value by reducing the currents by a suitable choice of diodes and of the pumping conditions. In this connection, mention should be made of the work of L. Becker and R. L. Ernst [4] in which they studied the effects of modulation of the Shottky noise (especially the effects of modulation of this noise by the pump) on the performance of non linear reactance and conductance converters.

5.2. Noise voltage at the input of the detector

Reverting to Fig. 12.6 and localizing in it the different noise sources we get Fig. 12.11. In this diagram,

$e_e\sqrt{\Delta f}$ represents the effective random amplitude of the noise produced by the resistance R_e in the frequency band $f_e, f_e + \Delta$

AMPLIFICATION OF LOW-FREQUENCY SIGNALS

$i_+\sqrt{\Delta f}$ ($i_-\sqrt{\Delta f}$) represents the effective random amplitude of the current produced by the conductance G in the frequency band $f_+, f_+ + \Delta f$ ($f_-, f_- + \Delta f$).

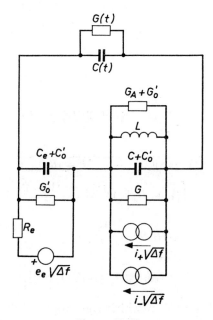

Figure 12.11

From the general equations established in section 2, we obtain:
Frequency:

$$\begin{array}{ll}
\omega_e & e_e = (1 + R_e Y'_e)V_e + R_e(G''_1 + j\omega_e C'_1)(V_+ + V^*_-) \\
\omega_+ & i_+ = G_T\{j\beta V_e + (1 + 2jQx)V_+ + j\alpha V^*_-\} \\
-\omega_- & i^*_- = G_T\{-j\beta V_e - j\alpha V_+ + (1 + 2jQx)V^*_-\}
\end{array} \right\} (12.75)$$

It should be noted that we have:

$$\langle e_e\, e^*_e \rangle = 4kT_0 R_e \tag{12.76}$$

$$\left.\begin{array}{l}
\langle i_+\, i^*_+ \rangle = 4kT_G G \\
\langle i_-\, i^*_- \rangle = 4kT_G G
\end{array}\right\} \tag{12.77}$$

$$\langle e_e\, i_+ \rangle = \langle e_e\, i_- \rangle = \langle i_+\, i_- \rangle = 0$$

$$Y'_e = G''_0 + j\omega_e[C''_0 + C_e]$$

Next, we will calculate the noise components V_+, V_- at the input of the detector. (12.75) gives us:

$$\left.\begin{aligned}
V_+ &= \frac{1}{[(1 + 2Qjx)^2 - \alpha^2][1 + R_e Y_{\text{in}}]} \\
&\quad \times \Bigg\{ (1 + R_e Y'_e)\frac{(1 + 2jQx)i_+ - j\alpha i_-^*}{G_T} \\
&\qquad + jR_e\beta(\gamma_1 + jx)\omega_p C'_1 \frac{(i_+ + i_-^*)}{G_T} \\
&\qquad + e_e\beta[-j(1 + 2jQx) + \alpha] \Bigg\} \\
V_-^* &= \frac{1}{[(1 + 2jQx)^2 - \alpha^2][1 + R_e Y_{\text{in}}]} \\
&\quad \times \Bigg\{ (1 + R_e Y'_e)\frac{j\alpha i_+ + (1 + 2jQx)i_-^*}{G_T} \\
&\qquad - jR_e\beta(\gamma_1 + jx)\omega_p C'_1 \frac{(i_+ + i_-^*)}{G_T} \\
&\qquad + e_e\beta[+j(1 + 2jQx) + \alpha] \Bigg\}
\end{aligned}\right\} \quad (12.78)$$

Let us consider the case in which the input admittance Y_{in} of the amplifier is much lower than the admittance $1/R_e$ of the source; we then have:

$$R_e Y_{\text{in}} \ll 1$$
$$R_e Y'_e \ll 1$$
$$R_e\beta(\gamma_1 + jx)\omega_p C'_1 \ll 1$$

Under these conditions, relations (12.78) may be written approximately,

$$\left.\begin{aligned}
V_+ &= \frac{1}{(1 + 2jQx)^2 - \alpha^2} \\
&\quad \times \left[\beta e_e[-j(1 + 2jQx) + \alpha] + \frac{(1 + 2jQx)i_+ - j\alpha i_-^*}{G_T} \right] \\
V_-^* &= \frac{1}{(1 + 2jQx)^2 - \alpha^2} \\
&\quad \times \left[\beta e_e[j(1 + 2jQx) + \alpha] + \frac{(1 + 2jQx)i_-^* + j\alpha i_+}{G_T} \right]
\end{aligned}\right\} \quad (12.79)$$

Since voltages V_+ and V_- are at different frequencies, they are not

AMPLIFICATION OF LOW-FREQUENCY SIGNALS

correlated, but, as we saw in Chapter 1, these components may give rise after synchronous detection to correlated noise voltages. The quantity we are interested in is then:

$$n_d = \langle v_d\, v_d^* \rangle$$

where $v_d\sqrt{\Delta f}$ is the effective random amplitude of the noise voltage in the frequency band $f_e, f_e + \Delta f$ at the output of the LF filter which follows the synchronous detector.

5.3. Spectral density n_d of the parasitic signal at the output of the detector

5.3.1. SPECTRAL DENSITY n_{ed} OF THE PARASITIC SIGNAL DUE TO THE GENERATOR e_e

In order to determine the spectral density n_{ed} due to the generator e_e, the currents i_+ and i_-^* in relations (14.79) will be equated to zero; we then obtain:

$$V_{+e} = \frac{-j + \alpha + 2Qx}{1 - \alpha^2 - 4Q^2x^2 + 4jQx}\beta e_e$$

$$V_{-e}^* = \frac{j + \alpha - 2Qx}{1 - \alpha^2 - 4Q^2x^2 + 4jQx}\beta e_e$$

As these relations are identical with expressions (12.53) we find, by virtue of (12.65) and (12.76)

$$n_{ed} = 4kT_0R_e\frac{4n^2K^2\beta^2[(1+\alpha^2)^2 + 4Q^2x^2]}{(1+\alpha^2)[(1-\alpha^2-4Q^2x^2)^2 + 16Q^2x^2]} \quad (12.80)$$

5.3.2. SPECTRAL DENSITY n_{+d} OF THE PARASITIC SIGNAL DUE TO THE GENERATOR i_+

Equating e_e and i_-^* in relations (12.79) to zero, we have:

$$V_{++} = \frac{(1+2jQx)i_+}{[(1+2jQx)^2 - \alpha^2]G_T}$$

$$V_{-+}^* = \frac{j\alpha i_+}{[(1+2jQx)^2 - \alpha^2]G_T}$$

Using the same reasoning as in section 4.3, we obtain:

$$n_{+d} = n^2K^2[1 + 4Q^2x^2 + \alpha^2 + 2\alpha\sqrt{1+4Q^2x^2}\sin(\theta_1+2\phi)]\frac{\langle i_+\, i_+^*\rangle}{G_T^2}$$

which, by virtue of (12.76), gives:

$$n_{+d} = n^2 K^2 [1 + 4Q^2 x^2 + \alpha^2 + 2\alpha \sqrt{1 + 4Q^2 x^2} \sin(\theta_1 + 2\phi)]$$

$$\times \frac{4kT_G G}{G_T^2} \quad (12.81)$$

with a phase angle of ϕ of the demodulation wave and where θ_1 is defined by the relations:

$$\left. \begin{array}{l} \sin \theta_1 = \dfrac{2Qx}{\sqrt{1 + 4Q^2 x^2}} \\[2mm] \cos \theta_1 = \dfrac{1}{\sqrt{1 + 4Q^2 x^2}} \end{array} \right\} \quad (12.82)$$

5.3.3. SPECTRAL DENSITY n_d OF THE PARASITIC SIGNAL DUE TO THE GENERATOR i_-

Equating e_e and i_+ to zero in relations (12.79), we obtain:

$$V_{+-} = \frac{-j\alpha i_-^*}{[(1 + 2jQx)^2 - \alpha^2] G_T}$$

$$V_{--}^* = \frac{(1 + 2jQx) i_-^*}{[(1 + 2jQx)^2 - \alpha^2] G_T}$$

which gives:

$$n_{-d} = n^2 K^2 [1 + 4Q^2 x^2 + \alpha^2 + 2\alpha \cdot \sqrt{1 + 4Q^2 x^2} \sin(2\phi - \theta_1)]$$

$$\times \frac{4kT_G G}{G_T^2} \quad (12.83)$$

5.3.4. SPECTRAL DENSITY n_d OF THE TOTAL PARASITIC SIGNAL AT THE OUTPUT OF THE DETECTOR

As generators e_e, i and i_- are not correlated, we have

$$n_d = n_{ed} + n_{+d} + n_{-d}$$

which, by virtue of (12.80), (12.81), (12.82), (12.83) and taking account of (12.64), give

$$n_d = 4n^2 K^2 \left\{ \frac{4kT_0 R_e \beta^2 [(1 + \alpha^2)^2 + 4Q^2 x^2]}{(1 + \alpha)^2 [(1 - \alpha^2 - 4Q^2 x^2)^2 + 16Q^2 x^2]} \right.$$

$$\left. + \frac{2kT_G G[(1 + \alpha^2)^2 + 4(1 + \alpha^2) Q^2 x^2 + 4\alpha^2]}{G_T^2 (1 + \alpha^2)[(1 - \alpha^2 - 4Q^2 x^2)^2 + 16Q^2 x^2]} \right\} \quad (12.84)$$

5.4. Noise factor

By virtue of the formulae established in Chapter 1, the noise factor in the case under consideration is equal (after synchronous detection) to:

$$F = \frac{\int_0^B H^2(f_e) n_a(f_e) \, df}{\int_0^B H^2(f_e) n_{ed}(f_e) \, df} \qquad (12.85)$$

If we assume that the transfer function of the LF filter is unity in the frequency range f_e of the input signal (see note in section 4), (12.85) may be written in the form:

$$F = 1 + \frac{\int_0^B (n_{+d} + n_{-d}) \, df}{\int_0^B n_{ed} \, df}$$

Since the bandwidth B of the LF filter was assumed to be much wider than that of the parametric amplifier itself, we may write:

$$F = 1 + \frac{\int_0^\infty (n_{+d} + n_{-d}) \, df}{\int_0^\infty n_{ed} \, df}$$

hence, taking (12.80), (12.81) and (12.83) into account,

$$F = 1 + \frac{G}{\beta^2 G_T^2 R_e} \frac{(1 + 3\alpha^2)}{\alpha^4 + \alpha^2 + 2} \frac{T_G}{T_0}$$

This may be written, with a sufficient degree of approximation, as shown in Fig. 12.12, in the form:

$$F = 1 + \frac{1+\alpha}{2} \frac{G}{\beta^2 G_T^2 R_e} \frac{T_G}{T_0} \qquad (12.86)$$

or, taking account of (12.19)

$$F = 1 + \frac{1+\alpha}{2} \frac{G}{\omega_p^2 C_1'^2 R_e} \frac{T_G}{T_0} \qquad (12.87)$$

Note. Relation (12.87) differs from the result obtained by J. R. Biard [1]. This author introduces a supplementary noise source: the noise assumed to be of thermal origin produced by the input admittance G_i of the conventional amplifier (load of the parametric amplifier). Strictly speaking, however, to take into account the noise produced by the conventional amplifier, we must know not only the input admittance G_i but also the admittance matrix of this amplifier as well as the nature and location of its internal noise sources.

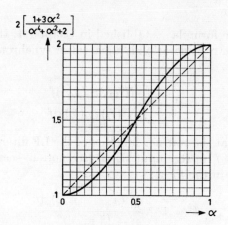

Figure 12.12

6. Special case in which $\alpha = 0$; tuned-load modulator

When the pumping conditions are such that the effect of the negative resistance due to the second harmonic of the Fourier series development of the differential capacitance is negligible or even non-existent,[†] the circuit of Fig. 12.3 operates as a classic tuned-load modulator. If α is equated to zero in the expressions obtained in the previous paragraphs, we obtain:

$$\eta = \frac{2\beta}{\sqrt{1 + 4Q^2 x^2}} = \frac{2\omega_p C_1'}{G_T} \cdot \frac{1}{\sqrt{1 + 4Q^2 x^2}} \quad (12.88)$$

$$\eta_0 = 2\beta = \frac{2\omega_p C_1'}{G_T}$$

$$B_{3 \text{ dB}} = \frac{1}{2Q} f_p$$

$$Y_{\text{in } 0} = G_0'$$

$$Y_{\text{in 3 dB}} \cong j\omega_p \frac{(C_0' + C_e)}{2Q}$$

$$F = 1 + \frac{G}{2\omega_p^2 C_1'^2 R_e} \cdot \frac{T_G}{T_0}$$

This type of operation takes place especially when the internal impedance of the source of the signal is very high. In order to reduce

[†] Pumping by means of a flat-topped wave (see Chapter 14).

the terms G_0'' and G_1'' (due to the losses in the diodes) to a negligible value with respect to $1/R_e$, we must apply to the diodes a suitable bias and a pumping voltage of relatively small amplitude. Under these conditions the effect of the curvature in the characteristic giving the differential capacitance as a function of the voltage is negligible, and the second harmonic of the Fourier series development of this capacitance is very small (cissoidal pumping).

The conversion efficiency may in this case be expressed very simply as a function of the pumping parameters. As will be seen later (Chapter 14 (relation (14.29))), the coefficient C_1 of the first harmonic of $C(t)$ may be written in the form:

$$C_1 = \frac{C_0 V_p}{n(V_0 + \phi)} \quad (12.89)$$

with C_0, the first term of the Fourier series development of $C(t)$,
V_0 the bias voltage,
$2V_p$, amplitude of the pumping voltage,
ϕ, the characteristic potential of the junction,†
n, a number depending on the type of junction.†

With diodes D_a and D_b properly paired, we have approximately

$$C_1' = C_{a_1} + C_{b_1} \sim 2C_1 \quad (12.90)$$

which, by virtue of (12.88), gives

$$\eta_0 = \frac{4\omega_p C_0 V_p}{n G_T (V_0 + \phi)} \quad (12.91)$$

References

[1] Biard, J. R., 'Low frequency reactance amplifier', *Proc. I.R.E.*, vol. 51 (No. 2), pp. 298–303, February 1963.
[2] Dachert, F., 'Théorie de l'amplification paramétrique à diode', *Annales de Radioélectricité*, vol. XV (No. 60), pp. 109–119, April 1960.
[3] Decroly, J. C., 'Note sur la dégénérescence des amplificateurs paramétriques', *Revue M.B.L.E.*, vol. V (No. 1), pp. 4–15, 1962.
[4] Becker, L. and Ernst, R. L., 'Nonlinear admittance mixers', *R.C.A. Review*, vol. XXV (No. 4), pp. 662–691, December 1964.

† See Chapters 8 and 13.

13

DECOUPLING AND METHODS OF ADJUSTMENT

by G. MARECHAL

1. Balancing of the bridge

In the theoretical study in Chapter 12 it was assumed that there was complete decoupling between the pump circuit and the output circuit. It was seen that this decoupling could not be obtained simply by filtering, and that a differential structure was required. It is proposed in this chapter to study the problems involved in the production of this decoupling, which will be referred to also as balancing of the bridge or of the input head. In principle only the output circuit having the filter tuned to ω_p must be in common to the two branches of the differential structure. The relative position, in every branch and in the combination of the branches of the varactor, of the pump circuit and of the input filter (which can also be tuned to ω_p) determine ten principal variants to the base circuit of Fig. 12.3, every one having its dual.

For this analysis we shall keep the base configuration studied in the previous chapter, although some of the other configurations have interesting advantages, the same technological problems being present in all variants.

1.1. Preliminary conditions for satisfactory balancing

The bridge is balanced if in the absence of a signal at the input ($e_e = 0$, $i_e = 0$) no voltage appears at the output ($v_s = 0$). Simple inspection of the circuit of Fig. 12.3, repeated in Fig. 13.1, and of equations (12.11), shows that the bridge is automatically balanced if the following conditions are satisfied:

(1) the two varactors are identical;
(2) the pumping voltages v_p and $-v_p$ delivered by the secondary of the transformer pass through zero simultaneously and have

DECOUPLING AND METHODS OF ADJUSTMENT 343

an identical form, symmetrical with respect to the centre point of the band $(0, 2\pi/\omega_p)$;
(3) the bias voltages are equal;
(4) the operation of the circuit is not disturbed by any parasitic element or effect.

These ideal conditions are adequate, although certain of them are not indispensable. However, in order to ensure maximum stability and regular operation it is desirable to satisfy these conditions as fully as possible.

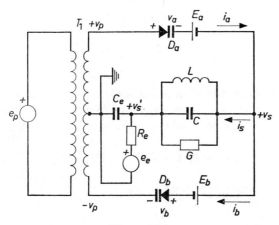

Figure 13.1

1.1.1. IDENTICAL VARACTORS

By the identity of varactors is meant the identity of the equivalent circuits which represent them. We will therefore start this chapter with the study of the equivalent circuit of the semiconductor diode.

(1) *Equivalent circuit and basic relations*

In all essentials, varactors used in parametric amplifiers for low frequencies are identical with those used at high frequencies and described in Chapter 8. All the theoretical conditions developed in Chapter 8 are applicable to low-frequency varactors. However, although in the high-frequency field the non-linear conductance of the semiconductor diode can be ignored, this does not apply to low frequencies, since the value of the differential conductance of the diode becomes comparable with its differential susceptance. It was for this reason that in Chapter 12 a semiconductor diode was represented by the equivalent circuit of Fig. 12.5. Strictly speaking, this circuit, which is reproduced in Fig. 13.2, should be completed by the elements R_s and L_c which were mentioned in Chapter 8. However, at the frequencies we are considering and for the diodes normally used at low frequencies, the admittance of R_s and L_c is very high with respect to the differential admittances of the junction. We will therefore omit R_s and L_c in this study.

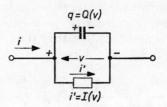

Figure 13.2

On the basis of the theoretical and experimental work on semiconductor diodes, functions $I(v)$ and $Q(v)$ (Fig. 13.2) may be expressed, to a good degree of approximation, by the relations:

$$i' = I(v) = I_s(1 - e^{-av}) + G_s v \qquad (13.1)$$

$$G = \frac{dI}{dv} = aI_s e^{-av} + G_s \qquad (13.2)$$

and

$$q = Q(v) = C_{0v}\phi \frac{(1 + v/\phi)^{1-(1/n)}}{1 - 1/n} + C_c v \qquad (13.3)$$

$$C = \frac{dQ}{dv} = \frac{C_{0v}}{\sqrt[n]{1 + v/\phi}} + C_c \qquad (13.4)$$

where v is the voltage applied to the diode, made positive in the cut-off direction.

I_s is the specific saturation current of the diode at a given temperature (positive in the cut-off direction). For low-frequency varactors, this current is usually between 10^{-14} and 10^{-11} A, at 25°C.

a is a parameter defined by:

$$a = \frac{q}{mkT} \qquad (13.5)$$

In this relation q is the charge of the electron, k is the Boltzmann constant, and m is a factor between 1 and 2.

The parameter a is expressed by the reciprocal of a voltage. At 25°C it is equal to:

$$a = \frac{39}{m} \quad \text{(volts)}^{-1} \qquad (13.6)$$

G_s is a conductance, the introduction of which in (13.1) takes account of the phenomena of which the effects become appreciable when the diode is biased in the cut-off direction. For good low-frequency varactors, G_s usually lies between 10^{-14} and 10^{-11} mho.

C_{0v} is the differential capacitance dQ/dv for $v = 0$. At low frequencies, C_{0v} usually lies between 10 and 10,000 pF. The most common values are from 50 to 500 pF.

ϕ is a characteristic parameter which is positive depending on the nature of the semiconductors used and the temperature T of the diode. The ϕ parameter has the dimensions of a voltage. It is often called the contact potential of the junction. ϕ usually lies between 0·4 and 1·5 volts. The most usual values are between 0·6 and 1·2 volts.

n is the power of the root shown in Fig. 13.4. The value of n depends on the method used to produce the junction. n lies between 2 and 3 for standard junctions.

C_c is the capacitance of the internal connections and of the diode cartridge. The mechanical configuration of low-frequency varactors is usually the same as that of standard diodes. The capacitance of the cartridge, which usually lies between 0·1 and 0·4 pF, may in general be ignored by comparison with C.

Finally, it should be noted that relations (13.1) to (13.4) are only valid in the voltage range defined by:

$$V_B \geqslant v \geqslant -\phi \qquad (13.7)$$

Relations (13.1) to (13.4), on which this study is based, are relatively well-established in practice. However, there are some minor differences and some remarks to be made:

Static characteristic. An analysis of the principle differences from the approximate theoretical law will be found in reference [1]. The author has endeavoured to make a compilation of the special effects encountered.

The first term of (13.1) corresponds to that given in the classical diffusion theory if we make:

$$a = \frac{q}{kT} \quad (m = 1) \tag{13.8}$$

$$I_s = i_s \, e^{-a\phi} \tag{13.9}$$

where i_s is a given current independent of temperature. It should be remembered that in the bandwidth zone ($v < 0$) the term e^{-av} rapidly becomes much larger than unity and that the characteristic curve tends towards an exponential. When we enter the cut-off region ($v > 0$), the term e^{-av} rapidly becomes much smaller than unity and the characteristic curve tends to become horizontal (I_s).

In accordance with more complete theories [1], confirmed by experience, the characteristic curve of the diode in the bandwidth zone will be the sum of several distinct exponentials whose relative contributions depend on the range of voltages considered. Thus at the beginning of the bandwidth zone, the predominating exponential is characterized by values of m generally lying between 1·1 and 1·45. The characteristic curves of the diode in the cut-off zone will be the sum of the first term and of a correcting term which, as a first approximation, may be taken as equal to $G_s V$. Once the static characteristics have been determined in this way, it is important to know the influence of temperature on these characteristics ([2], [3] and [4]) in order to estimate the effect of temperature on the operation of the amplifier and in this way to fix the pumping conditions. Usually, in the bandwidth zone, this influence is expressed by the variation, in degrees centigrade, to which the voltage applied to the diode must be subjected in order to maintain the current flowing through it at a constant value. Typical values of this variation are from 2 to 3 mV/°C.

Differential capacitance. The theoretical law (13.4) is confirmed in practice. Experience has shown that the effect of temperature is to produce a displacement of the characteristic curve $C = C(v)$ parallel to the axis of the v's by an amount equal to $(\partial \phi / \partial T) \, dT$ (Fig. 13.3).

We then have:

$$C + dC = C_{0v} \left(1 + \frac{V - (\partial \phi / \partial T) \, dT}{\phi} \right)^{-1/n} \tag{13.10}$$

which gives:

$$\frac{1}{C} \cdot \frac{dC}{dT} = \frac{\partial \phi / \partial T}{n(\phi + v)} \tag{13.11}$$

DECOUPLING AND METHODS OF ADJUSTMENT

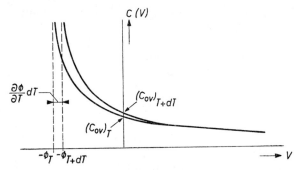

Figure 13.3

This gives, for $v = 0$:

$$\frac{1}{C_{0v}}\frac{dC_{0v}}{dT} = \frac{1}{n\phi}\cdot\frac{\partial\phi}{\partial T} \tag{13.12}$$

Relation (13.11) shows that C_{min} varies very little with temperature ($V_B \gg \phi$). The value of $(1/\phi)(\partial\phi/\partial T)$ for most junctions lies between:

at 25°C $10^{-3}/°C$ and $3 \times 10^{-3}/°C$
at 50°C $2 \times 10^{-3}/°C$ and $6 \times 10^{-3}/°C$
at $-20°C$ $0.4 \times 10^{-3}/°C$ and $2 \times 10^{-3}/°C$

Note. A new type of semiconductor varactor (M.O.S.) has recently been developed beyond the laboratory stage. Its performance makes it particularly suitable for amplifiers with a very high input impedance. The differential capacitance does not vary in accordance with the law (13.4) but with a more complex law, the study of which falls outside the scope of this work. The reader is referred to articles [5], [6], [7] and [8] amongst many others. In these will be found theoretical and experimental studies of the operation of these diodes, especially from the point of view of their differential capacitance. With these diodes, it is hoped that input impedances for the circuit studied in the previous chapter will be available of the order of 10^{17} ohms/150 pF, and a slope of 1200 pF/volt at the transition point (characteristic of the M.O.S. diode) [9].

(2) *Paired varactors*

In order to satisfy condition (1) as fully as possible, the choice of paired varactors should be made with special care. Generally, for varactors of the same type, the quantities ϕ and n differ by only several per cent. If, further, the differences between the static characteristics of these varactors are very small, it can be shown that the same applies to the values of $(\partial \phi/\partial T)(1/\phi)$. It follows, in accordance with (13.11), that the relative variation of the differential capacitance C, per degree centigrade, is the same for all these varactors. It will thus be seen that if amongst varactors of the same type we choose those which have the same values of C_{ov}, a and I_s, the difference between the temperature coefficient and the differential capacitance C is reduced at the same time. This choice results in a very stable balance, as shown by relations (13.29), (13.33) and (13.34) which will be considered later. For a series of very carefully constructed varactors, the value of a can be satisfactorily maintained constant for voltages lying between 0 and $-\phi$ volts.

It should be noted at this stage that certain manufacturers, instead of pairing the diodes in accordance with the static characteristic, try to give minimum values G_s and G_a. This method, which has certain advantages, involves the risk of increasing the deviation of the coefficient of T^0 of the C's.

Finally, varactors will be chosen in such a way that they meet at least the following practical criteria:

$$\left. \begin{aligned} \frac{\Delta C_{ov}}{C_{ov}} &< 5 \times 10^{-2} \\ \Delta \frac{1}{C_{ov}} \cdot \frac{\partial C_{ov}}{\partial T} &< 10^{-5}/°C \end{aligned} \right\} \quad (13.13)$$

$$\left. \begin{aligned} \frac{\Delta a}{a} &< 3 \times 10^{-2} \\ I_s &< 10^{-11} \text{ A} \\ G_s &< 2 \times 10^{-11} \text{ mho} \end{aligned} \right\} \quad (13.14)$$

1.1.2. SIMULTANEOUS PASSAGES THROUGH ZERO OF VOLTAGES v_p AND $-v_p$

The presence of parasitic elements and effects may give rise to a phase shift in the time interval Δt between the passages through zero of voltages v_p and $-v_p$, which we will assume to be of identical form, symmetrical with respect to the centre point of the interval $(0, 2\pi/\omega_p)$. If it is not essential that these voltages should pass through zero simultaneously when we are concerned only with the balancing of the circuit, it is nevertheless fundamental for correct operation of the amplifier.

If this simultaneousness is not respected, the initial phases of the harmonics of the same order of the Fourier series of the differential

DECOUPLING AND METHODS OF ADJUSTMENT

capacitances $C_a(v_a)$, $C_b(v_b)$ of each of the diodes are no longer identical. If we designate by θ the phase shift between the fundamentals of the voltages v_a and v_b, resulting from the time difference Δt, it can easily be shown that:

$$C_a(t) + C_b(t) = C_{a_0} + C_{b_0} + 2C_1'' \cos(\omega_p t - \theta_1) + 2C_2'' \cos(2\omega_p t - \theta_2) + [\text{harmonics of a higher order}]$$

with

$$\left. \begin{array}{l} C_1'' = \sqrt{(C_{1a} + C_{1b} \cos \theta)^2 + C_{1b}^2 \sin^2\theta} \leqslant C_{1a} + C_{1b} = C_1' \\ C_2'' = \sqrt{(C_{2a} + C_{2b} \cos 2\theta)^2 + C_{2b}^2 \sin^2 2\theta} \leqslant C_{2a} + C_{2b} = C_2' \end{array} \right\} \quad (13.15)$$

$$\tan \theta_1 = \frac{C_{1b} \sin \theta}{C_{1a} + C_{1b} \cos \theta}$$

$$\tan \theta_2 = \frac{C_{2b} \sin 2\theta}{C_{2a} + C_{2b} \cos 2\theta}$$

As shown by relations (13.15) a phase shift θ between the two voltages leads to a decrease in coefficients C_1' and C_2' occurring in expression (12.45), and consequently it causes a decrease in the conversion efficiency. Further, this phase shift θ causes an initial phase difference between the first and second harmonics, which may produce additional attenuation (12.42).

The time shift Δt introduces some further disadvantages. As will be seen later, one of the conditions for balance is that the currents I_a and I_b, of frequency ω_p flowing through the diodes should be equal. The balance of the circuit is then represented by the vectorial diagram of Fig. 13.4.

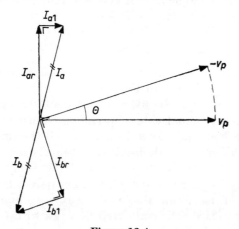

Figure 13.4

This diagram shows that the existence of the phase shift θ makes it impossible to obtain the value of the active currents I_{a_1} and I_{b_1} to be at a minimum simultaneously (in phase with the fundamental of voltages v_a and v_b respectively). It will be seen in Chapter 14 that the quantities G_0, G_1, G_2, which it is most desirable to make as low as possible (section 3, Chapter 12), increase with the active current I_1.

On the other hand, this current contributes very largely to the instability of the balance of the bridge, as I_1 depends exponentially on the bias voltage and the temperature. We see then that from these two points of view it is desirable to reduce the value of I_1 to as low a value as possible and consequently to cause the phase shift θ to be zero. Finally, a trimming component should be provided in order to ensure compensation for any phase difference between the pumping voltages.

Shape of pumping voltages

The presence of parasitic elements in series with the varactors may modify the shape of the voltages applied to varactors. Voltages of different shapes lead to Fourier series of the differential capacitances for which the initial amplitudes and phases of the harmonics of the same order are different. This, as we have seen, causes a deterioration in the performance of the amplifier. Alteration in the shape of the voltages applied to the varactors can be rendered negligible only by reducing as much as possible the influence of the parasitic elements causing these effects. The absolute values of these voltages will be as nearly equal as possible, but the effect of a small difference between them on the balance of the circuit may be compensated for by adjusting the bias voltages and, if necessary, introducing a trimming component into the circuit.

1.1.3. BIAS VOLTAGES

Basically the effect of a d.c. input signal is the same as that of the bias voltages; however, thanks to the small-signal theory, small variations in these biases may be likened to a small d.c. signal, the C_k's maintaining their nominal value. Special care must be taken therefore that these bias variations do not disturb the operation of the amplifier. To this end, it is desirable to obtain the bias voltages from the same source, since the fluctuations of this source, causing the biasings to vary in the same sense, will have only a negligible effect on the balance of the circuit. In certain cases, varactors can be biased by means of standard cells connected in series with them, but it is then necessary by a suitable choice of pumping voltage to suppress,

DECOUPLING AND METHODS OF ADJUSTMENT 351

or at least to reduce to a negligible value, the currents I_0 flowing through them. When possible, the bias circuit will be dropped ($V_0 = 0$).

1.1.4. PARASITIC ELEMENTS AND EFFECTS

(1) The differential phase shift of the transformer. This phase shift may be reduced to less than 10^{-2} degrees by the use of bifilar windings which have a very low resistance. The coupling of this transformer should be very close. 'Closed pots' of ferroxcube are well-adapted for this equipment.

(2) The series resistances of the varactors and the internal impedance of the pump generator.

(3) The parasitic capacitances of the transformer windings.

(4) The parasitic currents from various sources injected into the circuit when the shielding of the 'hot points' is insufficient.

Other parasitic elements, such as the base of the printed circuit, the insulation and the capacitances between the transformer windings and the parasitic capacitances placed in parallel with the varactors may upset the balance of the circuit. After this review of the preliminary requirements for a satisfactory balance, we will next deal briefly with the choice of the trimming components and of certain conditions affecting the pumping.

1.1.5. CHOICE OF TRIMMING COMPONENTS

Since the conditions set out at the beginning of this section cannot be fully satisfied, the use of trimming components becomes necessary. There are numerous parameters which can be varied in order to adjust the balance of the input head, but for practical reasons it should be possible to make the various adjustments independently, at least if they are carried out in a given sequence. The trimming components chosen for introduction into the circuit should, if possible, be of the same nature as the parasitic elements the effects of which they are designed to compensate. However, it is desirable to avoid the use of dissipative components which degrade the noise factor and cause a decrease in the gain and the input impedance.

1.1.6. CHOICE OF PUMPING CONDITIONS

As shown by relation (12.44), which gives the conversion efficiency of the parametric amplifier, the efficiency will increase as the quality factor Q of the output circuit increases:

$$Q = \frac{\omega_p(C_0' + C)}{G_T} = \frac{\omega_p(C_0' + C)}{G + G_A + G_0}$$

C is generally small by comparison with C_0' and is the known fraction

($k - 1$) of it (14.86). It follows that in order to obtain a high conversion efficiency we should have:

$$k_1 \omega_p C_0' \gg G + G_A + G_0' \qquad (13.16)$$

This inequality will be satisfied *a fortiori* if we have:

$$k_1 \omega_p C_0' \gg G_0'$$

or:

$$G_0 \ll k_1 \omega_p C_0$$

In most cases in practice, $aV_p > 1$ and this enables us to write

$$G_0 \ll k_1 a \omega_p C_0 V_p \qquad (13.17)$$

On the other hand, as will be seen in Chapter 14 (14.38) and (14.45), we have, if we ignore G_s,

$$G_1 = \alpha I_1$$

and

$$\frac{G_0}{G_1} = \frac{G_0}{\alpha I_1} = \frac{\mathscr{I}_0(2aV_p)}{\mathscr{I}_1(2aV_p)}$$

where $\mathscr{I}_K$ represents the modified Bessel function of order K. (13.17) is therefore equal to:

$$I_K \ll \frac{\mathscr{I}_1}{\mathscr{I}_0} k_1 \omega_p C_0 V_p$$

For usual values of aV_p, $\mathscr{I}_0(2aV_p)/\mathscr{I}_1(2aV_p)$ is near to 1.

The above inequality may then be written: $I_1 \ll k_1 \omega_p C_0 V_p$. Experience has shown that, in order to ensure satisfactory operation of the amplifier in a temperature range of $\pm 20°C$, the pumping conditions should be such that at the mean temperature we have:

$$I_1 \leqslant 10^{-4} \omega_p C_0 V_p \qquad (13.18)$$

Reasoning in a same way, trying to minimize the noise, we find again (13.18) and we estimate the value given in (13.14).

1.2. Conditions for equilibrium

1.2.1. GENERAL

In this discussion of the conditions for equilibrium we will deal with cissoidal and rectangular pumping only. We saw in Chapter 12 that in the absence of parasitic elements, the following three conditions were required for the equilibrium of the circuit.

Cissoidal pumping

$$I_{a_0} - I_{b_0} = 0$$
$$I_{a_1} - I_{b_1} = 0$$
$$C_{a_0} - C_{b_0} - C_{a_2} + C_{b_2} = 0$$

Rectangular pumping

$$I_{a_0} - I_{b_0} = 0$$
$$I_{a_1} - I_{b_1} = 0$$
$$C_{a_0} - C_{b_0} = 0$$

In order to satisfy these conditions, three pumping parameters are available; the amplitude $2V_p$ of the pumping voltage and the bias voltages V_{a_0} and V_{b_0} of each of the varactors.

However, in practice, it is always easier to adjust the amplitude of a d.c. voltage than that of an a.c. voltage. On the other hand, as the three above parameters cause a modification of the first order of the quantities I_n and C_n, it is not possible to devise a simple method of adjustment. For these reasons, the amplitude of the pumping voltage will be maintained at a uniform value and a trimming component will be introduced into the circuit. Finally, as we saw in section 1, a supplementary trimming component should be provided in order to compensate for any phase difference between the voltages v_a and v_b.

We will now proceed to determine, for cissoidal pumping, the conditions for equilibrium of the low-frequency parametric amplifier whose circuit is shown in Fig. 13.5. This circuit differs from that shown in Fig. 13.5 only by the addition of trimming components. That is the differential capacitor $C_{a_d} + C_{b_d}$ the components of which are connected in parallel across the varactors D_a and D_b respectively, and the phase changers θ_a and θ_b. Using the notations adopted in Fig. 13.5, the voltages v_a and v_b applied to the diodes are (for cissoidal pumping):

$$v_a = v'_a - v_s \qquad (13.19)$$

$$v_b = v'_b + v_s \qquad (13.20)$$

with

$$\begin{cases} v'_a = V_p[e^{j(\omega_p t + \theta_a)} + e^{-j(\omega_p t + \theta_a)}] - V_{a_0} \\ v'_b = V_p[e^{j(\omega_p t + \theta_b)} + e^{-j(\omega_p t + \theta_b)}] - V_{b_0} \end{cases}$$

PARAMETRIC AMPLIFIERS

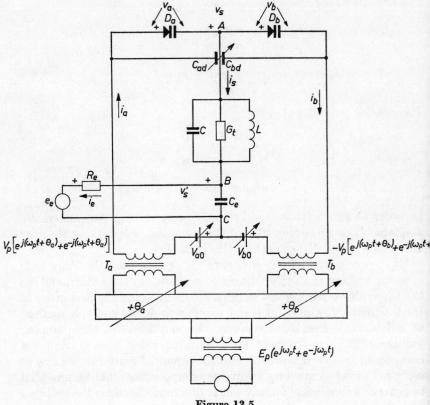

Figure 13.5

According to the small-signal hypothesis, the currents i_a and i_b flowing through the varactors are (see (12.3) and (12.4)):

$$i_a = I_a(v'_a) + C_a(v'_a)\frac{dv'_a}{dt} - G_a(v'_a)v_s - \frac{d}{dt}C_a(v'_a)v_s \quad (13.21)$$

$$i_b = I_b(v'_b) + C_b(v'_b)\frac{dv'_b}{dt} + G_b(v'_b)v_s + \frac{d}{dt}C_b(v'_b)v_s \quad (13.22)$$

with

$$C_a(v'_a) = C_{a_d} + \sum_{-\infty}^{+\infty} C_{a_k} e^{jk(\omega_p t + \theta_a)} \quad (13.23)$$

$$C_b(v'_b) = C_{b_d} + \sum_{-\infty}^{+\infty} C_{b_k} e^{jk(\omega_p t + \theta_b)} \quad (13.24)$$

$$G_a(v'_a) = \sum_{-\infty}^{+\infty} G_{a_k} e^{jk(\omega_p t + \theta_a)} \quad (13.25)$$

$$G_b(v_b') = \sum_{-\infty}^{+\infty} G_{b_k} e^{jk(\omega_p t + \theta_b)} \qquad (13.26)$$

$$I_a(v_a') = \sum_{-\infty}^{+\infty} I_{a_k} e^{jk(\omega_p t + \theta_a)} \qquad (13.27)$$

$$I_b(v_b') = \sum_{-\infty}^{+\infty} I_{b_k} e^{jk(\omega_p t + \theta_b)} \qquad (13.28)$$

It should be remembered that by definition the input head is said to be in equilibrium if for a zero input signal ($e_e = i_e = 0$) the output voltage is zero ($v_s = 0$).

Taking relations (13.23) to (13.28) into account, it can be shown by a reasoning similar to that in section 2 of Chapter 12, that in the case under consideration the conditions for equilibrium are:

d.c. component
$$I_{a_0} - I_{b_0} = 0 \qquad (13.29)$$

Component of frequency ω_p
$$I_{a_1} e^{j\theta_a} - I_{b_1} e^{j\theta_b} + j[I_{a_r} e^{j\theta_a} - I_{b_r} e^{j\theta_b}] = 0 \qquad (13.30)$$
with
$$\left.\begin{array}{l} I_{a_r} = [C_{a_d} + C_{a_0} - C_{a_2}]\omega_p V_p \\ I_{b_r} = [C_{b_d} + C_{b_0} - C_{b_2}]\omega_p V_p \end{array}\right\} \qquad (13.31)$$

Putting $\theta = \theta_a - \theta_b$, (13.30) may be written in the form
$$e^{j\theta_b}\{I_{a_1} e^{j\theta} - I_{b_1} + j(I_{a_r} e^{j\theta} - I_{b_r})\} = 0 \qquad (13.32)$$

Since $e^{j\theta_b}$ is never zero, (13.32), after separation of the real and imaginary parts, gives us the relations:

$$\tan \frac{\theta}{2} = \frac{I_{b_r} - I_{a_r}}{I_{a_1} + I_{b_1}} \qquad (13.33)$$

$$\tan \frac{\theta}{2} = \frac{I_{a_1} - I_{b_1}}{I_{a_r} + I_{b_r}} \qquad (13.34)$$

(13.29), (13.33), (13.34) which express the conditions for equilibrium of the input head for $\theta \neq 0$.

However as we saw in section 1.1.2, in order to ensure satisfactory operation of the amplifier, θ must be made as small as possible. For $\theta = 0$, equations (13.29), (13.33), (13.34) reduce to:

$$I_{a_0} - I_{b_0} = 0 \qquad (13.35)$$
$$I_{a_1} - I_{b_1} = 0 \qquad (13.36)$$
$$I_{a_r} - I_{b_r} = 0 \qquad (13.37)$$

1.2.2. ADJUSTMENT OF THE PHASE SHIFT θ

As we have seen, different parasitic effects may cause a slight phase shift between voltages v_a and v_b; amongst these are the parasitic capacitances which charge the secondaries of the transformer if the internal impedance of the pump generator is resistive. This effect provides us with a particularly simple method for the differential adjustment of the phases θ_a and θ_b of the pumping voltages. A method which has a second-order influence on the amplitude of these voltages is a second differential capacitor of low capacity C_a connected in the circuit as shown in Fig. 13.6 which will enable the phase θ to be adjusted (by several tenths of a degree).

1.2.3. METHOD OF OBTAINING EQUILIBRIUM

If we designate by $2V_p$ the amplitude of the pumping voltage, it is possible theoretically to satisfy the three conditions (13.35), (13.36), (13.37). Three adjustment parameters are available; the bias voltages V_{a_0} and V_{b_0} and the differential capacitance $C_{a_d} + C_{b_d}$ incorporated in the circuit for this purpose. By adjusting these bias voltages it is possible to satisfy the two conditions (13.35), (13.36), and the differential capacitor enables condition (13.37) to be satisfied. However, in order to ensure temperature stability, it is preferable to adjust the bias voltages in such a way as to satisfy the following two conditions separately:

$$\left. \begin{array}{l} I_{a_0} = 0 \\ I_{b_0} = 0 \end{array} \right\} \qquad (13.38)$$

these conditions giving, in accordance with the relations established in Chapter 14, very low currents I_{a_1} and I_{b_1} and satisfying the working conditions (13.18). It should be mentioned that the value of the Schottky noise decreases as I_{a_0} and I_{b_0} approach zero. In this case however, condition (13.36) is not necessarily satisfied and there is no remaining adjustment parameter enabling it to be satisfied.

It should be noted that in principle this condition is not strictly necessary. In fact, if we assume that this is the only condition which is not satisfied, it can be shown from equations (12.11) (ignoring the terms G_1, G_2) that an unbalance of the active currents I_{a_1}, I_{b_1} gives rise, at the output of the parametric amplifier, to *signals of frequency ω_p in quadrature* with the signals resulting from the injection of a useful d.c. signal at the input. These signals in quadrature with the useful signals are not, consequently, detected by synchronous demodulation. In practice, however, this unbalance of the active

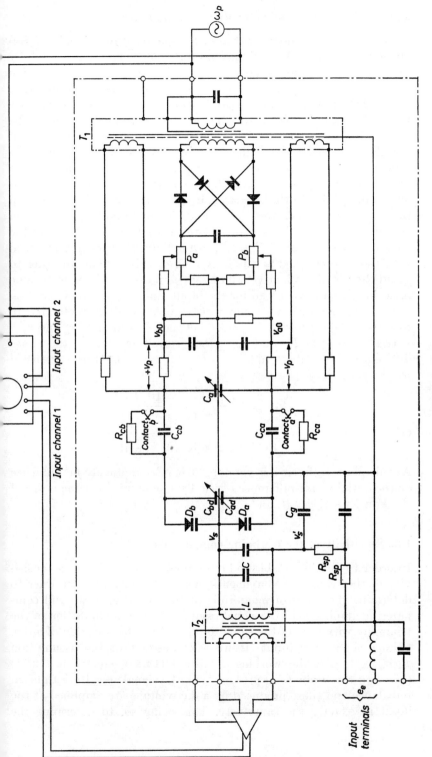

Figure 13.6

currents can have no importance, as the resulting signals of frequency ω_p on the one hand may saturate the conventional amplifier following the parametric device, and on the other hand a suppression greater than 100 to 1 cannot be hoped for from the demodulator.

It follows that in certain cases, it will be necessary to compensate for the inequality of the active currents I_{a_1}, I_{b_1}. For this purpose the two diodes can be connected to the terminals of a potentiometer R of very low resistance, the sliding contact of which is connected to the centre branch (see Fig. 13.7). This method, although having certain drawbacks, possesses one advantage; as the potentiometer is connected in series with the varactors, it does not cause any reduction in the input impedance. This simple and cheap method can be used in certain cases. It should be noted, however, that the conditions for the balance of the circuit of Fig. 13.7 are not strictly speaking identical with those established at the beginning of this section. In order to avoid the use of dissipative trimming components, which on the one hand involve an extra load for the input and output circuits and on the other hand introduce new sources of noise, it is preferable to tolerate a slight phase difference θ between the voltages v_a and v_b.

If the currents I_{a_1} and I_{b_1} satisfy condition (13.18), the phase difference which allows conditions (13.33) and (13.34) to be satisfied will be such that:

$$\tan \frac{\theta}{2} = \frac{I_{a_1} - I_{b_1}}{I_{a_r} + I_{b_r}} \leqslant \frac{10^{-4}}{2}$$

or:

$$\theta < 10^{-4} \text{ rad} \approx 5 \times 10^{-3} \text{ degree}$$

As this phase difference is very slight it has practically no influence on the initial phase difference of the harmonics of the same order of the Fourier series $C_a(t)$ and $C_b(t)$.

1.3. Sensitivity of the trimming components

In order to obtain a stable and reproducible equilibrium, the sensitivity of the trimming components must be ascertained. In order to determine the order of magnitude of this sensitivity, we will compare the amplitude of the signal resulting from a modification of the trimming with that of the signal produced by the injection of a d.c. voltage E_e at the input. If we take account of the assumption $R_e Y_{\text{in}} \ll 1$, and the working condition (13.18), equations (12.11) (ignoring the terms G_1, G_2) show that the signals resulting from an unbalance and those produced by a d.c. voltage are amplified at the input in exactly the same way. This being so, to determine the

DECOUPLING AND METHODS OF ADJUSTMENT

equivalent input voltage, we will take the simple case in which the pumping parameter β_v is small. On the other hand, if condition (13.18) is satisfied, we can ignore the variations in the balance due to the currents I_{a_1}, I_{b_1} with respect to those produced by the currents I_{a_r} and I_{b_r}.

1.3.1. VARIATION OF THE BIASING

For a pumping voltage of a given amplitude $2V_p$ and in accordance with relation (14.25), we have a variation dV_0 of the biasing:

$$dC_0 \approx -\frac{C_0 \, dV_0}{n(\phi + V_0)}$$

Assuming that the biasing changes only in one of the branches, we obtain, if we take (13.31) and (13.38) into account and noting that $C_2 \ll C_0$:

$$(I_{a_r} - I_{b_r}) = \omega_p \, dC_0 V_p = \frac{-\omega_p C_0 V_p}{n(\phi + V_0)} \, dV_0 = \omega_p C_1 \, dV_0 \quad (13.39)$$

It follows from equations (12.11) that the term $I_{a_r} - I_{b_r}$ is amplified in the same manner as the term $-j\omega_p C_1' \cdot E_e$, and consequently that the value E_e of the input signal equivalent to a variation dV_0 is equal to:

$$E_e = \tfrac{1}{2}|dV_0| \quad (C_1' \approx 2C_1) \quad (13.40)$$

By the variations of the biasing voltage, independently in every branch (aleatory noise for instance), we must take into account accompanying variations. Let us suppose that the two biasing voltages are obtained through potentiometric division from the same voltage (Fig. 13.6) and that the varactors are paired, we can write:

$$V_{a_0} = k_1 E$$

$$V_{b_0} = k_2 E$$

$$V_{a_0} + V_{b_0} = 2V_0$$

$$V_{a_0} - V_{b_0} = \Delta V_0$$

$$C_{a_0}(V_0) + C_{b_0}(V_0) = 2C_0$$

$$C_{a_0}(V_0) - C_{b_0}(V_0) = \Delta C_0$$

$$\frac{dE}{E} = \frac{dV_{a_0}}{V_{a_0}} = \frac{dV_{b_0}}{V_{b_0}} = \frac{dV_0}{V_0}$$

For paired varactors, $\Delta V_0/V_0$ and $\Delta C_0/C_0$ are small and we shall suppose ϕ and n respectively identical. Three main cases can be considered:

(1) *Identical biasing*

The equilibrium condition (13.37) is obtained through the adjustment of the differential capacitor C_{a_d}, C_{b_d}. One can easily demonstrate that:

$$\Delta V_0 = 0$$

$$E_e \approx -\frac{\Delta C_0}{2C_0} V_0 \left(\frac{\mathrm{d}E}{E}\right)$$

(2) *No differential capacitor*

The equilibrium condition (13.37) is obtained through the adjustment of the biasing voltages. It is easy to demonstrate that:

$$\begin{cases} \Delta V_0 \approx n(\phi + V_0)\left(\frac{\Delta C_0}{C_0}\right) \\ E_e \approx -n\phi \frac{\Delta C_0}{2C_0} \left(\frac{\mathrm{d}E}{E}\right) \end{cases}$$

(3) *Joint adjustment ideal for the sensibility to* $\mathrm{d}E$

The equilibrium condition (13.37) is obtained through the adjustment of the differential capacitor C_{a_d}, C_{b_d} and of the biasing voltages such as (13.37), independently of the first order of $\mathrm{d}E/E$. This independence is obtained when:

$$\begin{cases} \Delta V_0 \approx -V_0 \frac{n(\phi + V_0)}{(n\phi - V_0)} \left(\frac{\Delta C_0}{C_0}\right) \\ E_e \approx 0 \end{cases}$$

One can easily see that case (1) is far better than case (2) and that the difference between the biasing corresponding to case (1) and (2) is small. In the same way the equilibrium condition (13.29) will be better fulfilled in case (1) than in case (2). If the average biasing voltage (V_0) of the varactors (choose those which are best paired) is chosen so as to satisfy the conditions of Chapter 14, section 4, V_0 will be small compared to ϕ; case (1) is about ten times more favourable than case (2).

1.3.2. MALADJUSTMENT OF THE DIFFERENTIAL CAPACITOR $C_{a_d} + C_{b_d}$

In accordance with (13.31) and (13.38), an unbalance is associated with a maladjustment ($dC_d = d(C_{a_d} - C_{b_d})$) of the differential capacitor.

$$(I_{a_r} - I_{b_r}) = \omega_p \, dC_d V_p$$

Hence, by a reasoning similar to that in the previous section:

$$E_e = n(\phi + V_0) \frac{dC_d}{2C_0} \qquad (13.42)$$

As shown by (13.42), special care should be taken in the choice of a trimming capacitor. A difference between the C_0 values of the varactors leads to a relation identical with (13.42). The stability with time and as a function of temperature of the variable capacitor should be better than the very high stability of the varactor. The value of C_{0v} for good varactors varies at the most by several 10^{-5} per month, or by several femtofarads for junctions of about 100 pF. For paired varactors, the difference between the relative variations of their capacitances as a function of temperature is several 10^{-6} per degree centigrade (13.10).

Further, the adjustment of the variable capacitor should be sufficiently fine for the residual signal due to imperfect balance to be completely masked by the noise. Further, the insulation between unit and base should be much higher than $1/G_0$. Finally, the maximum value of the total capacitance $C_{a_d} + C_{b_d}$ of the differential capacitor should be calculated, taking into account the maximum guaranteed divergence of the values of C_{0v}, n and ϕ and of any discrepancies which may exist between the pumping parameters. In practice (see (13.31)), we will always have $C_{a_d} + C_{b_d} \ll C_0 - C_2$. Note that the use of a differential capacitor causes the tuning of the output circuit to be independent of the adjustment of C_{a_d}, C_{b_d}.

1.3.3. MALADJUSTMENT OF THE PHASE CHANGER

We saw that if the working condition (13.18) is satisfied, we can compensate for the effect of an unbalance of the active currents I_{a_1}, I_{b_1} by creating a slight difference θ between the voltages v_a and v_b applied to the varactors. We will now proceed to determine the influence of maladjustment by $d\theta$ of the phase difference θ. In accordance with (13.32) and assuming that θ and $d\theta$ are small, we obtain for the components in phase dI_1 and in quadrature dI_k respectively with the pumping voltage:

$$dI_r \approx (I_{a_1} - I_{a_r}\theta) \, d\theta = I_{a_r}\left(\frac{I_{a_1}}{I_{a_r}} - \theta\right) d\theta \qquad (13.43)$$

$$dI_1 \approx -(I_{a_1}\theta + I_{a_r}) \, d\theta = I_{a_r}\left(\frac{I_{a_1}}{I_{a_r}}\theta + 1\right) d\theta \qquad (13.44)$$

If we take (13.18) into account, the factor $(I_1/I_r - \theta) \, d\theta$ is very low and we can consequently ignore the influence of dI_r. Conversely the current component dI_1 may be of the same order of magnitude or greater than the difference of the currents $I_{a_1} - I_{b_1}$ for which we are attempting to compensate. If in expression (13.44) we ignore the term $(I_{a_1}/I_{a_r})\theta$ by comparison with unity, we have:

$$dI_1 = I_{a_r} \, d\theta \qquad (13.45)$$

Equations (12.11) tell us that the term $I_{a_1} - I_{b_1}$ is amplified in the same way as the term $\omega_p C_1' E_e$. It follows that the value E_e of the input signal, which is equivalent to a variation $d\theta$, is equal to:

$$E_e = n(V_0 + \phi)\frac{d\theta}{2} \qquad (13.46)$$

The permissible value for $d\theta$ will be determined by taking account of the suppression factor of the demodulator and the saturation characteristics of the conventional amplifier (see section 1.2.3). It should be noted here that because of the slight effect that the adjustment of θ has on the value of dI_r, it is possible to suppress dI_1 without disturbing the value of dI_r, which may then be suppressed by another method, using the differential capacitor $C_{a_d} + C_{b_d}$ (independent first-order adjustments).

For specially carefully designed circuits and high-quality diodes, condition (13.18) is largely satisfied. The only effect of the adjustment of the phase is to compensate for the phase difference between the voltages v_a, v_b resulting from the presence of residual parasitic elements which it is difficult to obtain weak enough so as to be able to do without any adjustment device.

1.3.4. INFLUENCE OF TEMPERATURE

Suppose the same hypothesis and let us choose the same notations as in section 1.3.1 (equations (13.41)). Suppose that the variation of the contact potential with temperature $(\partial\phi/\partial T)$ is the same for the two varactors. Let us calculate E_e in the three cases considered in section 1.3.3. Let us try to find the value of ΔV_0 which will cancel E_e (case (4)):

(1) Identical biasing.
$$\begin{cases} \Delta V_0 = 0 \\ E_e \approx \dfrac{\Delta C_0}{2C_0} \left(\dfrac{\partial \phi}{\partial T}\right) \mathrm{d}T \end{cases}$$

(2) No differential capacitor.
$$\begin{cases} \Delta V_0 \approx n(\phi + V_0) \dfrac{\Delta C_0}{C_0} \\ E_e \approx -n \dfrac{\Delta C_0}{2C_0} \left(\dfrac{\partial \phi}{\partial T}\right) \mathrm{d}T \end{cases}$$

(3) Ideal combined adjustment for sensitivity to $\mathrm{d}E$.
$$\begin{cases} \Delta V_0 \approx -V_0 \dfrac{n(\phi + V_0)}{(n\phi - V_0)} \dfrac{\Delta C_0}{C_0} \\ E_0 \approx \left(\dfrac{n\phi}{n\phi - V_0}\right) \dfrac{\Delta C_0}{2C_0} \left(\dfrac{\partial \phi}{\partial T}\right) \mathrm{d}T \end{cases}$$

(4) Ideal combined adjustment for sensitivity to temperature.
$$\begin{cases} \Delta V_0 \approx \dfrac{n}{n+1} (\phi + V_0) \dfrac{\Delta C_0}{C_0} \\ E_e \approx 0 \end{cases}$$

As in section 1.3.1, case (1) is better than case (2). Sensitivity to temperature in cases (1) and (2) is nearly the same. The ideal adjustment (case 4) is somewhere between cases (1) and (2), but nearer to (1) than to (2). Sensitivity of the current equilibrium (equations (13.38)) is analysed in Chapter 14, section 4.3: the ideal case is, in principle, case (1).

Generally speaking, case (1) is the best compromise; often, so as to satisfy the equilibrium equations (13.38), biasing will be slightly modified, approaching either case (3) favourable for sensitivity to biasing variations, or case (4) favourable for the temperature coefficient. If the results wished for are not attained, the varactors will have to be paired with a greater care.

An interesting variation is to place a varactor in series with every branch of the differential capacitor. Biasing of these two varactors is achieved independently of the biasing of the main varactors. This shows that in this configuration one can simultaneously achieve the two equilibrium conditions (13.38), insensitivity to differential variations of the biasing voltages and insensitivity to temperature variations. This configuration can also be adopted through voltage adjustment, that is the state of equilibrium represented in (13.37). It is also possible, in this case, to replace the differential capacitor by a highly isolated capacitor.

1.4. Method of adjustment

A proposed circuit for simple and stable adjustment of the balance is shown in Fig. 13.6. The adjustments, which are carried out successively, are nearly independent. Further, the measuring apparatus is connected at points where its use does not affect the operation of the circuit.†

The adjustment is made as follows. With the input terminals short-circuited, the voltage at the output of the conventional amplifier is observed on an oscillograph whose time base is synchronized with the pump. A weak signal of frequency ω_p appears on the screen.

1. Tune the head filter to ω_p by means of T_2 (maximization of the signal). As the head is not balanced when tuned, the conventional amplifier usually becomes saturated.
2. Reduce the amplitude of the signal due to the unbalance by adjusting C_{a_d}, C_{b_d} and C_a. Complete the tuning.
3. When this signal is suppressed, open contact a.
4. Observe the signal and wait until the new state of balance is reached. At this instant, the diode is biased to a voltage V_{a_0} which brings the current I_{a_0} to zero.
5. Suppress the amplitude of the signal appearing on the screen by means of C_{a_d}, C_{b_d} (and if necessary of C_a).
6. Close contact a.
7. Suppress the signal by adjusting P_a. The bias voltage is then equal to the above-mentioned value V_{a_0}.
8. Repeat operations 3, 4, 5, 6, 7 for the branch b.
9. Open the input terminals: there should now be no signal on the screen. If necessary make a final adjustment.

Notes. (1) It is possible to adapt this method for the simultaneous adjustment of each of the two branches.

† Brevet No. 661722.

DECOUPLING AND METHODS OF ADJUSTMENT 365

(2) It is generally desirable to observe the output signal of the demodulator and if possible to record it (operation 4).

(3) The loops a, R_{ca}, C_{ca} and b, R_{cb}, C_{cb} are introduced into the circuit of the amplifier only to permit ready adjustment of the balance. The values of R_c and C_c should be chosen so as to have:

$$C_c \gg C_0$$

$$\frac{1}{\omega_p C_c} \ll R_c$$

$$\frac{1}{\omega_e C_0} \gg R_c$$

Capacitor C is specified to have as low a value as possible, and with the largest possible negative temperature coefficient, but with low tolerance so as to compensate exactly for the effect of temperature drift of C'_0 on the tuning of the output filter.

(4) The resistances R_{sp} introduced into the circuit have a double role: to prevent a slight detuning of the filter due to changes in the impedance of the source, and to filter off the signals of frequency close to $\omega_p/2$ and harmonics which might appear at the input.

1.5. Special type of amplifier for d.c. signals

When it is required to detect the difference of a d.c. signal from the preserved value, the amplifier circuit may be considerably simplified. Let us consider the diagram of Fig. 13.7, and find the value of the

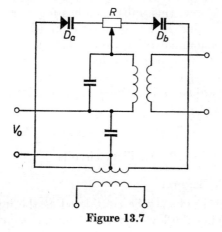

Figure 13.7

input voltage V_0 which makes it possible, by the use of potentiometer R, to obtain the balance of the bridge for components I_1 and I_r.

In the case being considered, if we ignore C_2 by comparison with C_0, equations (13.31) reduce to:

$$I_{a_r} = C_{a_0}\omega_p V_p$$
$$I_{b_r} = C_{b_0}\omega_p V_p$$

Consequently, the condition for balance as far as I_r is concerned may be written simply as:

$$C_{a_0} = C_{b_0} \qquad (13.47)$$

We saw that the deviation of ϕ and n is usually small for all varactors of the same type. Taking account of the accuracy required and that paired varactors are usually delivered with their capacitances C_{0v} equal to within -5%, we can take ϕ and n as equal for the two varactors.

If it is assumed that the value of the pumping parameter β_v is small, C_{a_0} and C_{b_0} depend only on V_0 (see sections 2.1.2 and 3.1.2 Chapter 14). Equation (13.47) becomes:

$$C_{a_{ov}}\left(1 + \frac{V_0}{\phi}\right)^{-1/n} = C_{b_{ov}}\left(1 - \frac{V_0}{\phi}\right)^{-1/n}$$

hence:

$$V_0 = \phi \frac{C_{a_{ov}}^n - C_{b_{ov}}^n}{C_{a_{ov}}^n + C_{b_{ov}}^n} \qquad (13.48)$$

The bridge is balanced for the I_1 components by means of potentiometer P.

It should be noted that the bridge is not necessarily balanced for the I_0 components. However, if the internal resistance of the input signal generator is low, this unbalance has no effect on the output signal. When the value V_0 has been determined, the input signal is brought, not to its nominal value, but to that value decreased by V_0. One can show that this type of amplifier, with varactors which were given a certain pairing quality, is less stable and noisier than the one from Fig. 13.5, adjusted according to the method described in section 1.4.

2. Auxiliary circuits

2.1. Overall block diagram

The parametric head may be considered as the input link of an assembly amplifying the low-frequency and d.c. signals. If, as seen from the input and output terminals, this assembly appears to resemble standard amplifiers, it has nevertheless certain special operational features, the consequences of which may sometimes be unexpected and against which measures have often to be taken. The assembly configuration is as shown in Fig. 13.8.

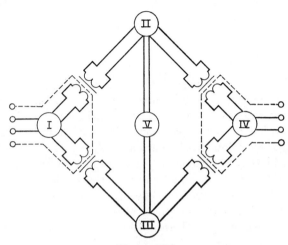

Figure 13.8

 I Parametric head.
 II Conventional amplifier.
 III Local oscillator (pump).
 IV Demodulator.
 V Power supply.

Signals from the amplifier and oscillator reach the demodulator. Their relative phase depends on the parametric gain, the phase changes between the signals delivered to the two outputs of the oscillator and on the transfer function of the conventional amplifier. For correct operation, the two signals arriving at the demodulator should have correctly related phases. An adjustable phase changer is generally used for this purpose. It is often preferable to locate the phase changer in the amplifier rather than in the oscillator where it usually causes a large loss of power.

2.2. Conventional amplifier

The conventional amplifier following the parametric head should be especially well designed from the point of view of noise, phase shift, and symmetry where saturation is concerned. The maximum gain that can be given to it is limited by the maximum amplitude of the input voltage. This voltage is due to the noise generated by the source of the useful signal, to the noise generated by the noise sources themselves and to faulty balancing of the bridge at the maximum useful signal. Further, there should be adequate and accurate decoupling between the local oscillator and the input of the conventional amplifier, in order to reduce as far as possible the parasitic signal, of frequency ω_p, at the input of the amplifier.

2.3. Pump oscillator

The oscillator should deliver alternating voltage of very stable frequency and amplitude. The chief effect of frequency instability of the pumping voltage is to detune the anti-resonant circuit, which, when the value of the α parameter is near to unity, causes very considerable variations in the gain and phase. The chief effect of an instability in the amplitude of the pumping voltage is to cause a variation in the gain due to a change in the values of C_0, C_1 and C_2. This instability may unbalance the bridge. It is clear that we must use a pump which is always identical to the two varactors. Fluctuations in the frequency and amplitude of the pump oscillator may contribute to the noise.

2.4. Demodulator

When the sign of the input voltage is known or when it is of no importance, a simple full-wave rectifier may be used as a demodulator. However, the use of a demodulator of this type causes the appearance of a d.c. component due to the parasitic signal entering the demodulator, a component which does not exist with synchronous demodulation. Phase-correlated demodulation results in a better noise factor.

Finally it should be noted that the output signal can be demodulated by multiplying it by a periodic signal of frequency ω_p, which need not be sinusoidal. This type of demodulator can very easily be constructed when the pumping is by means of crenellated signals.

2.5. Internal correction networks

If the amplifier is to be looped, correction networks to ensure its stability must be incorporated in the amplifier circuit. It is important to provide a filter at the input of the amplifier in order to reject all signals at frequencies equal to or greater than $\omega_p/2$, and a component at this frequency is present in the output signal as a product of intermodulation. If the frequency of the input signal is near to a multiple of $\omega_p/2$, the output signal, after synchronous demodulation, may, by a process identical with that just described, possess a component of frequency equal to the difference between the input frequency and the frequency of the multiple of the $\omega_p/2$ under consideration, that is the frequency lying within the bandwidth of the amplifier. These effects may produce oscillation in the looped amplifier.

2.6. Automatic gain control

A gain stabilization circuit is desirable in all parametric amplifiers. This may act on the pumping amplitude, the load, the tuning of the plug circuit or the bias. These different methods affect the noise factor and the output phase, the input differential and the bandwidth to different extents. The reference signal at the fringe of the transmitted spectrum may enable a fault in a highly reliable amplification chain to be detected.

2.7. Anti-saturation circuit

If the signals arriving at the input are not weak, the values of C_k are modified, the small-signal hypothesis is no longer valid, and the phase of the output signal (45° to 90° depending on the gain) is changed. This effect can be prevented by placing two diodes back-to-back at the input of the amplifier, biased if necessary, and a series resistance to limit the input current. In addition, a circuit to increase the load on the output filter when the output of the conventional amplifier exceeds a certain level may also be used.

3. Technical limitations

The criteria for the appraisal of a parametric amplifier depend essentially on the purpose for which it is intended. For example, the conversion efficiency may be required to be as high as possible, or to be not very sensitive to fluctuations in the load, the pumping or the temperature. Other criteria, the input impedance, the bandwidth or the noise factor at very low frequency may also have to be taken into consideration. In every case the choice of the components of the amplifier which satisfy as far as possible the criteria adopted, (optimization), should of course take account of the technical limitations.

Whilst the choice for certain components is wide, for others, it is very restricted. This applies especially to high-quality paired varactors which are now on the market. It is for this reason that the process of optimization of the essential parameters of the amplifier is usually not started until the actual type of varactor has been chosen, even if it may have to be repeated for another type.

The procedure to be followed when it is required to obtain the maximum voltage gain A_1 of a chain formed by connecting in cascade a parametric amplifier, a transformer, a conventional amplifier and a synchronous detector (see Fig. 12.4) is as follows.

Given:

(1) The type of varactor and the pumping conditions, that is the values of C'_0, C'_1, C'_2, R_s.
(2) The voltage gain of the conventional amplifier: $A(\omega_p)$.
(3) The input conductance of this amplifier: $G_i = G_i(\omega_p)$.
(4) The quality factor of the inductance in the anti-resonant circuit: $Q' = f(\omega_p, L)$.
(5) The parameters to be determined are the frequency ω_p and the transformation ratio n of the transformer.

According to the definition of the conversion efficiency we have:

$$A_1 = nA\eta_0 \tag{13.49}$$

Taking account of (12.45) this gives:

$$A_1 = nQA\left\{d\,\frac{\sqrt{1 + bQ^2}}{(1 - bQ^2)}\right\} \tag{13.50}$$

with

$$d = \frac{2C'_1}{C'_0 + C} \tag{13.51}$$

$$b = \left(\frac{C'_2}{C'_0 + C}\right)^2 \tag{13.52}$$

$$Q = \frac{\omega_p(C'_0 + C)}{G_A + G + G_0} \tag{13.53}$$

The ratio between the differential capacitance C'_0 and the capacitance C, intended to compensate for the effect of the temperature on C'_0, is usually known ($k_1 C'_0 = C'_0 + C$). Ignoring G_0 by comparison with $G_A + G$ we have:

$$\frac{1}{Q} = \frac{G_A + G}{\omega_p k_1 C'_0} \tag{13.54}$$

DECOUPLING AND METHODS OF ADJUSTMENT

On the other hand, we may write:

$$G_A = n^2 G_i$$

$$L = \frac{1}{k_1 \omega_p^2 C_0'}$$

$$G = \frac{k_1 \omega_p C_0'}{Q'} = \frac{\omega_p C_0'}{f(\omega_p, (1/k_1 \omega_p^2 C_0'))}$$

As the value of $k_1 C_0'$ is given, G and Q' depend only on ω_p.

$$\left. \begin{array}{c} G = G(\omega_p) \\ Q' = Q'(\omega_p) \end{array} \right\} \qquad (13.55)$$

(13.45) may be written:

$$\frac{1}{Q} = \frac{1}{Q'(\omega_p)} + \frac{n^2 G_i(\omega_p)}{k_1 \omega_p C_0'} = F(n, \omega_p) \qquad (13.56)$$

which, after substitution in (13.50), gives:

$$A_1 = \frac{n}{F(n, \omega_p)} A \left\{ d \, \frac{\sqrt{1 + b[F(n, \omega_p)]^{-2}}}{1 - b[F(n, \omega_p)]^{-2}} \right\} \qquad (13.57)$$

This relation makes it possible, in principle, to find those values of ω_p and n which give the gain a maximum value A_1. At this stage it is desirable to check whether G_0' is sufficiently low compared to $G_A + G$. The value of ω_p should not be so high as to cause great difficulties in preserving the phase for a synchronous demodulation, and on the other hand for the series resistance R_s to have a considerable influence on the admittance of the diode. The results are summarized below; they will illustrate the various technical limitations.

Given:

(1) The varactors are assumed to be identical, and the pumping conditions used as starting point are characterized by:

$C_{0v} = 77 \text{ pF}$
$\phi = 0.7 \text{ V}$
$n = 3$ (index of the root in (13.4))
$a = 29.5 \text{ V}^{-1}$
$I_s = 2.5 \times 10^{-13} \text{ A}$
$G_s = 37 \times 10^{-13} \text{ mho}$
$R_s = 0.3 \, \Omega$
$2V_p = 0.130 \text{ V}$
$V_0 = 0.058 \text{ V}$

where V_0 is the value of the bias voltage which must be applied to suppress the current I_0 when $2V_p = 0{\cdot}130$ V. This value is given by curve 1 in Fig. 14.20. β_v is given by (14.17):

$$\beta_v = 0{\cdot}170$$

From Fig. 14.8 we obtain:

$$C(V_0) = 0{\cdot}98, \qquad C_{v_0} = 75{\cdot}5 \text{ pF}$$

Figures 14.3, 14.4 and 14.5, respectively, give us C_0, C_1 and C_2.

$$C_0 = 76 \text{ pF}$$
$$C_1 = 2{\cdot}18 \text{ pF}$$
$$C_2 = 0{\cdot}125 \text{ pF}$$

Taking account of (13.11) and of the negative temperature-coefficient capacitors which are available, we find, for example, $k_1 = 1{\cdot}32$. As the two varactors were assumed to be identical, we have:

$$C'_0 = 152 \text{ pF} \qquad (k_1 C'_0 = 200 \text{ pF})$$
$$C'_1 = 4{\cdot}36 \text{ pF}$$
$$C'_2 = 0{\cdot}25 \text{ pF}$$

And this, taking (13.51) and (13.52) into account, gives:

$$\begin{cases} d = 4{\cdot}36 \times 10^{-2} \\ b = 1{\cdot}56 \times 10^{-6} \end{cases}$$

(2) The transfer function of the conventional amplifier is given in Figs. 13.9 and 13.10. The upper cut-off frequency of the

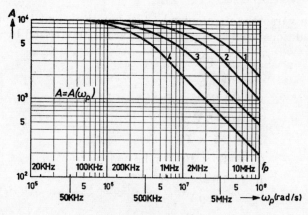

Figure 13.9

amplifier may be chosen between 2×10^7 rad/sec and 2×10^6 rad/sec (curves 1, 2, 3, 4). At the price of a slight loss of gain, the transfer phase of the conventional amplifier may be adjusted by the choice of its upper cut-off frequency.

(3) The input conductance of the amplifier is equal to:

$$G_i = 2 \times 10^{-3} \text{ mho}$$

(4) The appropriate value of the quality factor of the inductance coil is determined by the value of the inductance, by the frequency and by the methods used to produce the coil. Taking account of the frequency band of the pump in question,

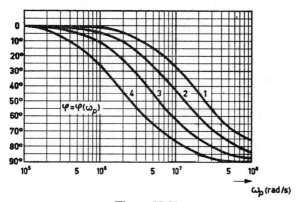

Figure 13.10

the rejection of the common mode usually required, the convenience in use, and above all of the fact that the tuning frequency of the output filter is that of the pump oscillator, the use of 'closed pots' of ferroxcube is usually advisable (low leakage flux, low sensitivity to external fields). By varying all the available parameters of the 'closed pots' (material of the base, dimensions, air gap, number of turns, type of wire ...), a trace is made of the maximum 'iso-Q' diagram that can be obtained with variable inductance and frequency. For the M.B.L.E. series 'P' pots (the dimensions of which vary from (11/7 mm) to (42/29 mm)) this diagram $Q' = f(\omega_p, L)$ is given by Fig. 13.11.

Now that the various parameters have been determined, we are in a position to determine n and ω_p. For this purpose, it should be noted that the expression (13.57) giving the voltage gain A_1 of the chain under consideration depends on two factors:

$$\frac{n}{F(n, \omega_p)} \quad \text{and} \quad d \cdot \frac{\sqrt{1 + b[F(n, \omega_p)]^{-2}}}{1 - b[F(n, \omega_p)]^{-2}}$$

In order to simplify the investigation of the values n and ω_p these two factors will be considered separately:

Note that in (13.57) the factor $n/(F(n, \omega_p))$ is almost independent of the pumping conditions. To determine this factor

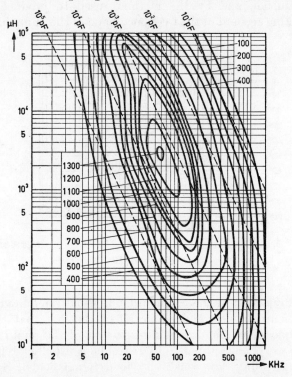

Figure 13.11. Maximum experimental values of 'iso Q' M.B.L.E. 'P' pots

from Fig. 13.12 we find the value of $1/(Q'(\omega_p))$ defined in (13.56) and the value of $(n^2 G_i)/(k_1 \omega_p C_0')$ for different values of n. From Fig. 13.13 we find the value of $F(\omega_p, n)$ defined in (13.56). The factor $n/(F(n, \omega_p))$ is given in Fig. 13.14. If, when n varies, the maximum values of this factor remain almost exactly equal over a wide range of frequencies, it is important to note that for a given value of n the choice of an optimum frequency is quite critical.

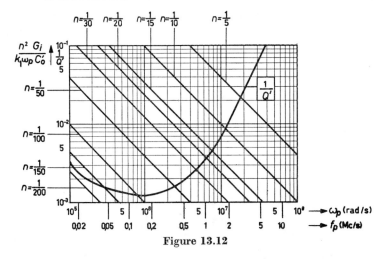

Figure 13.12

Note that if $2 \cdot 5 k_1$, $F(n, \omega_{p_2})$ is smaller than 1 (the maximum value of b being equal to $(0 \cdot 4/k_1)^2$, (14.32)), the factor

$$d \cdot \frac{\sqrt{1 + bF(n, \omega_p)^2}}{1 - bF(n, \omega_p)^{-2}}$$

which occurs in (13.57) may be made as large as required by an appropriate choice of the pumping conditions. In this, factor a increases, as a first approximation, linearly with β_v. It corresponds to the simple modulation gain due to C_1.

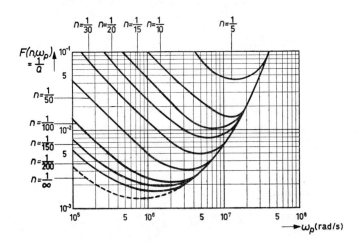

Figure 13.13

Figure 13.15 shows the increased gain due to C_2 as a function of ω_p for several values of n. This increased gain is unfortunately small except for transformation ratios unobtainable in practice. The immediate remedy is to increase the pumping amplitude slightly in order to obtain this high gain for the lower values of the transformation ratio. If we take Figs. 13.9, 13.14 and 13.15 into account we can choose n and ω_p. The best compromise from the points of view of the bandwidth, the noise factor of the conventional amplifier and the

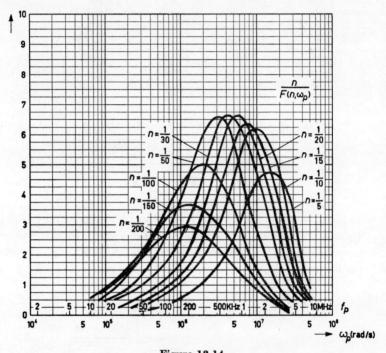

Figure 13.14

stability is: $n = 1/30$, $\omega_p = 4 \times 10^6$. It is possible to obtain a transformation ratio of 30 at 650 kHz. Let us calculate G_0'. (14.74) gives us:

$$G_0' = 3 \cdot 5 \times 10^{-11} \text{ mho}$$

For $n = 1/30$ and $\omega_p = 4 \times 10^6$, Fig. 13.12 gives us:

$$G_A = 2 \cdot 3 \times 10^{-6} \text{ mho}$$
$$G = 1 \cdot 7 \times 10^{-6} \text{ mho}$$

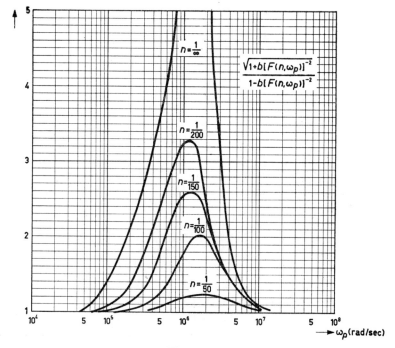

Figure 13.15

The working condition (13.18) is largely satisfied. The working condition (15.8) gives the following value of R_s:

$$R_s \ll 430 \ \Omega$$

with R_s equal to 0·3 Ω, condition (15.8) is satisfied. To take account of Q scatter, an additional adjustable load may be connected in parallel with G_i, thus enabling the parametric gain to be adjusted. The following value can therefore be taken for G_T:

$$G_T = 4\cdot 4 \times 10^{-6} \text{ mho}$$

hence $Q = 180$. When

$$\left(\frac{C_2'}{C_0'}\right)\left(\frac{Q}{k_1}\right) = 1$$

the gain becomes infinite and β_v is then equal to (14.28):

$$\beta_v = 0\cdot 36$$

The values of Q, n and ω_p being known, we can calculate, as functions of β_v, the quantities A_1, $B_{3 \text{ dB}}$, α, β. The results are shown in Fig. 13.16,

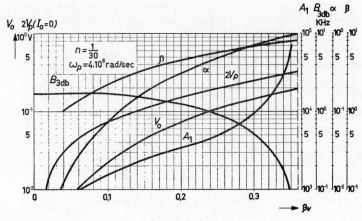

Figure 13.16

in which will be found also the values of V_0 and $2V_p$ giving I_0 zero as a function of β_v. Figure 13.17 shows the value of the noise factor (F) as a function of β_v for various values of the input resistance R_e.

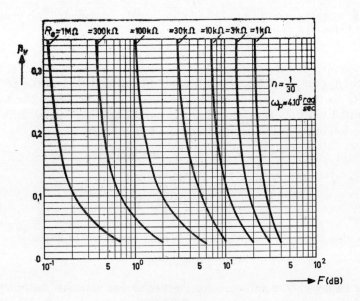

Figure 13.17

4. Examples

The results of Chapter 12 concerning the conversion efficiency are shown in the diagrams of Figs. 13.18, 13.19 and 13.20. Figure 13.18 shows the theoretical conversion efficiency as a function of the *total* load, or of Q, for various pumping voltages. The increase in the conversion efficiency for a given V_p when α approaches unity can be clearly seen. It should be noted, however, that for low values of V_p the value of G_1, for $\alpha = 1$, may become less than $G + G_0$, but this

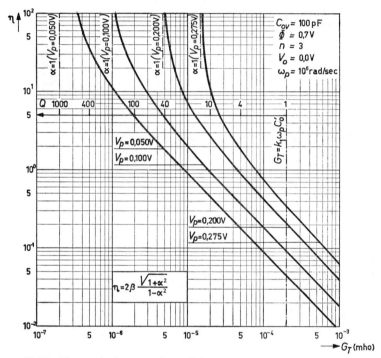

Figure 13.18. Theoretical conversion efficiency as a function of the total load for various pumping voltages

relates to a case which cannot occur in practice ($G_A < 0$). On the other hand, for high values of V_p it is G_0 which may exceed the value of G_T for $\alpha = 1$, which also relates to a case which cannot occur in practice ($G_A + G < 0$).

These considerations are illustrated in Fig. 13.19, in which is shown the conversion efficiency, as a function of the useful load, for an experimental amplifier in which the inductance of the output filter had an intrinsic Q of 370. This figure clearly shows that, even for low

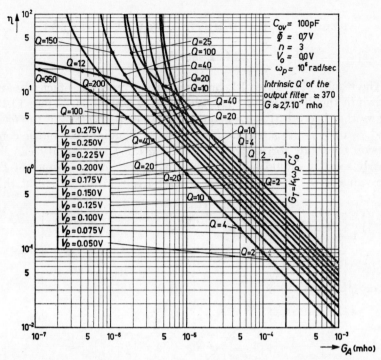

Figure 13.19. Experimental conversion efficiency as a function of the total useful load G_A for various pumping frequencies

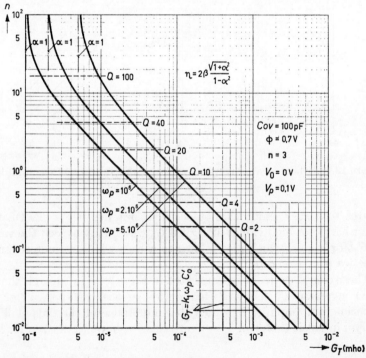

Figure 13.20. Theoretical conversion efficiency as a function of the total load for various pumping frequencies

values of G_A, the $\alpha = 1$ regions are not reached for extreme values of V_p. Fig. 13.20 shows the theoretical conversion efficiency as a function of the total load for various pumping frequencies. It will be seen that it is advantageous to increase the pumping frequency.

5. Details of amplification of d.c. and low-frequency signals by the parametric varactor method

It is a matter for surprise that nearly six years have elapsed between the first important publications on the properties of parametric amplification and the first development of parametric amplifiers for d.c. and low-frequency signals, especially when we realize that the required qualities for d.c. amplifiers are precisely those offered by a parametric amplifier. These qualities are:

(1) Very high input impedance.
(2) Very low background noise.
(3) High power gain.
(4) Transformation of d.c. or low-frequency signals into high-frequency a.c. signals.
(5) The possibility of a total rejection of the common mode.
(6) The possibility of obtaining the required bandwidth.

5.1. Input impedance

The impedance of the source of certain pick-ups is higher than $10^{13}\,\Omega$ for an effective spectrum extending from 10^{-4} Hz to several Hz. In certain cases the actual spectrum extends from 10^{-8} Hz to 10^2 Hz. The impedance of the load and the input in balance current presented by the low-frequency parametric amplifier are determined by the static characteristic $I = f(V)$ of the varactors and by the pumping parameters. Thanks to new techniques, the most modern varactors have a static characteristic which is nearly in accordance with the theoretical law together with a very low saturation current. A judicious choice of the pumping parameters enables the unbalance current to be suppressed and a high input impedance to be obtained ($> 10^{12}\,\Omega$). The input capacitance is unfortunately, necessarily rather high; an important reduction of this capacitance can be obtained by replacing C_e by an anti-resonant circuit tuned on ω_p. Operation becomes notably more complicated and the equations from Chapters 12 and 13 are valid only for a small ω_e. From (12.70) we get the input impedance as:

$$Y_{in} = \frac{1}{R_{in}} = G'_0 + j\omega_p(C'_0 + C_e)x$$
$$+ 2(G'_1 + j\omega_p C'_1 x)\beta \frac{[-jx(1 + 2jQx) + \alpha]}{(1 + 2jQx)^2 - \alpha^2} \quad (13.58)$$

5.2. Noise

In most standard amplifiers the noise factor, which may be excellent over a wide band, becomes less favourable below a certain frequency f_f. For d.c. amplifiers with a given bandwidth, the noise factor becomes less favourable as f_f increases. Standard amplifiers are therefore particularly unsuitable for the analysis of signals of small amplitude and of very low frequencies. By the use of high-quality diodes and by a judicious choice of the pumping parameters, the parametric amplifier may have an excellent noise factor. One of the essential properties of these amplifiers is the striking absence of noise at $1/f$, which is well borne out by experience (exhaustive tests made down to frequencies of 10^{-3} Hz demonstrated the total absence of noise at $1/f$) [11], [9].

It seems unlikely that from this point of view the standard type of amplifier will ever be superior to the parametric type. As the parametric amplifier transposes d.c. or low-frequency signals in a narrow band each side of the pumping frequency, these signals can be amplified by standard methods over a frequency band where these methods have the lowest noise factor. From (12.86) we get the noise factor as:

$$F \simeq 1 + \frac{1 + \alpha}{2} \frac{G}{\beta^2 R_e G_T^2} \frac{T_G}{T_0} \quad (13.59)$$

5.3. Gain

By adjusting C_1 and C_2, the amplification effect due to C_2 enables a high voltage gain to be obtained. As the conversion efficiency for ω_e approaches zero (12.66) gives:

$$\eta_0 = 2\beta \frac{\sqrt{1 + \alpha^2}}{1 - \alpha^2} \quad (13.60)$$

5.4. Transformation of d.c. or low-frequency signals into a.c. high-frequency signals

There is no necessity to point out the great importance of the transformation of d.c signals into a.c. signals to enable them to be amplified without drift, the only cause of which after conversion is the variation of the gain of the conventional a.c. amplifier.

Another advantage of the use of parametric amplifiers to transform d.c. signals into a.c. signals is that there is a very high rate of common mode rejection when making differential measurements. This property is due to the fact that the power to be supplied to the parametric head in order to effect the transformation is delivered by an a.c. source and may consequently be transmitted by means of magnetic coupling. It should, however, be noted that this rejection cannot be complete for a.c. signals as, with the circuit envisaged, it is technically impossible to have the same parasitic capacitance between the frame and each of the input terminals. However some variants of the basic diagram have a symmetrical structure at the input ω_e and the common mode rejection into the alternative is very good. A more complex structure with four varactors is then chosen.

5.5. Bandwidth

The required bandwidth will be obtained by a judicious choice of the pumping frequency and of the parametric gain. Bandwidth at 3 dB (12.71) is:

$$B_{3\ \text{dB}} \simeq \frac{f_p}{2Q} (1 - \alpha) \qquad (13.61)$$

If the desired bandwidth cannot be achieved, use of a 'Goldberg' structure for the whole amplifier should be made [35].

5.6. Remarks

(1) The use of a parametric amplifier may very often be justified at low frequencies. Several examples will be given. Operational amplifiers for general purposes; control amplifiers for specialized uses; pH meters; miniaturized amplifiers used as radio transmitters for the transmission of encephalogram and cardiogram readings; various measuring instruments (error amplifier of the differential voltmeter, picoammeter); special apparatus for nuclear physics (measurement of nuclear radiation by its direct action on the varactors); measurement of the current of ionization chambers; (logarithmic amplifiers, etc.); seismographs and oceanographic probes; amplifiers for very high fidelity sound reproduction equipment.

(2) Use for teaching purposes. The parametric amplifier makes possible the showing of parametric effects on standard measuring instruments (oscilloscope, etc.).

(3) One can also envisage the use of the parametric amplifier for the amplitude-modulation of a high-frequency signal. This signal would then constitute the pumping signal and the modulation voltage the input signal. The amplitude of the carrier is proportional to the d.c. voltage and the amplitude of the side bands to that of the low-frequency voltage. If necessary the two side bands could be separated as follows. In a hybrid structure, two identical parametric bridges are pumped by signals of the same frequency ω_e but in phase quadrature. The input signals (ω_e) are injected into them also in phase quadrature. The addition or subtraction of the output signals supplies the sum-frequency or difference-frequency components.

The low-frequency parametric amplifier studied in the previous chapters has, when oscillating (when α is higher than 1), some remarkable properties which have not, so far as we are aware, been mentioned in any technical publications. The operation recalls that of the parametron, but in which each of the two stable states of oscillation may be imposed by the d.c. input voltage. The use of this operation [37] is outside the scope of this work which is confined to the small-signal analysis of parametric devices.

References

[1] Kennedy, D. P., 'Semiconductor device evaluation', in *Advances in Electronics and Electron Physics*, Academic Press.
[2] Izumi, I. and Okano, M., 'An improved solid-state logarithmic amplifier', *I.E.E.E. Trans. on Nuclear Science*, July 1964.
[3] San, C. T., Noyce, T. N. and Shockley, W., 'Carrier generation and recombination in *P–N* junction and *P–N* junction characteristics', *Proc. I.R.E.*, vol. 45 (No. 5), p. 1228, September 1957.
[4] Goulding, F. S., Lennox, C. G. and Robinson, L. B., 'Analysis of the non-linear behaviour of semiconductor junctions and its application in nucleonics', *AECL-801*, p. 77, April 1959.
[5] Terman, L. M., 'An investigation of surface states at a silicon/silicon oxide interface employing metal-oxide silicon diodes', in *Solid-State Electronics*, vol. 5, pp. 285–299, Pergamon Press, 1962.
[6] Nicollian, E. H. and Goetzberger, A., 'Laterial a.c. current flow model for metal insulator semiconductor capacitor', *I.E.E.E. Trans. on Electronic Devices*, p. 108, March 1965.
[7] Pfann, W. G. and Garret, C. G. S., 'Semiconductor varactors using surface space-charge layers', *Proc. I.R.E.*, vol. 47, pp. 2011–2012, November 1959.

[8] Lindner, R., 'Semiconductor surface varactor', *Bell System Technical Journal*, vol. XLI (No. 3), May 1962.
[9] Lindner, R., 'M.O.S. diodes', *Electronics*, 9 August 1963.
[10] Brouwn, C. M. and MacFarland, H. T., 'Low-frequency parametric amplifiers', *U.S. Government Research Report, AD 601378*.
[11] Biard, J. R., 'Low-frequency reactance amplifier', *Proc. I.E.E.E.*, pp. 298–303, February 1963.
[12] Schneider, B. and Strutt, M. J. O., 'Theory and experiments on shot noise in silicon p–n junction diodes and transistors', *Proc. I.E.E.E.*, vol. 47, pp. 546–554, April 1959.
[13] Sudland, K., 'Transistorisierte neutronen Flußverstärker', *Kerntechnik*, vol. 6, pp. 17–19, 1964.
[14] 'Electrical resistance of insulating materials', *A.S.T.M. D257–61*.
[15] Schoninger, E. and Seidt, F., 'Halbleiter-Bauelemente für hochempfindliche Gleichspannungsverstärker', *Siemens Zeitschrift*, vol. 11, pp. 837–843, November 1964.
[16] Keller, H., Lehmann, E. and Micle, L., 'Diffundierte Silizium Kapazitasdiode', *Radio Mentor*, p. 661, August 1962.
[17] Bourdel et al., 'Caractéristiques des modules destinés aux ensembles de mesures pour le contrôle neutronique des piles automatiques', *Onde Électrique*, No. 448–449, p. 74C.
[18] Davis and Ezell, 'Sub-audio parametric amplifier', *Electronics*, pp. 28–31, 1 March 1963.
[19] Izumi and Kokubu, 'A transistorized high voltage regulator using a.c. control', *Electronic Engineering*, August 1964.
[20] El-Ibiary, M. Y., 'Semiconductor logarithmic d.c. amplifier', *I.E.E.E. Trans. on Nuclear Science*, April 1963.
[21] 'How to measure differential amplifier, common mode rejection', *E.E.E.*, vol. 12 (No. 7), July 1964.
[22] 'Varactors in voltage tuning application', *Microwave Journal International*, p. 60, July 1964.
[23] Josephs, H. C., 'A d.c. reactance amplifier', *Proc. I.E.E.E.*, pp. 1669–1670, October 1965.
[24] Goldberg, E. A., 'Stabilization of wide-band direct-current amplifiers for zero and gain', *R.C.A. review*, pp. 296–300, June 1950.
[25] Cantraine, G., 'Usages speciaux et calcul approfondi de circuits lineaires', comportant un élément non dissipatif périodiquement variable.
[26] *Bulletin Scientifique*, 1ère partie (No. 3), p. 217, 1964; *A.I.M.*, 2ème partie (No. 1), p. 31, 1965.
[27] Gustin, P. and Van Halle, M., 'Representation analogique de phenomenes parametriques', *Rapport d'Études Université Catholique de Louvain, Faculté des Sciences Appliquées*, 1963.
[28] Van Reepinghen, 'Bruit superpose et bruit de mode commun', *Evolution Electronique* (No. 126), p. 47, July/August 1964.
[29] Arlowe, D., 'Electrical noise in instrumentation systems', *U.S. Office of Technical Services Report, TID 17749*.

[30] Hoge, R. R., 'A sensitive parametric modulator for d.c. measurements', *I.R.E. International Convention Record*, Pt. 9, pp. 34–42 1960.
[31] Hyde, F. J., 'Varactor-diode parametric amplifiers', *Proc. I.E.E.* vol. III (No. 6), June 1964.
[32] De Bolt, H. E., 'A high-sensitivity semiconductor diode modulator for d.c. current measurement', *I.R.E. Trans. on Nuclear Science* December 1960.
[33] Loos, 'Een ingangsschakeling met zeer hoge ingangsweestand', *NAT LAB Report*, No. 113/61.
[34] Danloux and Dumesnils, *Calcul and logique par courants continus* Dunod.
[35] Gorelik, C., 'Phénomènes de résonance dans les systèmes linéaires et paramètres périodiques', *Tech. Phys. U.S.S.R.*, vol. 2, pp. 81–134 1935.
[36] Mandeltam, L., Papalexi, M., Andronov, A., Chaikin, S. and Witt, A. 'Exposé de recherches récentes sur les oscillations non linéaires' *Tech. Phys. U.S.S.R.*, vol. 2, 1935.
[37] Decroly, J. C., 'Phénomènes d'Hystérèse dans les régimes oscillatoires d'un amplificateur paramétrique basse frequence', *Revue H.F.* vol. 6, No. 8, 1965.

14

THE PUMPING OF PARAMETRIC DIODES FOR LOW FREQUENCIES

by G. MARECHAL

1. General

The object of this chapter is the calculation of the quantities C_k, G_k, I_k of a parametric diode as functions of the characteristic pumping parameters. The relations connecting the differential capacitance C, the differential conductance G and the static current I with the voltage v applied to the varactor, may be expressed with good approximation for standard diodes by (Chapter 13):

$$C = C_{0v}\left(1 + \frac{v}{\phi}\right)^{-\nu} + C_C \qquad (14.1)$$

$$G = G_a\, e^{-av} + G_s \qquad (14.2)$$

$$I = I_s(1 - e^{-av}) + G_s v \qquad (14.3)$$

with

$$\nu = \frac{1}{n} \qquad (14.4)$$

and

$$G_a = aI_s \qquad (14.5)$$

It should be noted that relations (14.1), (14.2) and (14.3) will be used only in the region defined by:

$$V_B \geqslant v \geqslant -\phi \qquad (14.6)$$

If the voltage v is a periodic function of fundamental frequency ω_p, the quantities C, G, I may be developed as Fourier series, and quite generally we have:

$$C(t) = \sum_{-\infty}^{+\infty} C_k \, e^{+jk\omega_p t} \qquad (14.7)$$

$$G(t) = \sum_{-\infty}^{+\infty} G_k \, e^{jk\omega_p t} \qquad (14.8)$$

$$I(t) = \sum_{-\infty}^{+\infty} I_k \, e^{jk\omega_p t} \qquad (14.9)$$

As in Chapter 8, it will be assumed that:
(1) The pumping voltage v_p varies from $V_{\min}$ to $V_{\max}$.
(2) The time origin is chosen so that $v_p(t)$ is at its maximum.
(3) The function $v_p(t)$ in the interval $(0, 2\pi/\omega_p)$ has as its axis of symmetry the centre point of this interval.

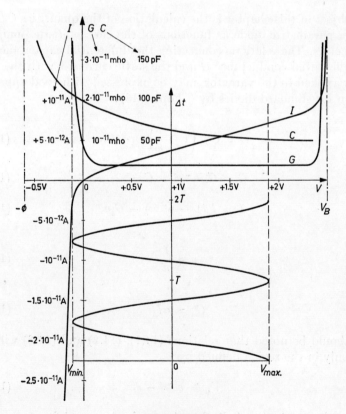

Figure 14.1

Since there is a biunivocal and monotonic relation between C and v, G and v, I and v, the functions $C(t)$, $G(t)$ and $I(t)$ have the same axis of symmetry as $v_p(t)$. We then have:

$$2C_k = C_k + C_k^* = C_k + C_{-k} \quad (14.10)$$

$$2G_k = G_k + G_k^* = G_k + G_{-k} \quad (14.11)$$

$$2I_k = I_k + I_k^* = I_k + I_{-k} \quad (14.12)$$

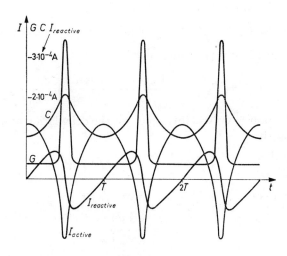

$C(v) \longrightarrow C(t)$
$G(v) \longrightarrow G(t)$
$I(v)_{active} \longrightarrow I(t)_{active}$
$I(v)_{reactive} \longrightarrow I(t)_{reactive}$

$a = 29.5 \ (V^{-1})$
$I_S = 2.5 \cdot 10^{-13} (A)$
$G_S = 37 \cdot 10^{-13} (mho)$
$V_0 = 0.9 \ (V)$
$\alpha_v = 0.645$
$\beta_v = 0.666$
$n = 2$
$\phi = 0.6 \ (V)$
$V_B = 2.5 \ (V)$
$T = 5 \cdot 10^{-6} \ sec$

Figure 14.2

It has been shown that the optimum pumping voltage is a square wave if it is required to give C_1 a maximum value. Since in the microwave field it is practically impossible to obtain this waveform at low frequencies, a voltage pumping with any given waveform is possible. We will deal here with cissoidal and rectangular voltage pumping. The d.c. bias voltage of the varactor will be designated by V_0. Figures 14.1 and 14.2 show C, G and I as functions of time for cissoidal pumping.

2. Cissoidal voltage pumping

2.1. Calculation of the C_k's

2.1.1. GENERAL FORMULAE OF THE C_k'S

Chapter 8, the notations of which will be recalled, gives the details of the developments and of the discussions applicable to the calculation of the parameters of cissoidal voltage pumping. The voltage applied to the varactor is:

$$v_p = V_0 + V_p(e^{j\omega_p t} + e^{-j\omega_p t}) = V_0 + 2V_p \cos \omega_p t \quad (14.13)$$

After substitution in (14.1) we obtain

$$C(t) = C_{0v}\left(1 + \frac{V_0 + 2V_p \cos \omega_p t}{\phi}\right)^{-\nu} + C_c \quad (14.14)$$

As in Chapter 8, we will introduce for the calculation and discussion of the quantities

$$C_k = \frac{1}{T}\int_0^T C(t)\, e^{-jk\omega_p t}\, dt \quad (14.15)$$

the parameters α_v and β_v defined by:

$$\alpha_v = \frac{4V_p}{\phi + V_B} \quad (14.16)$$

$$\beta_v = \frac{2V_p}{\phi + V_0} \quad (14.17)$$

Conditions (14.6) give us immediately

$$0 < \alpha_v \leqslant 1 \quad (14.18)$$

$$0 < \beta_v \leqslant 1 \quad (14.19)$$

$$\frac{\alpha_v}{2 - \alpha_v} \leqslant \beta_v \leqslant 1 \quad (14.20)$$

The reader is referred back to formulae (8.58), (8.87) and (8.91) in Chapter 8.

Partial pumping with non-minimum bias ($\alpha_v < 1, \beta_v < 1$)

$$\frac{|C_k|}{C_{\min}} = \frac{4^\nu \sin \pi\nu}{2\pi \cdot \sqrt{\pi}} \left(\frac{\beta_v}{\alpha_v}\right)^\nu \beta_v^k \frac{\Gamma(1-\nu)\Gamma((\nu+k)/2)\Gamma((\nu+k+1)/2)}{k!}$$

$$\times F\left(\frac{\nu+k}{2}, \frac{\nu+k+1}{2}, 1+k, \beta_v^2\right) \quad (14.21)$$

PUMPING OF DIODES FOR LOW FREQUENCIES

Full pumping ($\alpha_v = 1$, $\beta_v = 1$)

$$\frac{|C_k|}{C_{\min}} = 4^\nu \left| \frac{\Gamma(1-2\nu)}{\Gamma(1-\nu+k)\cdot\Gamma(1-\nu-k)} \right| \qquad (14.22)$$

Partial pumping with minimum bias ($\alpha_v < 1$, $\beta_v = 1$)

$$\frac{|C_k|}{C_{\min}} = \frac{1}{\alpha_v^\nu} \cdot 4^\nu \left| \frac{\Gamma(1-2\nu)}{\Gamma(1-\nu+k)\Gamma(1-\nu-k)} \right| \qquad (14.23)$$

2.1.2. SIMPLIFIED FORMULAE OF THE C_k's

In most low-frequency parametric amplifiers the type of pumping used is neither full ($\alpha_v = 1$, $\beta_v = 1$), nor partial with minimum bias ($\alpha_v < 1$, $\beta_v = 1$). Partial pumping with high bias ($\beta_v \ll 1$) often occurs.

(1) For $\beta_v \ll 1$, the series (14.21) giving the C_k's converges rapidly. The general formulae can then be simplified. Relating the C_k to C_0 and C_0 to $C(V_0)$:

$$C(V_0) = C_{\min} \cdot 2^\nu \left(\frac{\beta_v}{\alpha_v}\right)^\nu = C_{0v}\left(1 + \frac{V_0}{\phi}\right)^{-\nu} \qquad (14.24)$$

If we consider only the first term of the development of (14.21) we at once find:

$$\frac{C_o}{C(V_0)} = f_1(n, \beta_v) \approx 1 + \frac{(\nu+1)\nu}{4}\beta_v^2 = 1 + \frac{n+1}{4n^2}\beta_v^2 \approx 1 \qquad (14.25)$$

$$\frac{|C_k|}{C_0} = f_k(n, \beta_v) \approx \left| \frac{\Gamma(1-\nu)}{2^k \cdot k! \cdot \Gamma(1-\nu-k)} \beta_v^k \right| \qquad (14.26)$$

which gives:

$$\frac{1}{\beta_v} \cdot \frac{C_1}{C_0} = \frac{1}{2n} = \frac{\nu}{2} \qquad (14.27)$$

$$\frac{1}{\beta_v^2} \cdot \frac{C_2}{C_0} = \frac{n+1}{8n^2} = \frac{\nu(\nu+1)}{8} \qquad (14.28)$$

(14.27) and (14.28) give the interesting ratio:

$$\mathscr{L} = \frac{C_0 C_2}{C_1^2} \approx \frac{n+1}{2} \qquad (14.29)$$

It is seen thus that as a first approximation $\mathscr{L}$ is independent of the pumping parameters (α_v and β_v). It is a function only of the power n of the root, that is of the type of diode. This statement can be found in [2]. Relation (14.29) and Fig. 14.6 enable us to calculate the value of this interesting ratio. If β_v is not considerably smaller than 1, the series giving the C_k's converges very slowly, and it might be thought that the approximate relations (14.25), (14.26), (14.27), (14.28), (14.29) would become rapidly unusable. However, as β_v increases, the C_k's and C_0's also increase. Their ratio changes more slowly than each of the terms. The independence of $\mathscr{L}$ of the pumping parameters is still more marked. It follows from the nature of the non-linear relation which unites the differential capacitance of the diode with the voltage at its terminals.

(2) *For* $\beta_v = 1$, relation (14.23) allows us to calculate the exact values corresponding to relations (14.25), (14.26), (14.27) and (14.28).

$$\frac{C_0}{C(V_0)} = 2^\nu \frac{\Gamma(1 - 2\nu)}{[\Gamma(1 - \nu)]^2} \tag{14.30}$$

$$\frac{C_1}{C_0} = \frac{1}{n - 1} = \frac{\nu}{1 - \nu} \tag{14.31}$$

$$\frac{C_2}{C_0} = \frac{n + 1}{(n - 1)(2n - 1)} = \frac{(1 + \nu)\nu}{(1 - \nu)(2 - \nu)} \tag{14.32}$$

$$\mathscr{L} = \frac{n^2 - 1}{2n - 1} = \frac{1 - \nu^2}{(2 - \nu)\nu} \tag{14.33}$$

2.1.3. NOMOGRAPHS OF THE C_k'S

The nomographs of Figs. 14.3, 14.4, 14.5, 14.6, 14.7, 14.9 show the exact values of C_0, C_1, C_2, $\mathscr{L}$, as functions of the pumping parameters, and they enable us to estimate the degree of approximation obtained when we use the approximate formulae for important values of β_v.

Table 14.1. $C_0/C(V_0) = f(\beta_v)$ for cissoidal pumping

β_v	n					
	2	2·2	2·4	2·6	2·8	3
0·100	1·002	1·002	1·001	1·001	1·001	1·001
0·200	1·008	1·007	1·006	1·005	1·005	1·005
0·300	1·018	1·016	1·014	1·013	1·011	1·010
0·400	1·033	1·029	1·026	1·023	1·021	1·019
0·500	1·055	1·048	1·043	1·038	1·034	1·031
0·600	1·085	1·075	1·066	1·059	1·053	1·049
0·700	1·130	1·113	1·100	1·090	1·081	1·074
0·800	1·200	1·173	1·152	1·136	1·122	1·111
0·850	1·253	1·218	1·191	1·170	1·152	1·138
0·875	1·288	1·247	1·216	1·192	1·172	1·156
0·900	1·332	1·284	1·247	1·219	1·196	1·177
0·910	1·353	1·301	1·262	1·231	1·207	1·187
0·920	1·377	1·321	1·279	1·246	1·219	1·199
0·930	1·404	1·343	1·297	1·262	1·233	1·210
0·940	1·436	1·369	1·319	1·280	1·249	1·225
0·950	1·474	1·400	1·345	1·302	1·269	1·241
0·960	1·521	1·438	1·376	1·329	1·292	1·261
0·970	1·582	1·487	1·416	1·363	1·321	1·287
0·980	1·670	1·556	1·472	1·409	1·360	1·321
1·000		5·421	3·180	2·432	2·060	1·857

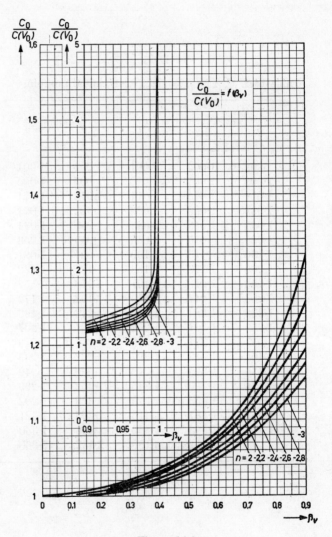

Figure 14.3

Table 14.2. $C_1/\beta_v C_0 = f(\beta_v)$ with cissoidal pumping

B_v	\multicolumn{6}{c}{n}
	2	2·2	2·4	2·6	2·8	3
0·100	0·2507	0·2279	0·2089	0·1928	0·1791	0·1671
0·200	0·2529	0·2299	0·2107	0·1945	0·1806	0·1686
0·300	0·2566	0·2333	0·1974	0·1974	0·1836	0·1710
0·400	0·2623	0·2384	0·2186	0·2017	0·1873	0·1748
0·500	0·2704	0·2458	0·2253	0·2079	0·1930	0·1801
0·600	0·2818	0·2561	0·2347	0·2166	0·2010	0·1876
0·700	0·2983	0·2711	0·2484	0·2291	0·2126	0·1983
0·800	0·3236	0·2942	0·2695	0·2485	0·2305	0·2149
0·850	0·3426	0·3113	0·2851	0·2629	0·2437	0·2272
0·875	0·3548	0·3224	0·2952	0·2721	0·2523	0·2351
0·900	0·3698	0·3360	0·3076	0·2835	0·2628	0·2448
0·910	0·3769	0·3424	0·3135	0·2889	0·2677	0·2494
0·920	0·3848	0·3496	0·3200	0·2949	0·2732	0·2545
0·930	0·3938	0·3577	0·3274	0·3016	0·2795	0·2602
0·940	0·4040	0·3670	0·3359	0·3094	0·2866	0·2668
0·950	0·4161	0·3779	0·3458	0·3184	0·2949	0·2745
0·960	0·4306	0·3911	0·3577	0·3293	0·3049	0·2837
0·970	0·4488	0·4076	0·3728	0·3430	0·3174	0·2952
0·980	0·4736	0·4300	0·3930	0·3615	0·3343	0·3108
1·000	1·0000	0·8333	0·7142	0·6250	0·5555	0·5000

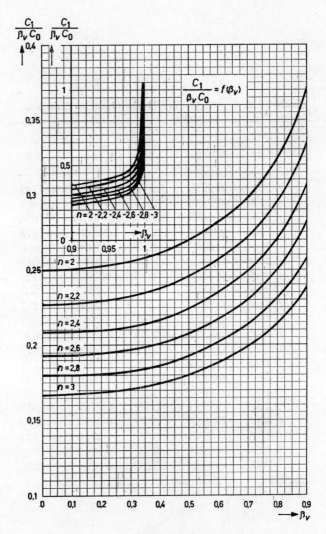

Figure 14.4

Table 14.3. $C_2/\beta_v^2 C_0 = f(\beta_v)$ with cissoidal pumping

β_v	n					
	2	2·2	2·4	2·6	2·8	3
0·100	0·0943	0·0831	0·0742	0·0669	0·0609	0·0559
0·200	0·0958	0·0844	0·0754	0·0680	0·0619	0·0567
0·300	0·0986	0·0869	0·0776	0·0700	0·0637	0·0584
0·400	0·1028	0·0905	0·0808	0·0729	0·0663	0·0608
0·500	0·1089	0·0960	0·0857	0·0773	0·0703	0·0644
0·600	0·1179	0·1040	0·0928	0·0837	0·0761	0·0697
0·700	0·1314	0·1158	0·1034	0·0932	0·0847	0·0776
0·800	0·1535	0·1352	0·1206	0·1087	0·0988	0·0905
0·850	0·1709	0·1506	0·1342	0·1209	0·1099	0·1005
0·875	0·1825	0·1608	0·1433	0·1291	0·1172	0·1073
0·900	0·1972	0·1737	0·1548	0·1394	0·1266	0·1158
0·910	0·2043	0·1800	0·1604	0·1444	0·1310	0·1199
0·920	0·2124	0·1871	0·1667	0·1500	0·1362	0·1245
0·930	0·2216	0·1952	0·1739	0·1565	0·1420	0·1298
0·940	0·2325	0·2047	0·1824	0·1640	0·1488	0·1360
0·950	0·2454	0·2161	0·1924	0·1730	0·1569	0·1433
0·960	0·2613	0·2300	0·2048	0·1841	0·1669	0·1524
0·970	0·2818	0·2480	0·2207	0·1983	0·1797	0·1640
0·980	0·3104	0·2732	0·2429	0·2181	0·1975	0·1800
1·000	1·0000	0·7843	0·6391	0·5357	0·4589	0·4000

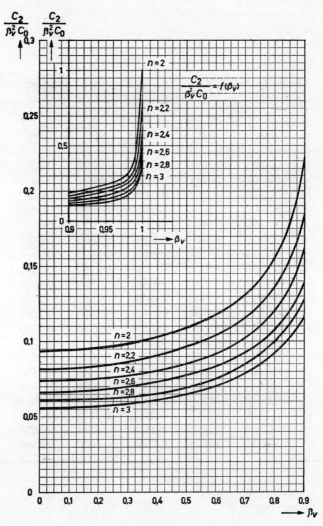

Figure 14.5

Table 14.4. $\mathscr{L} = C_0 C_2 / C_1^2 = f(\beta_v)$ with cissoidal pumping

β_v	\multicolumn{6}{c}{n}					
	2	2·2	2·4	2·6	2·8	3
0·100	1·500	1·600	1·700	1·800	1·900	2·000
0·200	1·498	1·597	1·697	1·797	1·897	1·997
0·300	1·497	1·597	1·697	1·797	1·897	1·996
0·400	1·495	1·592	1·691	1·791	1·891	1·991
0·500	1·489	1·588	1·688	1·788	1·887	1·987
0·600	1·486	1·585	1·684	1·783	1·883	1·981
0·700	1·477	1·576	1·675	1·775	1·874	1·974
0·800	1·466	1·563	1·661	1·761	1·860	1·959
0·850	1·456	1·554	1·651	1·750	1·850	1·948
0·875	1·450	1·547	1·645	1·743	1·842	1·941
0·900	1·442	1·539	1·636	1·734	1·833	1·932
0·910	1·439	1·535	1·632	1·730	1·829	1·927
0·920	1·434	1·531	1·628	1·726	1·824	1·923
0·930	1·429	1·525	1·623	1·720	1·818	1·917
0·940	1·424	1·520	1·616	1·714	1·812	1·910
0·950	1·417	1·513	1·609	1·706	1·804	1·903
0·960	1·409	1·504	1·600	1·697	1·795	1·893
0·970	1·399	1·493	1·589	1·686	1·783	1·881
0·980	1·384	1·477	1·573	1·669	1·766	1·864
1·000	1·000	1·1294	1·2526	1·3714	1·4869	1·6000

400 PARAMETRIC AMPLIFIERS

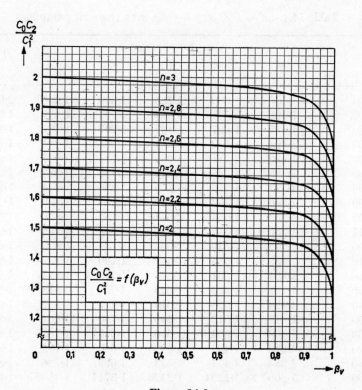

Figure 14.6

Table 14.5. $\dfrac{C(V_0)}{C_{0v}} = \left(1 + \dfrac{V_0}{\phi}\right)^{-\nu}$

$\dfrac{V_0}{\phi}$	\multicolumn{6}{c}{n}					
	2	2·2	2·4	2·6	2·8	3
−0·99	10·00	8·111	6·813	5·878	5·179	4·641
−0·98	7·071	5·919	5·104	4·502	4·044	3·684
−0·97	5·774	4·923	4·311	3·852	3·499	3·218
−0·96	5·000	4·319	3·824	3·449	3·157	2·924
−0·95	4·472	3·903	3·484	3·165	2·915	2·714
−0·93	3·780	3·349	3·028	2·781	2·585	2·426
−0·92	3·536	3·152	2·864	2·642	2·465	2·321
−0·90	3·162	2·848	2·610	2·424	2·276	2·154
−0·85	2·582	3·369	2·204	2·074	1·970	1·882
−0·80	2·236	2·078	1·955	1·857	1·777	1·710
−0·70	1·826	1·729	1·651	1·589	1·537	1·494
−0·60	1·581	1·517	1·465	1·423	1·387	1·357
−0·50	1·414	1·370	1·335	1·306	1·281	1·260
−0·30	1·120	1·176	1·160	1·147	1·136	1·126
−0·10	1·054	1·049	1·045	1·041	1·038	1·036
+0·00	1·000	1·000	1·000	1·000	1·000	1·000
+0·10	0·953	0·958	0·961	0·964	0·967	0·969
+0·20	0·913	0·920	0·927	0·932	0·937	0·941
+0·30	0·877	0·888	0·896	0·904	0·911	0·916
+0·50	0·816	0·832	0·845	0·856	0·865	0·874
+0·80	0·745	0·766	0·783	0·798	0·811	0·822
+1·00	0·707	0·730	0·749	0·766	0·781	0·794
+2·00	0·577	0·607	0·633	0·655	0·675	0·693
+3·00	0·500	0·533	0·561	0·587	0·610	0·630
+5·00	0·408	0·443	0·474	0·502	0·527	0·550
+10·00	0·301	0·336	0·368	0·398	0·425	0·450
+20·00	0·218	0·251	0·281	0·310	0·337	0·362
+30·00	0·180	0·210	0·239	0·267	0·293	0·318
+50·00	0·140	0·167	0·194	0·220	0·246	0·270
+100·00	0·099	0·122	0·146	0·169	0·192	0·215

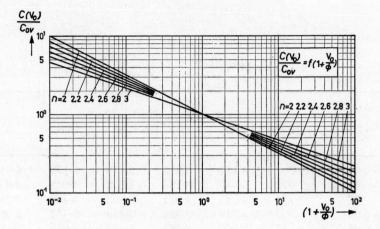

Figure 14.7

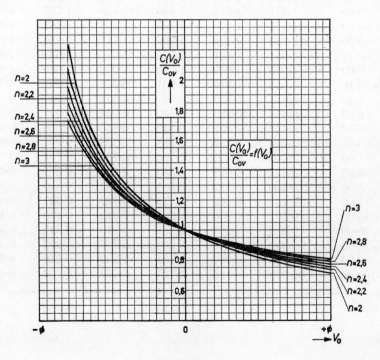

Figure 14.8

PUMPING OF DIODES FOR LOW FREQUENCIES

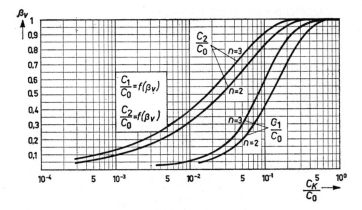

Figure 14.9. Cissoidal pumping

2.1.4. RECURRENCE FORMULA OF THE C_k's

The very important recurrence formula established in section 2.2, Chapter 8, is repeated here

$$|C_{k+2}| = -\frac{k+\nu}{k-\nu+2}|C_k| + \frac{2}{\beta_v} \cdot \frac{k+1}{k-\nu+2}|C_{k+1}| \quad (14.34)$$

2.2. Calculation of the G_k's

2.2.1. GENERAL FORMULAE OF THE G_k's

Reverting to relation (14.2), after expression (14.13) for V_p has been introduced into it:

$$G(t) = G_a e^{-a(V_0 + 2V_p \cos \omega_p t)} + G_s \quad (14.35)$$

The quantities G_k are given by:

$$G_k = \frac{G_k + G^*}{2} = \frac{1}{T} \int_0^T G(t) \cdot \cos(k\omega_p t) \, dt \quad (14.36)$$

hence, by virtue of (14.35):

$$G_0 = G_s + G_a e^{-aV_0} \cdot \mathscr{J}_0(2aV_p) \quad (14.37)$$

$$G_k = G_a e^{-aV_0} \cdot \mathscr{J}_k(2aV_p) \cdot (-1)^k \quad (14.38)$$

In formulae (14.37) and (14.38) $\mathscr{J}_k(x)$ represents the modified first-order Bessel function of the index k. Titles of important works on the functions used in this chapter will be found under references [3] to [8].

2.2.2. SIMPLIFIED FORMULAE OF THE G_k'S

Depending on the value of $2aV_p$, either one or other of the following developments may be used:

aV_p large (>1), ($1V_p > 0.1$ volt)

$$G_k \approx \frac{G_a \, e^{a(2V_p - V_0)}}{\sqrt{4\pi a V_p}}$$

$$\times \left\{ 1 - \frac{4k^2 - 1^2}{1!\,16aV_p} + \frac{(4k^2 - 1^2)(4k^2 - 3^2)}{2!\,(16aV_p)^2} \ldots \right\} \cdot (-1)^k \quad (14.39)$$

aV_p small (<1), ($2V_p < 0.1$ volt),

$$G_k = G_a \, e^{-aV_0} \cdot \sum_{p=0}^{\infty} \frac{(aV_p)^{k+2p}}{p!\,(k+p)!} \cdot (-1)^k \quad (14.40)$$

It should be noted, however, that these formulae are not convenient for use.

2.2.3. NOMOGRAPHS FOR THE DETERMINATION OF THE G_k'S

As the quantities G_k are functions of the three parameters a, V_0, $2V_p$, only the following nomographs will be given here:

$$\mathcal{J}_0(2aV_p), \quad \mathcal{J}_1(2aV_p), \quad \mathcal{J}_2(2aV_p), \quad e^{-2aV_p}\mathcal{J}_0(2aV_p),$$

$$e^{-2aV_p}\mathcal{J}_1(2aV_p), \quad \frac{\mathcal{J}_1(2aV_p)}{\mathcal{J}_0(2aV_p)}, \quad \text{and} \quad \frac{\mathcal{J}_2(2aV_p)}{\mathcal{J}_0(2aV_p)}.$$

Figures 14.10, 14.11 and 14.12 enable the G_k's to be calculated easily with the aid of formulae (14.37) and (14.38).

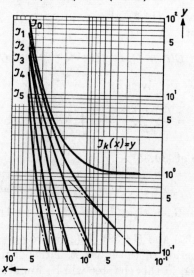

Figure 14.10. Bessel function of $\mathcal{J}_k(x) = y$

PUMPING OF DIODES FOR LOW FREQUENCIES 405

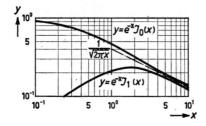

Figure 14.11

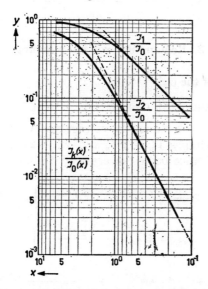

Figure 14.12. Bessel function of $\mathcal{J}_k(x)/\mathcal{J}_0(x)$

2.2.4. RECURRENCE FORMULA OF THE G_k'S

For $k > 0$,

$$G_{k+2} = G_k + \frac{k+1}{aV_p} \cdot G_{k+1} \quad (14.41)$$

For $k = 0$,

$$G_2 = G_0 - G_s + \frac{1}{aV_s} \cdot G_1 \quad (14.42)$$

2.3. Calculation of the I_k

2.3.1. GENERAL FORMULAE OF THE I_k

$$I_k = \frac{1}{T} \int_0^T \{I_s(1 - e^{-a(V_0 + 2V_p \cos \omega_p t)}) \\ + G_s(V_0 + 2V_p \cos \omega_p t)\} \cos(k\omega_p t)\, dt \quad (14.43)$$

Hence:

$$I_0 = I_s[1 - e^{-aV_0}\mathscr{J}_0(2aV_p)]$$

$$+G_sV_0 = \frac{G_a + G_s - G_0}{a} + G_sV_0 \quad (14.44)$$

$$I_1 = I_s e^{-aV_0}\mathscr{J}_1(2aV_p) + V_pG_s \quad (14.45)$$

For $k > 1$,

$$I_k = (-1)^{k+1}I_s e^{-aV_0}\mathscr{J}_k(2aV_p) \quad (14.46)$$

3. Rectangular voltage pumping

3.1. Calculation of the C_k's

3.1.1. GENERAL FORMULAE OF THE C_k

When pumped by a rectangular voltage the varactor is subjected successively to voltages $V_0 + 2V_p$ and $V_0 - 2V_p$: the value of the differential capacitance passes successively from C_min to C_max.

$$C_\text{min} = C_{0v}\left(1 + \frac{V_0 + 2V_p}{\phi}\right)^{-\nu} \quad (14.47)$$

$$C_\text{max} = C_{0v}\left(1 + \frac{V_0 - 2V_p}{\phi}\right)^{-\nu} \quad (14.48)$$

Putting again:

$$\beta_v = \frac{2V_p}{\phi + V_0} \quad (14.49)$$

and noting that:

$$C(V_0) = C_{0v}\left(1 + \frac{V_0}{\phi}\right)^{-\nu} \quad (14.50)$$

We obtain:

$$C_0 = C_{0v}\left(1 + \frac{V_0}{\phi}\right)^{-\nu}\left\{\frac{(1-\beta_v)^{-\nu} + (1+\beta_v)^{-\nu}}{2}\right\} \quad (14.51)$$

$$C_{2k+1} = \frac{2}{\pi}$$

$$\times \frac{(-1)^k}{2k+1}C_{0v}\left(1 + \frac{V_0}{\phi}\right)^{-\nu}\left\{\frac{(1-\beta_v)^{-\nu} - (1+\beta_v)^{-\nu}}{2}\right\} \quad (14.52)$$

PUMPING OF DIODES FOR LOW FREQUENCIES 407

These last two relations may also be written in the form

$$C_0 = C(V_0) \sum_{p=0}^{p=\infty} \frac{\Gamma(\nu + 2p)}{(2p)!\,\Gamma(\nu)} \beta_v^{2p} + C_C \quad (14.53)$$

$$C_{2k+1} = \frac{2}{\pi} \frac{(-1)^k}{2k+1} C(V_0) \sum_{p=0}^{p=\infty} \frac{\Gamma(\nu + 2p + 1)}{(2p+1)!\,\Gamma(\nu)} \beta_v^{2p+1} \quad (14.54)$$

It will be seen that there are no *even harmonics*.

Full pumping: $\alpha_v = 1, \beta_v = 1$.

$$\left.\begin{array}{l} C_0 = \infty \\ C_{2k+1} = \infty \end{array}\right\} \text{ whatever the value of } \nu$$

$$\frac{C_{2k+1}}{C_0} = \frac{2}{\pi} \frac{(-1)^k}{2k+1} \quad (14.55)$$

Partial pumping with minimum bias: $\alpha_v < 1, \beta_v = 1$.

$$C_0 = \infty$$

$$C_{2k+1} = \infty$$

$$\frac{C_{2k+1}}{C_0} = \frac{(-1)^k}{2k+1} \quad (14.56)$$

3.1.2. SIMPLIFIED FORMULAE OF THE C_k'S

Within the limits of the discussion in 2.1.2, we may write:

$$\frac{C_0}{C(V_0)} \approx 1$$

$$\frac{C_{2k+1}}{C_0} \approx \frac{2}{\pi} \frac{(-1)^k}{2k+1} \frac{\Gamma(\nu + 2p + 1)}{\Gamma(\nu)} \beta_v \quad (14.57)$$

which gives

$$\frac{1}{\beta_v} \frac{C_1}{C_0} \approx \frac{2}{\pi} \nu = \frac{2}{\pi n} \quad (14.58)$$

3.1.3. NOMOGRAPHS OF C_0 AND C_1

The nomographs of Figs. 14.13, 14.14, 14.15 give the exact values of C_0 and C_1 as functions of the pumping parameters. It should be noted that Figs. 14.7 and 14.8 and Table 14.5 are valid for rectangular pumping.

Table 14.6. $C_0/C(V_0) = f(\beta_v)$ rectangular pumping

β_v	n					
	2	2·2	2·4	2·6	2·8	3
0·100	1·004	1·003	1·003	1·003	1·002	1·002
0·200	1·015	1·014	1·012	1·011	1·010	1·009
0·300	1·036	1·032	1·028	1·026	1·023	1·021
0·400	1·068	1·060	1·053	1·048	1·043	1·040
0·500	1·115	1·101	1·090	1·081	1·073	1·067
0·600	1·186	1·162	1·144	1·129	1·116	1·106
0·700	1·296	1·257	1·227	1·202	1·182	1·166
0·800	1·491	1·422	1·369	1·327	1·294	1·266
0·850	1·659	1·562	1·489	1·432	1·386	1·348
0·875	1·779	1·662	1·574	1·505	1·450	1·405
0·900	1·944	1·797	1·688	1·603	1·535	1·481
0·910	2·028	1·866	1·745	1·652	1·578	1·519
0·920	2·129	1·948	1·813	1·710	1·628	1·563
0·930	2·250	2·045	1·894	1·779	1·688	1·615
0·940	2·400	2·166	1·994	1·863	1·760	1·678
0·950	2·594	2·320	2·121	1·969	1·851	1·757
0·960	2·857	2·528	2·290	2·110	1·972	1·862
0·970	3·243	2·829	2·532	2·311	2·142	2·008
0·980	3·891	3·326	2·928	2·636	2·414	2·240
1·000						

PUMPING OF DIODES FOR LOW FREQUENCIES

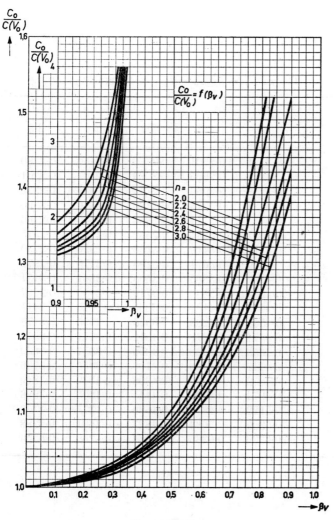

Figure 14.13

Table 14.7. $C_1/\beta_v C_0 = f(\beta_v)$ rectangular pumping

β_v	n					
	2	2·2	2·4	2·6	2·8	3
0·100	0·3191	0·2901	0·2660	0·2456	0·2280	0·2128
0·200	0·3216	0·2925	0·2683	0·2477	0·2301	0·2148
0·300	0·3258	0·2966	0·2722	0·2514	0·2336	0·2182
0·400	0·3323	0·3028	0·2780	0·2570	0·2390	0·2233
0·500	0·3412	0·3114	0·2864	0·2650	0·2466	0·2306
0·600	0·3537	0·3237	0·2982	0·2763	0·2575	0·2408
0·700	0·3714	0·3410	0·3150	0·2926	0·2730	0·2559
0·800	0·3979	0·3674	0·3408	0·3176	0·2971	0·2790
0·850	0·4169	0·3865	0·3598	0·3361	0·3151	0·2965
0·875	0·4289	0·3986	0·3719	0·3480	0·3267	0·3078
0·900	0·4433	0·4134	0·3866	0·3626	0·3411	0·3217
0·910	0·4500	0·4203	0·3935	0·3695	0·3478	0·3284
0·920	0·4574	0·4278	0·4012	0·3771	0·3553	0·3357
0·930	0·4655	0·4363	0·4098	0·3857	0·3639	0·3441
0·940	0·4747	0·4460	0·4196	0·3955	0·3736	0·3536
0·950	0·4852	0·4569	0·4308	0·4069	0·3849	0·3649
0·960	0·4974	0·4700	0·4444	0·4206	0·3986	0·3785
0·970	0·5122	0·4859	0·4609	0·4375	0·4158	0·3956
0·980	0·5309	0·5064	0·4827	0·4601	0·4388	0·4187
1·000	0·6366	0·6366	0·6366	0·6366	0·6366	0·6366

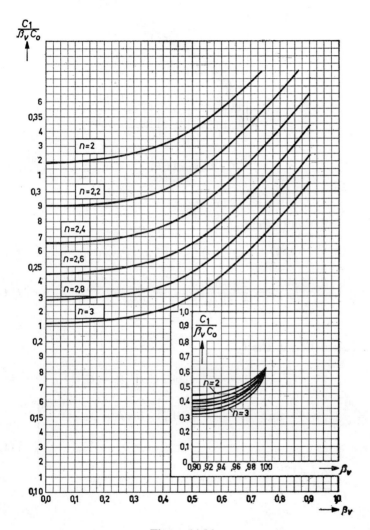

Figure 14.14

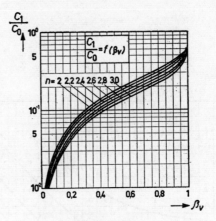

Figure 14.15

3.2. Calculation of the G_K's

$$G_0 = G_s + G_a \cdot e^{-aV_0} \cdot \cosh(2aV_p) \tag{14.59}$$

$$G_{2k+1} = \frac{2}{\pi} \frac{(-1)^{k+1}}{2k+1} \cdot G_a \cdot e^{-aV_0} \sinh(2aV_p) \tag{14.60}$$

3.3. Calculation of the I_K's

$$I_0 = I_s\{1 - e^{-aV_0}\cosh(2aV_p)\} + G_s V_0 \tag{14.61}$$

$$\left.\begin{array}{l} I_1 = \dfrac{2}{\pi}(I_s e^{-aV_0}\sinh(2aV_p) + 2G_s V_p) \quad \text{for } k = 1 \\[4pt] \text{For } k \geqslant 2: \\[4pt] I_{2k+1} = \dfrac{2}{\pi}\dfrac{(-1)^k}{2k+1}[I_s e^{-aV_0}\sinh(2aV_p) + 2G_s V_p] \end{array}\right\} \tag{14.62}$$

3.4. Recurrence formulae of C_{2k+1}, G_{2k+1}, I_{2k+1} ($k \geqslant 2$)

$$\frac{C_{2k+1}}{C_{2k-1}} = \frac{G_{2k+1}}{G_{2k-1}} = \frac{I_{2k+1}}{I_{2k-1}} = -\left(\frac{2k-1}{2k+1}\right) \tag{14.63}$$

$$\frac{C_{2k+1}}{C_1} = \frac{G_{2k+1}}{G_1} = \frac{I_{2k+1}}{I_1} = \frac{(-1)^k}{2k+1} \tag{14.64}$$

4. Specially important case: $I_0 = 0$

It is desirable for numerous applications that I_0 should be zero.

4.1. Relation between V_0 and $2V_p$ for $I_0 = 0$

Case in which $G_s V_0 \ll I_s$. (14.44) and (14.61) give respectively:

$$e^{-aV_0} \mathscr{J}_0(2aV_p) = 1 \quad \text{(cissoidal pumping)} \quad (14.65)$$

$$e^{-aV_0} \cosh(2aV_p) = 1 \quad \text{(rectangular pumping)} \quad (14.66)$$

For $2aV_p \gg 1$, relations (14.65) and (14.66) become respectively:

$$e^{-a(2V_p - V_0)} \approx 2\sqrt{\pi a V_p} \quad (14.67)$$

$$e^{-a(2V_p - V_0)} \approx 2 \quad (14.68)$$

or:

$$2V_p - V_0 \approx \frac{1}{2a} \log_n (4\pi a V_p) \quad (14.69)$$

$$2V_p - V_0 \approx \frac{1}{a} \log_n 2 \quad (14.70)$$

These relations, however, are not valid except for measurements in which the value of V_p is not too large. In fact, for increasing values of V_p, the corresponding value of V_0 increases and the condition $G_s V_0 \ll I_s$ can no longer be satisfied.

Case in which $G_s V_0$ has any given value

$$1 + \frac{G_s V_0}{I_s} = e^{-aV_0} \mathscr{J}_0(2aV_p) \quad \text{(cissoidal pumping)} \quad (14.71)$$

$$1 + \frac{G_s V_0}{I_s} = e^{-aV_0} \cosh(2aV_p) \quad \text{(rectangular pumping)} \quad (14.72)$$

4.2. Value of G_0 for $I_0 = 0$

Case in which $G_s V_0 \ll I_s$. Relations (14.37), (14.59), (14.65) and (14.66) give:

$$G_0 = G_a + G_s \quad (14.73)$$

Case in which $G_s V_0$ has any given value. Relations (14.37), (14.59), (14.71) and (14.72) give:

$$G_0 = G_a + G_s(1 + aV_0) \quad (14.74)$$

In practice, these conditions can be satisfied if voltage V_0 is obtained by the rectification of a voltage proportional to the amplitude of the pumping voltage. A judicious choice of this amplitude, of the rectifier diode and the resistance shunting the filtering capacitance, enables these conditions to be satisfied over a wide range of voltage $2V_p$ and temperature.

In order to guide this choice, especially concerning the effect of temperature (parameter a), it is useful to know the value of the differential $\partial V_0/\partial a$ for $I_0 = 0$, for a given V_p.

4.3. Influence of temperature

$V_0 G_s \ll I_s$. Relations (14.65) and (14.66) give respectively

$$V_0 = \frac{1}{a} \log_n \{\mathscr{I}_0(2aV_p)\} \qquad \text{(cissoidal pumping)} \qquad (14.75)$$

$$V_0 = \frac{1}{a} \log_n \{\cosh(2aV_p)\} \qquad \text{(rectangular pumping)} \quad (14.76)$$

hence:

$$\left(\frac{\partial V_0}{\partial a}\right)_{I_0=0} = \frac{1}{a}\left\{2V_p \frac{\mathscr{I}_1(2aV_p)}{\mathscr{I}_0(2aV_p)} - V_0\right\} \qquad (14.77)$$

$$\left(\frac{\partial V_0}{\partial a}\right)_{I_0=0} = \frac{1}{a}\left\{2V_p \frac{\sinh(2aV_p)}{\cosh(2aV_p)} - V_0\right\} \qquad (14.78)$$

Any given value of $V_0 G_s$

G_s varies with temperature in accordance with an unknown law, but its variation is slower than that of I_s, and the effect of G_s on the stability of the current balance I_0 is in general beneficial. In fact, as can be seen from Fig. 14.20, if $I_s \gg G_s V_0$ curve 1 merges with curve 3; on the other hand, for decreasing values of a, the shape of curve 3 follows that of 4, 3, 2. Since I_s and a vary in opposite senses with temperature, their effects on the stability of the equilibrium may be partially compensated.

For certain diodes, the compensation is almost exact over a temperature range of 50°C and a decade of bias voltage. At high temperatures, the expressions (14.93) and (14.94) remain, in practice, valid, since the influence of G_s then quickly becomes negligible. For the lower temperatures, the influence of G_s becomes dominant. However, in this case, I_0 becomes low enough to be ignored.

4.4. Nomographs

For easy solution of equations (14.65), (14.66), (14.71), (14.72), Figs. 14.16 and 14.17 have been prepared giving the functions:

$$Z = e^{-x} \mathscr{J}_0(y) \qquad (14.79)$$

$$Z = e^{-x} \cosh(y) \qquad (14.80)$$

for individual values of $Z = 1, 2, 5, 10$.

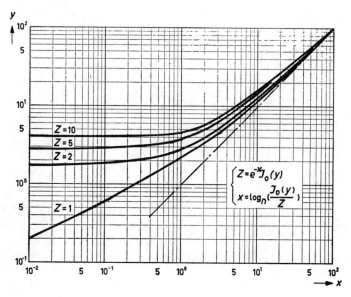

Figure 14.16

Relations (14.65) and (14.66) are given directly by Figs. 14.16 and 14.17 for $z = 1$. The plotting of (14.71) and (14.72) is a step-by-step process taking into account the particular value of G_s.

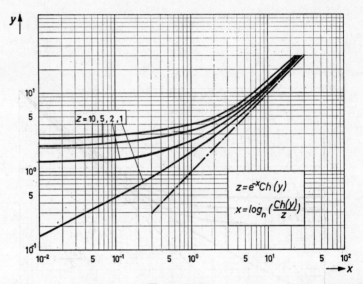

Figure 14.17

4.5 Practical example

In Figs. 14.18 and 14.19 are plotted the values of G_0, I_0 as functions of V_0 for several values of V_p; in Fig. 14.20 the function $f(V_o, V_p) = 0$ for $I_o = 0$ for several values of a with cissoidal and rectangular pumping. I_s was taken as $2 \cdot 5 \times 10^{-13}$ A and G_s as 37×10^{-13} mho.

These few examples reveal a striking agreement between theory and practice.

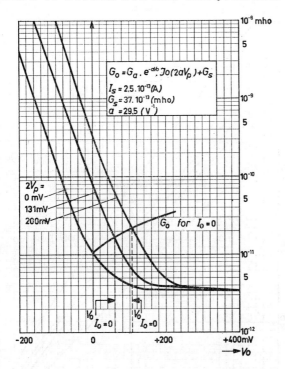

Figure 14.18

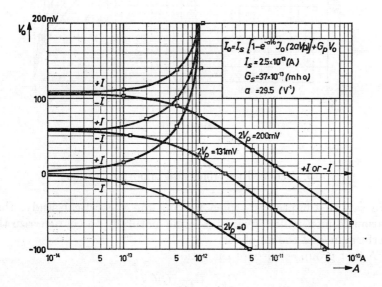

Figure 14.19

418   PARAMETRIC AMPLIFIERS

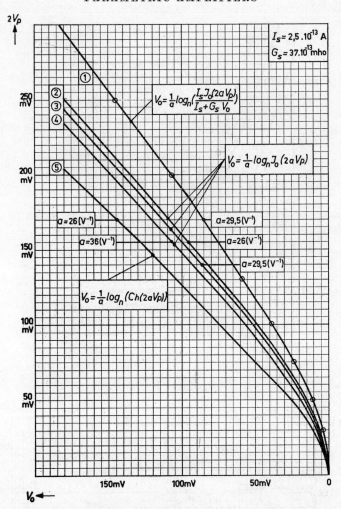

Figure 14.20

5. Reduced variables and important ratios

We saw in Chapter 12, when making the theoretical study, that various dimensionless quantities occur naturally to characterize the performance of the amplifier.

A first group consists of the ratios:

$$\frac{C_1}{C_0}; \quad \frac{C_2}{C_0}; \quad \frac{C_0 C_2}{C_1^2}$$

PUMPING OF DIODES FOR LOW FREQUENCIES

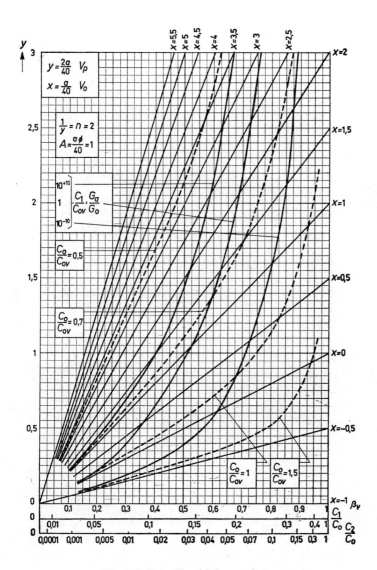

Figure 14.21. Cissoidal pumping

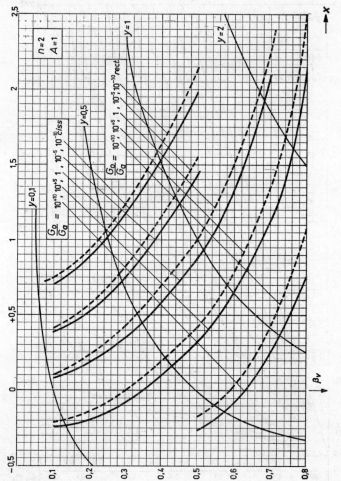

Figure 14.22

PUMPING OF DIODES FOR LOW FREQUENCIES   421

A second group consists of the ratios:

$$\frac{\omega_p C_0'}{G_T}; \quad \frac{\omega_p C_1'}{G_T}; \quad \frac{\omega_p C_2'}{G_T};$$

We made:

$$\alpha = \frac{\omega_p[C_{a_2} + C_{b_2}]}{G_T} = \frac{\omega_p C_2'}{G_T} = \frac{C_2'}{C_0'} \frac{Q}{k_1} \quad (14.81)$$

$$\beta = \frac{\omega_p[C_{a_1} + C_{b_1}]}{G_T} = \frac{\omega_p C_1'}{G_T} = \frac{C_1'}{C_0'} \frac{Q}{k_1} \quad (14.82)$$

$$Q = \frac{k_1 \omega_p[C_{a_0} + C_{b_0}]}{G_T} = \frac{k_1 \omega_p C_0'}{G_T} \quad (14.83)$$

$$\gamma_1 = \frac{G_{a_1} + G_{b_1}}{\omega_p[C_{a_1} + C_{b_1}]} = \frac{G_1'}{\omega_p C_1'} \quad (14.84)$$

$$\gamma_2 = \frac{G_{a_2} + G_{b_2}}{\omega_p[C_{a_2} + C_{b_2}]} = \frac{G_2'}{\omega_p C_2'} \quad (14.85)$$

with

$$k_1 = \frac{C_0' + C}{C_0'} \quad (14.86)$$

For the discussion of the choice of pumping parameters, reduced values must be introduced, related to the two basic characteristic parameters of the junction C_{0v} and G_a. In order to characterize the actual performance of the varactor for the various ratios, G_T will be replaced by $2G_0$. The reduced variables are then:

$$\frac{C_0}{C_{0v}} \quad (14.87)$$

$$\frac{G_a}{G_0} \quad (14.88)$$

$$\frac{C_0 G_a}{C_{0v} G_0} \quad (14.89)$$

$$\frac{C_1}{C_{0v}} \quad (14.90)$$

$$\frac{C_1 G_a}{C_{0v} G_0} \quad (14.91)$$

$$\frac{C_2}{C_{0v}} \quad (14.92)$$

$$\frac{C_2}{C_{0v}} \cdot \frac{G_a}{G_0} \quad (14.93)$$

Let

$$x = \frac{aV_0}{40} \qquad (14.94)$$

where 40 is the approximate numerical value of a, given by the theory of diffusion, at the temperature 25°C.

$$y = \frac{2aV_p}{40} \qquad (14.95)$$

$$A = \frac{a\phi}{40} \qquad (14.96)$$

When the rectangular voltage pumping, the functions

$$\frac{C_2}{C_{0v}} \quad \text{and} \quad \frac{C_2}{C_{0v}} \cdot \frac{G_a}{G_0}$$

are always zero. Taking (14.94), (14.95), (14.96) into account we have:

$$\beta_v = \frac{y}{x + A} \qquad (14.97)$$

Figures 14.21 and 14.22 show an example of the use of the previous relations (varactors characterized by $A = 1$ and $n = 2$). These show very clearly the particularly critical nature of the choice of the amplitude of the pumping voltage and of the bias, since the losses in the varactor increase exponentially. The figures do not take G_A into account. It is, however, very easy to introduce this conductance by relating the whole network to the curves

$$\frac{C_K}{C_{0v}} \cdot \frac{G_A}{G_s} = f(\beta_v V_0)$$

(almost merged and horizontal), which can be plotted directly from (14.91). Similarly G_0/G_a is related to G_s/G_a. The relation will be similar to that shown in Fig. 14.18.

6. Comparison between cissoidal and rectangular pumping

The most important characteristic for comparison purposes between cissoidal and rectangular pumping is the ratio of the values of C_1/C_0 for equal losses in the varactor. In order to make clearer the advantages and disadvantages of the two types of pumping, the comparison will first be made ignoring the losses.

6.1. Comparison of the C_1/C_0's with β_v given

The quotient of the ratios C_1/C_0 obtained for rectangular and cissoidal pumping respectively is shown in Fig. 14.23 as a function of β_v, for several values of n. When β_v is small, this ratio is almost independent of β_v and of n: the simplified formulae (14.27) and (14.58) are applicable and give:

$$\frac{(C_1/C_0)_{\text{rect}}}{(C_1/C_0)_{\text{ciss}}} = \frac{4}{\pi} \approx 1\cdot27 \qquad (14.98)$$

When β_v increases, Fig. 14.23 shows that the ratio of the C_1/C_0's is higher for $n = 3$ than for $n = 2$. It will be seen that the curves for a given n remain practically horizontal up to $\beta_v = 0\cdot9$.

6.2. Comparison for equal losses

For a given applied voltage, the losses in the varactor are proportional to G_0. They will be represented by the ratio G_0/G_a which is a dimensionless quantity. For a given type of pumping, G_0/G_a is a certain function of β_v and V_0 or of β_v and $2V_p$:

$$\left.\begin{array}{l}\left(\dfrac{G_0}{G_a}\right)_r = f_r(\beta_{v_r}, V_{0_r}) \quad \text{(rectangular pumping)} \\[6pt] \left(\dfrac{G_0}{G_a}\right)_c = f_c(\beta_{v_c}, V_{0_c}) \quad \text{(cissoidal pumping)}\end{array}\right\} \qquad (14.99)$$

or:

$$\left.\begin{array}{l}\left(\dfrac{G_0}{G_a}\right)_r = g_r(\beta_{v_r}, 2V_{p_r}) \\[6pt] \left(\dfrac{G_0}{G_a}\right)_c = g_c(\beta_{v_c}, 2V_{p_c})\end{array}\right\} \qquad (14.100)$$

The comparison for equal losses is expressed by:

$$\left(\frac{G_0}{G_a}\right)_r = \left(\frac{G_0}{G_a}\right)_c = \frac{G_0}{G_a} \qquad (14.101)$$

There are two cases to be considered.

Case 1:

$$V_{0_r} = V_{0_c} \qquad (14.102)$$

Let:

$$\beta' = \frac{\beta_{v_r} + \beta_{v_c}}{2}$$

$$\gamma = \frac{\beta_{v_r}}{\beta_{v_c}}$$

hence:

$$\beta_{v_c} = \frac{2\beta'}{1+\gamma}$$

$$\beta_{v_r} = \frac{2\gamma\beta'}{1+\gamma}$$

After substitution in (14.99) and taking (14.101) into account, we find:

$$f_c\left(\frac{2\beta'}{1+\gamma}, V_0\right) = f_r\left(\frac{2\gamma\beta'}{1+\gamma}, V_0\right)$$

Solving this last equation with respect to V_0, we obtain:

$$V_0 = \phi(\gamma, \beta')$$

which, in accordance with (14.101) gives:

$$\frac{G_0}{G_a} = f_c\left(\frac{2\beta'}{1+\gamma}, \phi(\gamma, \beta')\right)$$

By means of this relation we can determine γ as a function of β' with G_0/G_a as parameter (Fig. 14.24).

Case 2:

$$2V_{p_r} = 2V_{p_c} = 2V_p$$

We find successively

$$g_r(\beta_{v_r}, 2V_p) = g_c(\beta_{v_c}, 2V_p)$$

$$g_r\left(\frac{2\gamma\beta'}{1+\gamma}, 2V_p\right) = g_c\left(\frac{2\beta'}{1+\gamma}, 2V_p\right)$$

$$2V_p = \psi(\gamma, \beta')$$

$$\frac{G_0}{G_a} = g_c\left(\frac{2\beta'}{1+\gamma}, \psi(\gamma, \beta')\right)$$

By means of this last relation we can determine γ as a function of β' with G_0/G_a as parameter (Fig. 14.25). If we refer to equations (14.27) and (14.58) we see that we have:

$$\left(\frac{C_1}{C_0}\right)_r = \frac{2}{\pi}\frac{\beta_{v_r}}{n}$$

$$\left(\frac{C_1}{C_0}\right)_c = \frac{\beta_{v_c}}{2n}$$

hence:

$$\frac{(C_1/C_0)_r}{(C_1/C_0)_c} = \frac{4}{\pi}\frac{\beta_{v_r}}{\beta_{v_c}}$$

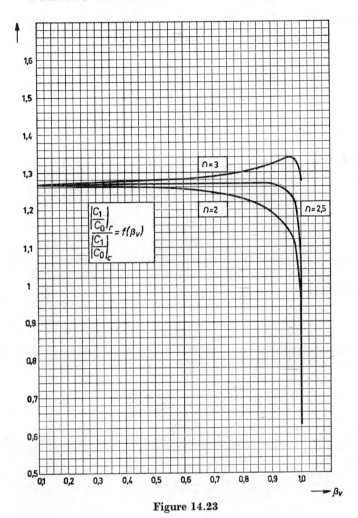

Figure 14.23

By means of this formula we can determine the quotient of the C_1/C_0's as a function of β' for a given G_0/G_a:

$$\frac{(C_1/C_0)_r}{(C_1/C_0)_c} = \frac{4}{\pi}\gamma(\beta')$$

Conclusion

Reference to Figs. 14.23, 14.24 and 14.25 shows that the quotient of the ratios C_1/C_0 with a given G_0/G_a is lower than that for the case when $\beta_{v_r} = \beta_{v_c} = \beta'$.

6.3. Comparison of the losses for given values of V_0 and $2V_p$

If we ignore G_s and assume that $2aV_p$ is larger than unity, and taking account of (14.37), (14.39) and (14.59):

$$\frac{(G_0)_r}{(G_0)_c} = \frac{\cosh(2aV_p)}{\mathscr{J}_0(2aV_p)} \approx \sqrt{2aV_p} \cdot \sqrt{\frac{\pi}{2}} = 1\cdot 253\sqrt{2aV_p} \quad (14.103)$$

6.4. Conclusion

It follows that where the value of C_1/C_0 is concerned, the advantage of rectangular pumping over cissoidal pumping, from the point of view of the losses, is quite small.

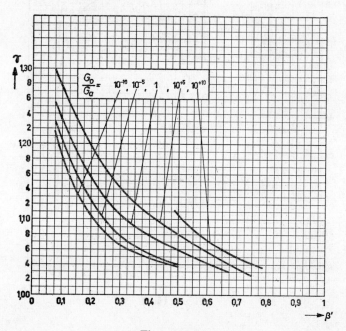

Figure 14.24

Amongst the disadvantages of rectangular pumping may be mentioned:

(1) The tuning of the output filter may be up to twice as sensitive to a variation of the amplitude of the pumping voltage (see the results recorded in Tables 1 and 6).
(2) The bias voltage to be applied in order to make $I_0 = 0$ is about 1·3 times greater than that required for cissoidal pumping.

(3) It leads to a slightly increased sensitivity to variations in temperature.
(4) It does not allow the appearance of 'parametric' gain with its advantages, namely on the noise factor.

Amongst the advantages of this type of pumping may be mentioned:

(1) An excellent stability of the gain and of the output phase (absence of C_2).
(2) Specially easy synchronous demodulation by means of sampling.

The type of circuit most suitable for rectangular pumping is the two-stage one, as this solution is very flexible. The phase of the output signal of the first amplifier is in quadrature with the pump; the pumping voltage of the second stage (frequency $2\omega_p$) will be held in

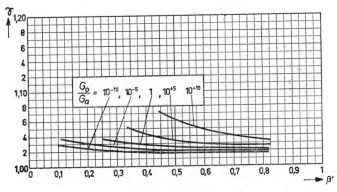

Figure 14.25

phase with the pumping voltage of the first stage (frequency ω_p). These two pumping voltages, with their identical amplitudes and phases adjustable, may be delivered by the standard chain consisting of: a quartz-synchronized multivibrator, an integrator and a two-to-one arrangement.

References

[1] Kennedy, P., 'Semiconductor device evaluation', *Advances in Electronics and Electron Physics*, vol. 18, 1963, Academic Press.
[2] Biard, J. R., 'Low-frequency reactance amplifier', *Proc. I.R.E.*, vol. 51 (No. 2), pp. 298–303, February 1963.

[3] *Sechsstellige Tafeln der Bessel'schen Funktionen imaginären Arguments*, Anding, Leipzig, 1911.
[4] Watson, G. N., 'A treatise on the theory of Bessel functions', at Cambridge University Press, 1958.
[5] Dwight, *Tables of Integration and other Mathematical Data*, The Macmillan Company, New York.
[6] Jahnke and Emde, *Tables of Functions*, Dover Publications, New York.
[7] Morse, P. M. and Feshbach, H., *Methods of Theoretical Physics*, McGraw-Hill, New York, 1953.
[8] Angot, A., 'Compléments de mathématiques à l'usage des ingénieurs de l'electrotechnique et des télécommunications', *Editions de la Revue d'Optique*, p. 735 (4th edn.), 1957.

15

MEASUREMENT OF THE PARAMETERS OF THE VARACTOR

by G. MARECHAL

The manufacturers of low-frequency varactors are often sparing with precise particulars: it is seldom possible to have available the values of the different parameters and their stability and tolerances. A knowledge of these data is essential for the production of a reliable and reproducible amplifier. This position, coupled with the simple desire to check the guaranteed performance, makes it essential to measure the different quantities I_s, a, G_s, R_s, C_{0v}, ϕ, n, C_c. We have seen in Chapter 11 that operation in low frequency of a given varactor is only approximately characterized by equations (13.2) and (13.4), or by (13.1) and (13.3).

For a given varactor, the value of each of these characteristic quantities varies slightly with the voltage applied to the varactor. However, it is usually sufficient to obtain the mean values corresponding to the range of operation expected. Methods of measurement must therefore be adopted which will supply the most reliable values of the parameters within the optimum theoretical approach to the problem in question. The most critical quantities from this point of view are a, I_s and G_s, which were discussed in section 1.1 (1) of Chapter 13. However, an exact knowledge of G_s is only necessary if it is essential for the condition of balance $I_0 = 0$ to be as accurate as possible. The group of measurements described below has been devised with a view to rapid and systematic tests: the reasonable price of varactors for low-frequency parametric amplifiers enables the latter to be manufactured in series.

PARAMETRIC AMPLIFIERS

1. Measurement of I_s, a, G_s, R_s

1.1. Measurement of the static characteristic

The static characteristic of varactors can be obtained rapidly by using it as the negative-feedback impedance of an operational amplifier. The general schematic of the measuring circuit is shown in Fig. 15.1. Since the positive input of the amplifier A is connected to earth, the negative input constitutes a virtual earth.

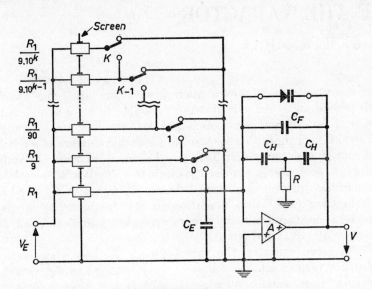

Figure 15.1

The current injector is shown with the series-parallel configuration. The injected current, obtained by closing the $K + 1$ first contact, is equal to:

$$i' = \frac{V_E 10^K}{R_1} = \frac{V_E}{R_k} \qquad (15.1)$$

By means of this structure, weak currents can be injected rapidly (a current of 10^{-12} A can be injected with a time constant of 10 msec). The contacts of the various inverters $0, 1, \ldots, K - 1, K$, will be of the break-before-make type in order to avoid any parasitic transients over the whole scale range. The insulation resistance of the open contact multiplied by the gain of the amplifier A is to be compared with the resistance R_K. The quality of this insulation may deteriorate with K without changing the accuracy. For weak currents, the measuring time is usually determined by the time constant of the varactor itself. To reduce the measuring times at low currents, one has to adopt the inverse structure where the output of the amplifier is connected to V_E in place of V. Measuring is then done in two times: rapid *loading* of the capacitance of the varactor (scale where $K \gg 0$); measuring of the current on the sensitive scale considered (scale where $K \approx 0$).

The *amplifier* A should have a static current error and a static voltage error which can be ignored by comparison with the injected current V_E/R_K and the output voltage V respectively. Its gain will be chosen as a function of the accuracy and speed of the required measurement. Very often, in order to prevent the circuit oscillating for certain values of current through the varactor, it is necessary to complete the amplifier circuit by the addition of an external phase advancer, or to make an appropriate choice of the internal time constants of the amplifier, so as to obtain, at the most, a step-by-step decrease in its gain with frequency [6], [7]. If account is taken of the fact that the differential conductance of the varactor may vary by more than 10^{10} times, there is always a current for which the time constant of the circuit formed by the varactor, shunted by the residual input capacitance C_E, is equal to the internal time constant of the amplifier. If the varactor is tested in an oven or a cryostat, the capacitance C_E cannot be small. The differential capacitance of the varactor is often great enough for a network such as C_H–R–C_H to be superfluous. It is important that the amplifier and the circuit should be well shielded from external effects due to parasitic pulses, which when detected by the diode, cause a blockage of the amplifier due to the appearance of voltages across C_E. Amplifier A will preferably be of the mechanical heteroparametric type (vibrating capacitance or 'Vibron') for d.c. signals, with an a.c. amplifier in parallel. With a satisfactory installation, the following performance may be expected: the value of the current injected will be ascertainable to better than 1% above 10^{-12} A; the value of the voltage at the terminals of the diode to be better than 200 p.p.m. ± 100 μV.

Note. Before use, the varactors should be carefully cleaned with absolute alcohol, then wiped and dried. They should be measured in the dark, at a constant and repeatable temperature.

1.2. Determination of a and I_s

When the static characteristic has been obtained, the determination of the optimum values of a and I_s should preferably be made by the method of least squares: the voltage region to be considered usually

extends from the most negative voltage of the pumped varactor to
−50 mV, so that in this way at least three current decades can be
covered. In this region, the characteristic curve is exponential. In
(13.1) the predominant term is $-I_s\,e^{-aV}$, and the measured static
characteristic should be compared with the law:

$$i' = -I_s\,e^{-aV} \tag{15.2}$$

The calculations to be performed are dealt with in numerous
articles [1].

Very often, in order to find a and I_s, a quicker method will be
adequate: in the pass zone of the static characteristic, the following
determinations will be made:

(a) Estimate the slope of the curve $V = \log I$, from which a is
obtained by:

$$a = \frac{\ln 10}{\text{volts per decade}} \approx \frac{2\cdot 305}{\text{volts per decade}} \tag{15.3}$$

(b) In the pass zone read the value of the current I_{CH} corresponding to a voltage of value $-V_{CH}$ which is usually chosen corresponding to a current of one decade lower than the maximum current of the pumped varactor.

$$I_s = -I_{CH}\,e^{-aV_{CH}} \tag{15.4}$$

1.3. Determination of G_s

If the values of I_s are known, G_s may be determined by the method
of least squares. The voltage region to be considered usually extends
from the most positive voltage of the pumped varactor to −50 mV.
The measured static characteristic should be compared with the law:

$$C_s = \frac{1}{V}[i' - I_s(1 - e^{-aV})] \tag{15.5}$$

If we are interested in an optimum balance of the parametric bridge
from the point of view of the d.c. component ($I_0 = 0$ in each branch),
a better evaluation of G_s will be obtained as follows:

(a) Determine experimentally one or two couples of values V_p, V_0
corresponding to $I_0 = 0$, by means of the circuit shown in Fig. 15.2
($C_E \gg C_0$).

In this determination, V_p is measured directly and, V_0 is measured at
the output of the operational amplifier ($V \approx V_0$).

(b) Calculate G_s by means of relations:

(1) for cissoidal pumping

$$G_s = \frac{I_s}{V_0} [e^{-aV_0} \mathscr{I}_0(2aV_p) - 1] \qquad (15.6)$$

(2) for rectangular pumping

$$G_s = \frac{I_s}{V_0} [e^{-aV_0} \cosh(2aV_p) - 1] \qquad (15.7)$$

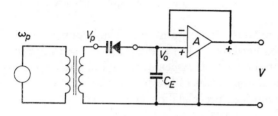

Figure 15.2

1.4. Determination of R_s

It is advisable to check that the series resistance R_s of the diode contributes only to a negligible extent to its admittance at the operational frequency.

$$R_s \omega_p^2 C_0^2 \ll G_A + G \qquad (15.8)$$

A rapid estimate of the value of R_s may be obtained by the pulse method of measurement (in order to avoid heating of the junction) of the static characteristic for voltages more negative than $-\phi$. In this region, the differential conductance of the junction is maintained at a practically constant value of $1/R_s$.

2. Measurement of C_{0v}, ϕ, n, C_c

2.1. Measurement of $C = C(v)$ and $dC(v)/dv$

This measurement will be made at the highest frequency at which the effect of L_c and R_s may be ignored. The approximate law (13.4) is strikingly verified by experience for diodes of standard geometry. The capacitance C_{0v} of the varactors for low frequencies is usually higher than 10 pF. A bridge measurement is required, especially as the bridge structure of the proposed parametric amplifier leads to a consideration of the variations of the two varactors. The measurement may be made by means of a parametric head in which one of the varactors has been replaced by a variable standard capacitance, fitted with a small additional adjustable capacitor for the measurement of the variations. The bandwidth of the amplifier from zero will be as narrow as required. The pump voltage will be low in order to make $(C(V_0))$ the differential capacitance of the varactor (biased by a voltage V_0), similar to the zero term of the series development of the differential capacitance $(C_0(V_0))$ of the varactor pumped by the test voltage and biased by voltage V_0. The measured differential capacitance is $C_0(V_0) + C_c = C_m(V_0)$. The quantities to be determined are $C(V_0)$ and C_c.

If the amplitude of the pump and the bias is such that $\beta_v \leqslant 0.2$, the difference between $C(V_0)$ and $C_0(V_0)$ is less than 10^{-3}, whatever the power of the root. If necessary, $C_0(V_0)$ could be converted into $C(V_0)$ by the nomographs of Chapter 14. If great care has been taken in the construction, the absolute error in the measurement of $C(V_0)$ can be made lower than $10^{-3} \pm 0.1$ pF.

Variations of 10 femtofarads (10×10^{-15} F) can then be detected. The measurement of the differences, made with the small additional variable standard capacitor, gives the value of the relative slope dC/dV_0 of the diode, a quantity which is often given by the manufacturer for $V_0 = 0$. The error on the slope of the diode will be less than 1% by default if ΔV is chosen as smaller than 0.1 $(V_0 + \phi)$, the measurement of the difference in capacitance ΔC being made at $\Delta V/2$ on each side of V_0.

2.2. Determination of C_{0v}, ϕ, n, C_c

2.2.1.

The method described in Chapter 9 may be used for this measurement; however, for most varactors C_{0v} varies from 10 pF to 10,000 pF whilst C_c may be estimated at 0.2 pF. It will therefore always be very difficult to find the value of C_c. For this reason it is preferable to determine C_{0v}, ϕ and n, taking an estimated value of C_c into account. If we know ϕ and n, a more exact value of C_c can be determined if necessary.

Differentiation of (13.4) gives us

$$C(V_0) = n(\phi + V_0)\left(-\frac{dC}{dV_0}\right)_{v_0} + C_c \approx C_m(V_0) \qquad (15.9)$$

We will introduce into (15.9) the results of the measurements made at a bias voltage V_1 close to $-\phi$, and proceed in the same way for the

MEASUREMENT OF PARAMETERS OF THE VARACTOR   435

measurements made at zero bias voltage. If n is eliminated by means of the two equations, we find:

$$\phi = V_1\left[\frac{\{C_m(V_1) - C_c\}(dC/dV_0)_{V_0=0}}{\{C_m(0) - C_c\}(dC/dV_0)_{V_0=V_1}} - 1\right]^{-1} \quad (15.10)$$

If we know ϕ, n will be determined by using either of the following two equations:

$$\left.\begin{aligned}\frac{1}{\nu} = n &= \frac{C_m(0) - C_c}{\phi(-dC/dV_0)_{V_0=0}} \quad \text{(Relation (15.9) for } V_0 = 0)\\ \frac{1}{\nu} = n &= \frac{C_m(V_1) - C_c}{(\phi + V_1)(-dC/dV_1)_{V_0=V_1}}\\ &\quad \text{(Relation (15.9) for } V_0 = V_1)\end{aligned}\right\} \quad (15.11)$$

On the other hand, from (13.4), for $V_0 = 0$ we have:

$$C_{0v} = C_m(0) - C_c \quad (15.12)$$

As ϕ and n are now known we may introduce into (15.9) the results of the measurements made at a bias voltage V_2 close to V_B:

$$C_c = C_m(V_2) - n(\phi + V_2)\left(-\frac{dC}{dV_0}\right)_{V_0=V_2} \quad (15.13)$$

(15.13) gives a more exact value of C_c than that which we introduced into (15.11) and (15.12).

It should be noted that (15.11) may be written,

$$\frac{C_{0v}}{n\phi} = \left(-\frac{dC}{dV_0}\right)_{V_0=0} \quad (15.14)$$

a relation independent of C_c. It will be seen that it is advantageous to choose V_1 and V_2 as far apart as possible.

2.2.2

The measurement of the differential capacitance is more accurate than the measurement of its slope. The determination of n and ϕ may be made without a knowledge of the slope if C_c is estimated. The expressions obtained give ϕ and n in the form of implicit equations which can be solved graphically by successive approximations.

$$V_2\left(\frac{C_m(0) - C_c}{C_m(V_1) - C_c}\right)^n - V_1\left(\frac{C_m(0) - C_c}{C_m(V_2) - C_c}\right)^n = V_2 - V_1 \quad (15.15)$$

$$\left(1 + \frac{V_1}{\phi}\right) = \left(1 + \frac{V_2}{\phi}\right)\left(\frac{\log\left(\frac{C_m(0) - C_c}{C_m(V_1) - C_c}\right)}{\log\left(\frac{C_m(0) - C_c}{C_m(V_2) - C_c}\right)}\right) \quad (15.16)$$

For a given type of varactor, at a given temperature ϕ and n have generally a low dispersion ($\pm 3\%$). Taking this into account a rapid method for the resolution of (15.15) and (15.16) could be elaborated.

3. Cut-off frequency—quality factor

In order to characterize their varactors, manufacturers very often give a diagram representing the quality factor (Q) of the varactor as a function of the frequency; very often they also guarantee one or several minimum values of Q for one or several frequency values; sometimes they introduce two cut-off frequencies. These diagrams and warrants, if correctly interpreted, can give much useful information. For frequencies lower than 100 MHz (where L_c does not yet interfere), and taking into account that R_s is clearly smaller than $1/G_0$, one can write:

$$Q \approx \frac{C_0 \omega}{G_0 + R_s C_0^2 \omega^2} \qquad (15.17)$$

which relation after writing

$$\omega_{c_L} = \frac{G_0}{C_0} \qquad (15.18)$$

and

$$\omega_{c_H} = \frac{1}{R_s C_0} \qquad (15.19)$$

becomes:

$$Q = \frac{1}{\dfrac{\omega_{c_L}}{\omega} + \dfrac{\omega}{\omega_{c_H}}} \qquad (15.20)$$

One remarks that these equations suppose the pumping conditions to be already defined. Very often these conditions correspond to $V_0 = 0$ and $\beta_v \leqslant 0 \cdot 2$; and we have then:

$$C_{0v} \approx C_0 \ (V_0 = 0)$$

$$G_0 \approx G_a + G_s$$

Representation of (15.20) is done easily in a double logarithmic diagram in ω and Q (or in f and Q). This diagram shows that the function has an axis of symmetry in

$$\omega_s = \sqrt{\omega_{c_H} \cdot \omega_{c_L}}$$

At this frequency, Q is maximum and is equal to:

$$Q_m = \frac{\omega_s}{2\omega_{c_L}} = \frac{\omega_{c_H}}{2\omega_s}$$

For $\omega \ll \omega_s$, (15.20) simplifies to

$$Q \approx \frac{\omega}{\omega_{c_L}}$$

An asymptotic line placed on the left side of the logarithmic diagram, crosses axis $Q = 1$ in ω_{c_L}.

For $\omega \gg \omega_s$ (15.20) simplifies to:

$$Q \approx \frac{\omega_{c_H}}{\omega}$$

An asymptotic line placed on the right side of the logarithmic diagram crosses axis $Q = 1$ in ω_{c_H}.

Typical values of the lower medium and higher cut-off frequencies and of a maximum Q are:

$$f_{c_L} = \frac{\omega_{c_L}}{2\pi} = 10^{-2} \text{ Hz}$$

$$f_s = \frac{\omega_s}{2\pi} = 5 \times 10^3 \text{ Hz}$$

$$f_{c_H} = \frac{\omega_{c_H}}{2\pi} = 2\cdot 5 \times 10^9 \text{ Hz}$$

$$Q_m = 2 \times 10^5$$

References

[1] Angot, A., 'Compléments de mathématiques à l'usage des ingénieurs de l'electronique et des télécommunications', *Edition de la Revue d'Optique*, p. 735 (4th edn.), 1957.
[2] Marton, L. (ed.), *Methods of Experimental Physics*, vols. 1–7, Academic Press.
[3] 'A.C. capacitance, dielectric constant, and loss characteristics of electrical insulating materials', *A.S.T.M. D150–59T*.
[4] 'Electrical resistance of insulating materials', *A.S.T.M. D257–61*.
[5] 'Cleaning plastic specimens for insulation resistance testing', *A.S.T.M. D1371–59*.
[6] Brookshier, W. K., 'Electrometer circuit design for extended band widths', *Nuclear Instruments and Methods*, vol. 25, pp. 317–327, 1964.
[7] Fowler, E. P., *A New Reactor Log. Power and Periodmeter*, Atomic Energy Establishment, Winfrith, Dorchester, Dorset.

INDEX

a.c. signals, 382
adjustment of a parametric amplifier, 258ff
  methods of, 342ff
admittance, 321
  input, 332
amplification of low-frequency signals, 311ff
  characteristics of, 381
amplifiers
  classification of, 53ff
  degenerate, 82, 157ff
  difference-frequency, 157
  double-band, 185ff
  for d.c. signals, 365
  four-frequency, 143ff
  isolators, use of, in, 114
  negative-resistance, 64
    transducer gain of, 109
    pumping of, 237
  noise temperature of, 19ff
  reflection, 81, 179
    adjusted for minimum noise, 258
    adjustment of, 260
    description of, 306
    deterioration in performance of, 155
    three-frequency, 101ff, 134
    with and without circulator, 112
amplitude and output signal, 320
anti-saturation circuit, 369
automatic gain control, 369
auxiliary circuits, 366

background noise
  in telecommunications, 3
  of low-frequency signals, 334
  sources of, 42ff
bandwidth
  3 dB, 331
  high gain, 133, 140
  of d.c. and low-frequency signals, 383

  of three-frequency converters, 98
bridge, balancing of, 342ff

capacitance of varactor, 264, 266
circle diagram, 247ff
circulators, 103
  with four and five ports, 112
cissoidal voltage pumping, 390
  comparison with rectangular pumping, 422
converters
  classification of, 53ff
  difference-frequency, 59, 64, 81, 99
    classification of, 157
    transducer gain of, 149
  four-frequency, 61, 67, 143ff
  frequency, noise temperature of, 32
  inverting, 101ff, 121
    adjusted for minimum intrinsic noise, 257
  sum-frequency, 57, 80, 153, 83ff
    adjusted for maximum gain, 255
    adjusted for minimum noise, 256
    description of, 303
    pumping of, 237
    three-frequency, 57, 83ff
    transducer gain of, 147ff
correction networks, internal, 368
coupling, directional, 291
current pumping, 210ff
  comparison with voltage pumping, 239
  tables and charts for, 228ff

d.c. signals
  amplifier for, 365
  transformation into a.c. signals, 382
degenerate amplifiers, 82, 157ff
demodulator, 368
detector
  instantaneous voltage of, 329
  noise voltage at input of, 334

INDEX

detector (*cont.*)
   spectral density of parasitic signal at output of, 337
difference-frequency converters, 59, 64, 81
   classification of, 157
   non-inverting, 99
   transducer gain of, 149
diodes, parametric, pumping of, 194ff
double-band reception, 184

elastance
   in current pumping, 228ff
   in voltage pumping, 219ff
equivalent circuits
   four-frequency converters, 143ff
   inverting converters, 101
   reflection amplifiers, 101
   sum-frequency converters, 83
   two-ports, 20, 267ff
   varactor, 261

filters, 301ff
four-frequency
   amplifiers, 143ff
   converters, 61, 67, 143ff
   parametric systems, classification of, 66
frequency converters, 57ff, 83ff, 143ff, 67
   effective spectral noise temperature of, 32
functions
   noise, 7
   transfer, for input signal and noise, 160

gain
   instability factor of, 85
   optimization of, 125, 138, 86
   transducer (*see* transducer gain)
gas discharge tube, measurement of noise temperature by, 285

harmonics of elastance, 222, 224, 231, 233

immittance, terminal, 75
impedance
   input, 144, 381
   seen by the varactor, 161
   transformers, location of, 303

input
   admittance, 332
   impedance, 144, 381
   signal and noise, transfer functions for, 160
instability
   coefficient, 122
   factor
      in optimization, 124ff
      of gain of sum-frequency converters, 85
      of reflection amplifiers, 136ff
internal noise sources, 42, 334
inverting converters, 101ff
   adjusted for minimum intrinsic noise, 257
isolators, 103, 114

junction, ideal, small-signal theory for, 70

losses, varactor, 77
low-frequency signals, amplification of, 311ff, 381

Manley and Rowe relations, 53
measurement
   of noise temperature, 276ff
   of parametric amplifiers, 246ff
   of response curves, 302
   of short-circuit planes, 302
modulator, tuned-load, 340
monochromatic signal generator, 278

narrow-band signal, 12
negative-resistance amplifier, 64, 109
noise
   background, 3ff, 334ff
   expressions for, 161
   intrinsic
      in inverting converters, 257
      in sum-frequency converter, 256
   optimization of
      in inverting converter, 126
      in reflection amplifier, 138ff
      in sum-frequency converter, 91, 94
   transfer functions for, 160
noise
   factor, 19, 25, 30
   functions, 7
   in amplification of d.c. signals, 382

sources, 20, 334
temperature
　automatic measurement of, 288
　of amplifier, 19ff, 276ff
　of frequency converter, 32
　of inverting converter, 122
　of reflection amplifier, 136
　　minimization of, 140
　of sum-frequency converter, 85
　of two-ports, 28ff
　voltage at input of detector, 334
nomographs of pumping, 218, 415

optimization
　instability factor in, 124ff, 138ff
　of gain, 86, 125, 138
　of noise temperature, 91ff, 126ff, 138ff
　pumping, 237
oscillator, pump, 368
output signal
　expression for, 161
　phase and amplitude of, 320
　synchronous detection of, 324

polarization
　current pumping, 233
　voltage pumping, 225
power gain, relation to transducer gain, 66
pump circle diagram, 249
pump circuit, adjustment of, 258
pumping
　current, 210ff
　　tables and charts for, 228
　　square wave, 238
　voltage, 198ff
　　cissoidal, 390
　　comparison with current pumping, 239
　　rectangular, 406ff
　　　comparison with cissoidal, 422
　　tables and nomographs for, 218ff
pumping of parametric diodes
　for low frequencies, 387ff
　for microwave frequencies, 194ff
pump oscillator, 368

quality factor of varactors, 264, 436

reactive networks, 297ff
reception
　single-band, 164ff
　double-band, 184ff
rectangular voltage pumping, 406ff
　comparison with cissoidal pumping, 422
recurrence formulae, 412
reflection amplifier, 81, 101ff, 134ff
　adjustment of, 258, 260
　description of, 306ff
　performance of, deterioration in, 155
　single-band, 179
response curves, measurement of, 302

signal generators, monochromatic, 278
signal-to-noise ratio, 12
single-band reception, 164ff
small-signal analysis, 70ff
Smith chart, 247
stability, 97, 151
sum-frequency converters
　adjustment
　　for maximum gain, 255
　　for minimum intrinsic noise, 256
　description of, 303
　four-frequency, 61, 153
　three-frequency, 57, 80
synchronous detection of output signal, 324, 329

telecommunications
　background noise in, 3
temperature
　influence of, 414
　of a system, 40
terminal immittances, 75
thermionic diode, measurement of noise temperature by, 281
three-frequency
　converters, 57, 83ff, 66, 80
　parametric systems, classification of, 66
　reflection amplifier, 101ff
three-port, 160
transducer gain
　of difference-frequency systems, 115, 149
　of four-frequency converters and amplifiers, 147ff

transducer gain (*cont.*)
 of inverting converters, 121
 of negative-resistance amplifiers, 109
 of reflection amplifiers, 134
 of sum-frequency converter, 84
 relation to power gain, 66
 transfer functions for input signal and noise, 160
transformer, impedance, location of, 303
trimming components, sensitivity of, 358
two-ports
 chain matrix of, 268
 effective spectral noise temperature of, 28
 equivalent circuit of, 20, 267
 noise factor of, 25, 30
 use of isolators in, 114

varactor
 characteristics of, 263
 comparison of, 242
 equivalent circuit of, 261
 impedances seen by, 161
 in reactive networks, 297
 losses of, 77
 measurement of, 261ff, 429ff
 parameters, 429ff
 pumped, description of, 194
 variables, reduced
  in four-frequency converters and amplifiers, 147
  in small-signal analysis, 79
voltage
 energizing the detector, 329
 output, 327
 pumping, 198ff
  cissoidal, 390
   comparison between cissoidal and rectangular, 422ff
  rectangular, 406ff
  tables and nomographs for, 218f
 waveform of pumping voltage, 198

TECHNICAL LIBRARY (MAIN)
U. S. NAVAL WEAPONS LAB.
DAHLGREN, VIRGINIA 22448